Networks and the Spread of Ideas in the Past

Networks and the Spread of Ideas in the Past: Strong Ties, Innovation and Knowledge Exchange gathers contributions from an international group of scholars to reconsider the role that strong social ties play in the transmission of new ideas, and their crucial place in network analyses of the past.

Drawing on case studies that range from the early Iron Age Mediterranean to medieval Britain, the contributing authors showcase the importance of looking at strong social ties in the transmission of complex information, which requires relationships structured through mutual trust, memory, and reciprocity. They highlight the importance of sanctuaries in the process of information transmission, the power of narrative in creating a sense of community even across geographical space, and the control of social systems in order to facilitate or stifle new information transfer.

Networks and the Spread of Ideas in the Past demonstrates the value of searching the past for powerful social connections, offers us the chance to tell more human stories through our analyses, and represents an essential new addition to the study and use of networks in archaeology and history. The book will be useful to academics and students working in the Digital Humanities, History, and Archaeology.

Anna Collar is a Lecturer in Roman Archaeology at the University of Southampton and a founder of *The Connected Past* international research network and conference series. Her research explores the material culture of religion; social networks; and landscape, mobility, and emotion.

Digital Research in the Arts and Humanities
Founding Series Editors: Marilyn Deegan, Lorna Hughes and Harold Short

Current Series Editors: Lorna Hughes, Nirmala Menon, Andrew Prescott, Isabel Galina Russell, Harold Short and Ray Siemens

Digital technologies are increasingly important to arts and humanities research, expanding the horizons of research methods in all aspects of data capture, investigation, analysis, modelling, presentation and dissemination. This important series covers a wide range of disciplines with each volume focusing on a particular area, identifying the ways in which technology impacts on specific subjects. The aim is to provide an authoritative reflection of the 'state of the art' in technology-enhanced research methods. The series is critical reading for those already engaged in the digital humanities, and of wider interest to all arts and humanities scholars.

The following list includes only the most-recent titles to publish within the series. A list of the full catalogue of titles is available at: https://www.routledge.com/Digital-Research-in-the-Arts-and-Humanities/book-series/DRAH

Transformative Digital Humanities
Challenges and Opportunities
Edited by Mary McAleer Balkun and Marta Mestrovic Deyrup

Medieval Manuscripts in the Digital Age
Edited by Benjamin Albritton, Georgia Henley and Elaine Treharne

Access and Control in Digital Humanities
Edited by Shane Hawkins

Information and Knowledge Organisation in Digital Humanities
Global Perspectives
Edited by Koraljka Golub and Ying-Hsang Liu

Networks and the Spread of Ideas in the Past
Strong Ties, Innovation and Knowledge Exchange
Edited by Anna Collar

Networks and the Spread of Ideas in the Past

Strong Ties, Innovation and Knowledge Exchange

Edited by Anna Collar

Routledge
Taylor & Francis Group
LONDON AND NEW YORK

First published 2022
by Routledge
4 Park Square, Milton Park, Abingdon, Oxon OX14 4RN

and by Routledge
605 Third Avenue, New York, NY 10158

Routledge is an imprint of the Taylor & Francis Group, an informa business

British Library Cataloguing-in-Publication Data
A catalogue record for this book is available from the British Library

Library of Congress Cataloging-in-Publication Data
A catalog record has been requested for this book

ISBN: 978-1-138-36847-7 (hbk)
ISBN: 978-1-032-26619-0 (pbk)
ISBN: 978-0-429-42921-7 (ebk)

DOI: 10.4324/9780429429217

Typeset in Times New Roman
by KnowledgeWorks Global Ltd.

Contents

Figures

Contributors

Nathanael Andrade has published extensively on the Roman and later Roman Near East and ancient world history. His books include *Syrian Identity in the Greco-Roman World* (Cambridge University Press, 2013); *The Journey of Christianity to India in Late Antiquity: Networks and the Movement of Culture* (Cambridge University Press, 2018); and *Zenobia: Shooting Star of Palmyra* (Oxford University Press, 2018). He is a Professor in the history department at Binghamton University (SUNY).

Sandra Blakely is an Associate Professor in the Department of Classics, Emory University. Her research interests include religion and magic of the Greek and Roman world, anthropological and comparative approaches to the ancient Mediterranean, archaeological theory, and digital approaches to antiquity including ArcGIS; social network analysis; and gaming, historiography, and archaeometallurgy.

Anna Collar is the author of *Religious Networks in the Roman Empire: The Spread of New Ideas* (Cambridge University Press, 2013); a finalist in the American Academy of Religion's Best First Book in the History of Religions 2014 award; and the co-editor of *The Connected Past: Challenges to Network Studies in Archaeology and History* (Oxford University Press, 2016), *Pilgrimage and Economy in the Ancient Mediterranean* (Brill, 2020), and *Pilgrims in Place, Pilgrims in Motion* (Aarhus University Press forthcoming). She is currently working on a monograph exploring Syrian mobility and place-making in the Roman world. She is the co-director of the Taseli-Karaman Archaeological Project in Cilicia, Turkey, and of the Roman Avebury Landscape Project in the United Kingdom.

Megan Daniels is an Assistant Professor of Greek Material Culture at the University of British Columbia. Her interests focus on cultural interactions in the eastern Mediterranean in the Late Bronze and Iron Ages. She is currently completing a monograph on the shared ideologies of divine kingship between the Aegean and western Asia through the figure of the Queen of Heaven. Her further interests include interdisciplinary approaches to ancient migration and the intersections of religion and economy in the ancient Mediterranean. She publishes mainly on religious syncretism in the contexts of economic and political expansion in the Mediterranean, and is also currently preparing publications on pottery from sites in Greece and Tunisia.

Esther Lewis held an Arts and Humanities Research Council studentship as part of the Midlands 3 Cities Doctoral Training Partnership. She was awarded her PhD in late medieval history in 2020 from the University of Nottingham, which focused on religious life in the fifteenth century. Her research interests include pious practice in late medieval England, social networks in urban settings, and questions of belonging and community.

Kilian Mallon is a Dean's Postdoctoral Fellow and Lecturer in Classics at Stanford University. A native of Dublin, Ireland, he completed his PhD in Classical Archaeology at Stanford University in 2019. He has worked on numerous archaeological projects in Spain, Greece, Italy, and Turkey. His research focuses on the social and political history of the Roman world through legal, historical, and archaeological sources.

John Mooring studied Ancient History and Ancient Studies at the Vrije Universiteit Amsterdam. He did his major in Hittite language and culture and in Museology. His research interests are the interaction and influence between cultures, and the classical past and its heritage. Currently, he is an external PhD Candidate at the Vrije Universiteit Amsterdam, working on the role of trust and integrity during the transition to fiduciary money. He is an associated PhD student of OIKOS, the National Research School in Classical Studies in the Netherlands.

Joanna C. Mundy is a Digital Projects Specialist, at the Emory Center for Digital Scholarship, Emory University. Joanna Mundy received her PhD from Emory University in Roman and Greek Art History in 2018. Her research explores the built domestic environment in ancient urban houses in Rome in the first to fifth century CE. She works on digital scholarship projects in database design, network analysis, and digital exhibitions.

Kevin Stoba received his doctorate from the University of Liverpool in 2022. His research interests include small-group religious activity in the Roman empire; ancient interactions with religious imagery; the psychophysiological and emotional aspects of ritual behaviour; the transmission of ideas and practices around the ancient world; and the impact of individuals, organisations, and institutions on religious forms. He is especially interested in inter- and multidisciplinary studies involving statistical or network analysis and cognitive approaches.

Rebecca Sweetman is a Professor of Ancient History and Archaeology at the University of St Andrews. Her research focuses on the archaeology of the Aegean in the Roman and Late Antique periods, and she is particularly interested in processes and impacts of change across the diachronic range. As such, she works on networks, resilience, and Christianisation processes. Having worked on mainland Greece and Crete, she has spent the last few years working on the Cyclades.

Christina G. Williamson is an Assistant Professor at the University of Groningen in the Netherlands. Her research crosses the boundaries of historical, archaeological, and spatial studies, with a focus on religious and urban dynamics. She is the author of *Urban Rituals in Sacred Landscapes in Hellenistic Asia Minor* (Leiden, 2021), and is co-director of the project 'Connecting the Greeks. Multi-scalar Festival Networks in the Hellenistic World' (2019–2023).

Acknowledgements

First and foremost, my thanks go to the contributors to this volume, for sharing with me their excellent work, for their enthusiasm for the project, and their support through the process of putting the volume together. I am very grateful to Rob Latham, Heidi Lowther, Manas Roy, Meeta Singh, the Routledge editorial board, and an anonymous reader for helpful advice and suggestions for improving the manuscript.

Anna Collar
Southampton, November 2021

1 Strong ties, social networks, and the diffusion of new ideas

Who do you trust?

Anna Collar

Introduction: the echo chamber

Donald Trump's victory in the US presidential elections in 2016 and the vote for Brexit by 17.4 million people in the United Kingdom in the same year are just two recent examples which demonstrate the unprecedented power of online social networks to effect political and ideological change. The role of social media was crucial in both these seismic political events, and the psychological and emotional micro-manipulation of individuals through social media channels has been revealed as the new, dark side of politics through the exposure of the practices of British consultancy firm Cambridge Analytica and their even shadier clients.[1] The rise of Facebook, Twitter, and Instagram has fundamentally shifted the way that politics is done, but also the way that human social life is done: we have moved into an era of global online social networks that have the capacity to incorporate widely geographically separate individuals into close-knit groups and communities that bolster and support them and their ideological stance on the world, while the companies that operate the system simultaneously gather and manipulate the rich data freely given by individuals into agglomerated information about the ideological proclivities of different social groups. From such a combination comes the power to market things—and more importantly, ideas—to certain individuals, through emotion: drawing on both an individual's feeling of belonging in a group of like-minded people and on the fears they have about the boundaries of those groups. We are witness to and enmired in our own personal echo chambers, whether this concerns fears about vaccinations, trans rights, or imminent ecosystem collapse.

In network terms, this is *homophily*, the tendency for like to attract like. We live in times dominated by homophilic online social networks, and by the powerful emotional perception of the edges of those networks—where 'our' group comes into contact, and conflict, with 'their' group. Such issues around group behaviour have always been at the heart of the study of human history; however, it is only in recent years that archaeologists and historians have had the technological capacity to analyse intra- and inter-group

DOI: 10.4324/9780429429217-1

behaviours as forming *through* the structures and dynamics of interconnected social networks. Partly as a result of the introduction of social media into our daily lives, archaeology and history have seen a huge increase in networks as part of the vocabulary used to talk about and think through the study of the past. Over the last fifteen years, network methods and language have increased in popularity and scope in archaeological and historical applications,[2] making network approaches to the past one of the strongest emerging research areas in our disciplines. Subjects of study range widely from the interactions of prehistoric Kuril islanders in the north of Japan to the Pueblo cultures of the American Southwest; and from the material remains of the Epipalaeolithic in the Near East to those of the late precolonial period in the Caribbean.[3]

As a central premise, network analysis is concerned with how entities relate to other entities. Such entities are most often called nodes, which are connected via edges, ties, or relationships. What is classified as a node ranges from brain cells to individual people to types of artefact to households to nation states and everything in between; and the ties that connect them range from neural, marriage, type of material, consumption practices to trade, and many more.[4] Archaeology and history have a more limited range of material than social science, but there is still considerable diversity of data under study using network methods by archaeologists and historians. This diversity reflects also the diversity of scholars with different disciplinary backgrounds: some scholarship has drawn on physics and mathematics to push forward the side of theory that derives from complexity science;[5] others have focused on the power of computing and agent-based models (ABM) to develop approaches which work to recreate possible past network scenarios *in silico*, and make inferences from similarities between their archaeological dataset and the computer-generated patterns.[6] Others have been inspired by the work of Bruno Latour in particular, adopting actor-network theory (ANT) to challenge the way we see relations between people, things, and institutions and to re-envision the world as created through these relations rather than existing *a priori*.[7] And others have derived their methods from sociology and social network analysis (SNA), where the relationships between people and how these relate to social structures are of primary importance. This last approach has perhaps been most effective for those studies with rich historical datasets, such as bodies of letters or papyrological data,[8] which have had the capacity to examine past social networks between individuals.

Although there is diversity in what is studied, most archaeological uses of network methods rely on similarities between material culture assemblages as proxies for the strength or frequency of social relationships, such as trade, thus elucidating mostly regional interconnections between sites.[9] Within this body of work and across network science more broadly, the 'small world network'[10] has had a huge impact: a network structure featuring strong social groups or clusters that are linked together by random 'weak ties'.[11] This network structure, and specifically the presence of these random weak ties between the strongly bound clusters, enables the easy transmission of certain kinds of information between separate groups, or clusters. These weak ties are therefore described as possessing strength, and the small world network has been observed in many different situations across time and space—ranging from the colonising world of the Archaic Greeks to the spread of disease epidemics in modern America.

However, there are issues with 'complex' information transfer—such as new religious beliefs, behaviours, or norms—across such weak social ties, precisely because of their weak social integration within a group structure. Weak ties possess little emotional strength, little power to persuade. For complex information transfer that requires trust or reciprocity in order to be accepted and spread, we need to look instead at the social ties that carry emotion—whether this is romantic or familial love, friendship, respect, admiration, or fear—and which bind individuals into meaningful relationships and communities: the strong ties that form human beings' closest and most important social bonds of family, friendship, and close association through geographical neighbourhood or occupation. Strong ties are central to our social understanding of ourselves. Today, we can have strong tie relationships that are global in scale—for example, family members on the other side of the world, and technology that enables us to maintain those relationships regularly—but in the past, strong tie relations were more localised. Of course, a sense of community or group identity was possible across longer distances, with some aspects of strong emotional social ties maintained by repeated trade or mobilities, the sending of letters, and by religious practice,[12] but at a general level, attachments to place, social norms and behaviours, language, profession, and so on which help to form and maintain strong tie clusters will have been more local in character.

How then, did new ideas or innovations that require introduction through trusted strong tie social bonds travel beyond the local to cross geographical distance? Because the transmission of complex behaviours across great distances can be seen materially in the archaeological record, in the case of certain aspects of 'globalisation',[13] or the adoption of new ways of doing things, such as technological innovations in the production of pottery or metallurgy, or the spread of new religious beliefs. Weak ties that connect local clusters are the obvious answer in network terms, but they cannot provide the whole answer in social terms. The chapters gathered together in this volume showcase the importance of looking beyond weak ties to focus on the role that strong social ties have in the transmission of complex information which, in order to be accepted and lead to social change, require individuals to be bound together by relationships that are structured through mutual trust, memory, and reciprocity. They highlight the vital role of places of gathering in the process of information transmission, in this case sanctuaries; the role of narrative and storytelling in creating a sense of community even across geographical space; and the role and control of social systems in order to facilitate or stifle new information transfer.

The strength of weak ties and the small world

To understand the importance of strong social ties more fully, we must start with weak ties, and why they have been seen as so socially powerful: why they are understood as structurally strong. Mark Granovetter coined the term 'the strength of weak ties' in the title of his seminal paper,[14] because sociologically, people's relationships or social ties are characterised by their strength or weakness: where strength is ascribed to close, repeated, and regular contact between individuals, with whom they form local groups or clusters; and conversely, weakness is ascribed to acquaintances or people that are not so well known, acquaintances who are part of different social groups. Strong ties exist between dense groups of friends and family, which can also include geographical neighbours and colleagues, binding them into a reciprocal social network, where social cohesion is marked by trust, repetition of interaction, shared memories, and shared emotions. In sociological terms, these individuals have been argued as forming 'closed triads', where all members of the triad know each other.[15]

By contrast, a 'weak tie' in a social network describes the connection that exists between people who are otherwise integrated into *different* tightly knit social networks, and despite the fact that they usually represent a relationship between the two people that is infrequent, casual, or temporary, research has shown time and again how effective these relationships are at spreading certain kinds of information. This is because the network structure of the tightly knit strong tie social groups of which people are generally part is localised: the people with whom we interact most are likely to have access to the same kind of information as ourselves.[16] A person who is a weak tie connector to another social cluster on the network is able to transfer information from their social group into our own. In network terms, these weak tie connections between different strong tie social clusters have a powerful cumulative effect on the structure of the network as a whole: they connect the separate, localised strong tie social groups together and enable information to transfer from one part of the network to the other with relative ease. These ties, although relationally weak, are structurally strong.[17] This phenomenon has been described as the 'small world'. The weak ties between clusters in a small world network enable local information to have global, that is, full network, reach. It is the weak ties which enable this network quality of a small world: they bring the capacity for 'short path length', the network term which describes the ability to directly access other disparate clusters.

The power of weak ties has been amply demonstrated through real-world examples. The spread of disease is a particularly good case in point, where the arrival of one contagious individual into a strongly tied social group can cause the infection of pretty much everybody. This is a 'simple' contagion, that is, the process of exposure, infection, and spread is not dependent on anybody having an opinion about the process, or a choice about whether they are infected. They just catch the virus (though wearing a mask and staying socially distant helps). Granovetter's example was more sophisticated, exploring instead the way in which information about the labour market was transferred, and where individuals in a group with fewer weak social ties were seen to be at a disadvantage in terms of knowing about job openings at the right moment. If we are part of a very strong, localised social network without weak tie connections to other people in other places or groups, then we are effectively insulated against new information, whether this be about jobs, fashions, news, or new technological or religious innovations.

Weak ties, and the small-world networks that they are instrumental in creating, have been important in the recent iterations of network analysis in archaeology and history, for example, in Irad Malkin's vision of the interconnected Mediterranean world of Archaic Greece;[18] or the distribution of Clovis lithic points in North America.[19] In these cases, the clustered communities made up of strong ties (often seen as fixed) are changed by agents from outside their group, with connectivity and change coming from weak ties acting as active *brokers* between clusters, bridging the 'structural holes'[20] that otherwise create disconnectedness across the network and impede the flow of information across it. Such people in brokerage positions across structural holes are noted by Burt to be able to manipulate and control the network for their own benefit.[21] Furthermore, Burt noted that the benefits of taking a position of brokerage between clusters on a network were maximised when that network involved strong relationships and trusted contacts.[22] Research has also shown that the strength of weak ties thesis has been overstated even with regard to novel information, and that stronger, more 'redundant' ties possess more 'bandwidth': that is, because they are more frequent contacts, they pass on larger and richer flows of information, of which a higher proportion will be new.[23]

Innovation, persuasion, strong ties, and community

So, it seems that there are a number of issues with the assumptions behind the discovery of small world networks in the past, including the idea that technological or social innovations always come from beyond a localised group or area, the equality of nodes, and even the very idea of a totally random connection. The way human beings work socially means that total equality in a relationship is quite unlikely, as are totally random ties—and Burt has shown that such bridging ties are in any case more effective if they connect into strong and trusted relationships. These asymmetries and imbalances in real human relationships need further consideration in the transfer of information, and innovation. Innovations in particular are an interesting case here. Job information or news can be passively heard, without requiring anything of the individual listening to this new information. Innovations are different, because they require the active adoption by the individual concerned, and the likelihood of a person's adoption of an innovation is very different from their vulnerability to infectious disease. Although there are different types of innovation—technological, material, ideological—general observations about the social character of their adoption remain: successful diffusion relies on trust; and trust requires strong social ties.[24]

The diffusion of innovation was first fully explored sociologically by Everett Rogers in the 1960s, when he outlined five major types of human response to innovations.[25] He observed broad characteristics of each, relating to social qualities such as status, financial position, and age. Innovators, those who actually do the innovating, are defined as being generally of higher social status, financially well off—which enables them to absorb losses associated with innovations, and are socially connected to 'scientists' which gives them access to new information. The next category of people defined by Rogers were those that are quick to take up an innovation, classed as Early Adopters. These people were defined again by their higher social status, their financial independence, and higher level of education. All these social qualities enable this group to shape the opinion of others regarding innovations; these people are the opinion leaders in society. The third category identified by Rogers was that of the Early Majority, who were observed to adopt innovations considerably later; these people are not opinion leaders, although they are of higher social status. The fourth category he outlined is the Late Majority, who are generally characterised by scepticism, lower social status, and less financial liquidity. Their main social contacts are with others in the Late Majority. The final category of people as discussed by Rogers were classed as Laggards. The key characteristics of this group are connected to their age, their lack of status, their lack of financial liquidity, and their aversion to new ideas, change agents, and the attachment to tradition.

These categorisations represent rather a blunt tool: diffusion processes are highly complex and may differ among different people depending on the nature of the innovation itself and the perception of it. In addition, there is an underlying assumption here that all innovations are positive, again, this is in the eye of the beholder. However, the essential premise of Rogers' work is that the process of diffusion and adoption of innovation is related to our social context and to the behaviour of those around us: if members of our social group that we admire, trust, or wish to emulate have adopted an innovation, then we ourselves are more likely to take up that innovation too. This is related to deeply social factors, such as reluctance to deviate from the social norm, respect for individuals with perceived social power or who are regarded as higher status, and attachment to 'the way things have always been done'. It is here that the sociological observations surrounding the diffusion of innovations start to intersect with the science of networks: social, crowd behaviour is also intimately bound up with connectivity.

Weak ties have been central to thinking about this connectivity and the spread of information or innovations across networks, and there is a continuing perception, for example, that weak ties are important in the spread of religious ideas,[26] but it is worth considering the *kinds* of information or the *content* of an innovation that might be adopted via a weak tie bridge to another part of the network, rather than across a trusted strong tie. It is also important to think about what was being transmitted and why, and what these long-distance connections enabled, and at what cost.[27] Recent studies, such as that by Valentine Roux et al.,[28] have nuanced both Rogers' categories and the links made by weak ties by accounting for expertise as part of this process of adoption. Their focus was on technological innovation, their case study was on the adoption of kiln technology, and their aim was to characterise conditions which led to technological borrowing.[29] They explore the likelihood of weak ties acting as 'bridges' for new information, a condition which they discovered to rest on the perceived expertise and social leadership possessed by the person adopting the innovation *from* that weak tie contact.

In addition to this nuancing of the early adopter category, innovation transmission is also marked by the notions of horizontal and vertical transmission. Horizontal transmission refers to the peer-to-peer diffusion process, often imitative in character, that is, an innovation will spread through observation, emulation, and copying; by contrast, vertical transmission requires depth of knowledge and experience, for instance through conditioned learning, the cultural traditions inherited from family and lived environment, usefully characterised as the individual's *habitus*.[30]

An individual's socio-religious and socio-political environment is generally vertically inherited, deeply ingrained in the habits and practices of family and neighbourhood, everyday existence, perception, and worldview. This is not to suggest that it cannot be changed, but that in general, a vertically inherited aspect of social identity will be less vulnerable to innovations brought by 'random' change agents—weak ties—in the social network. This is in part because in order to enact religious or ideological change, there needs to be a strong level of trust between the change agent bringing or advocating the innovation, and the individual doing the changing. Therefore, when social, political, or ideological change is introduced via such trusted means, it can have a very powerful effect indeed: for example, recent work has sought to examine the spread of religious violence in the Radical Reformation in Europe, modelling both the horizontal axis of transmission (travelling 'firebrand' preachers) and the vertical axis of transmission (through congregational lineages), and has shown that in this case, the vertical axis exerted more social power in disseminating the idea of religious violence.[31]

The ties that bind: the strength of strong ties

And so we return to strong ties. When it comes to thinking through changes in individual beliefs and practices which have profound social impacts—such as religious conversion or political radicalisation—the 'strength of weak ties' is no longer appropriate: these casual acquaintances or geographically distant friends do not exert the necessary social power to enable the adoption of new religious, political, or social ideas. It is certainly possible that a charismatic individual with a mission to convert and persuade may be able to influence people—the apostle Paul offers an example of the power of such methods—but, although he was technically a 'weak tie',[32] Paul operated within powerfully strong tie contexts. The people to whom he evangelised saw him as part of their strong tie social network: he stayed with them for long periods of time, worshipped with them, led them, motivated them, and communicated with them via letters. The weeping, embracing, and kissing as Paul leaves the Ephesians in Acts 20:36 underscore the strength of the emotional bond that the communities felt with Paul: Paul was felt by the Christian converts of Ephesus to be a strong tie.[33]

Strong ties feature certain important aspects: emotional intensity, the time spent with that other person, the intimacy (describing mutual confiding), and the reciprocity between the two people.[34] Memory and shared experience also have an important role in the creation and maintenance of strong tie bonds. However, strong ties are not all the same, and they are not simple, unidimensional connections.[35] There are multiple different ways in which a social tie can be strong—related to *frequency* of contact, *directionality* of the tie (i.e. although a bond between a mother and child may be strong, there is directionality involved in terms of certain aspects, such as duty of care; moreover, this directionality will also shift across a lifespan), and the *nature* of their strength, that is, whether they are multivalent—that is, they might be related, work together, and live nearby—or whether they represent a simple relationship, where they are only connected in one capacity (e.g. employer/employee).[36] Brashears and Quintane have sought to nuance the discussion of strength in social ties, taking on Aral and Van Alstyne's notion of 'bandwidth' to describe the quantity and quality of the relationship in question. They have also questioned the assumption of *redundancy* involved in strong tie (or high bandwidth) relationships, that is, the 'closed triad' quality that has long been attributed to strong ties, showing that the *capacity* of a tie to transmit new information is more important, and the *frequency* of the relationship is also vital to the spread of new ideas.[37]

Mathematically, however, strong ties are weak. In network terms, they do not have the capacity to spread information quickly across a whole network structure. Their path length (their ability to reach into neighbouring clusters) is very long, meaning that they cannot easily access new groups or new information—a point emphasised by Sweetman in her consideration of

Christianisation in the Mediterranean in this volume. However, in real human social terms, strong ties possess the capacity to persuade and to influence. Experiments have shown that when the weak ties in a test case social network were removed, the network shrank a little but did not fully disconnect: the strong ties continued to hold the network together. The authors of this paper concluded that 'a high-fidelity strong tie social network' of overlapping clusters was almost as efficient as a small world network at transferring information,[38] an observation also echoed by Damon Centola's work.[39]

So although weak ties connecting the small world present the most efficient method of spreading certain kinds of information or diseases, it is clear too that a network of overlapping clusters built on the fidelity of strong ties is more important as a model for thinking about the spread of complex new information. With the nuances regarding the nature of strong ties in mind, finding various kinds of strong ties in the past, the ties that bind communities into trusting relationships, will allow us to begin to understand how fundamental alterations in religious, political, and ideological behaviour spread through social networks and contribute to broader-scale systemic changes.

Wide bridges and the places, stories, and systems of information diffusion

Complexity is the key here. Centola and Macy argued that where the new idea or information carries behavioural change which involves social risk or cost, or is controversial, then adoption is not simple, and 'requires independent affirmation or reinforcement from various sources'.[40] 'Complex', in other words, refers to social innovations such as new religious beliefs, ideologies, or new social norms. Centola and Macy have adopted the language of disease—describing these as 'complex contagions'—to suggest that in order for behavioural innovations such as a new system of religious belief to be accepted and become the norm across a collective group, they require both trust in the source of the innovation—a strong tie to introduce them—and *also* need to be repeated, ideally from multiple sources in order to receive reinforcement and take hold.[41] Social networks based on trusted strong ties are *required* for complex contagion, to enable risky behavioural changes: they act to create the situation which Centola and Macy have termed 'wide bridges'. Multiple sources of exposure to the innovation are needed in order to support certain important social factors in the process of adoption, including strategic complementarity—that is, the more that adopt an innovation, the more beneficial it is; the credibility of the innovation, which is enhanced as it is introduced across more and more trusted strong ties; the legitimacy of an innovation—as more people participate in the new behaviour or adhere to the innovative idea, the idea moves away from a position of 'deviance' towards one of acceptance and legitimacy; and, as more adopt an idea, emotional contagion is also increased.[42] In other words, strong ties are what make 'complex contagions' possible.

The purpose of this volume is to bring together scholars using network methods to explore these more ephemeral aspects of the transmission of complex new ideas about religion, politics, or ideology in the past. Sometimes, these will be transmitted hand in hand with other more tangible aspects, such as technological changes, other times, perhaps not. Indeed, whether an idea is successful may in part depend on the connection the idea has with other technological elements, or with particular places of gathering and contact, or ideas may be more likely to take root if they are passed across multiple ties—in Centola and Macy's terms, 'wide bridges'. Can we identify occasions when ideas are transmitted in this way, and others when they have been passed through narrower ties? Also relevant to the discussion here will be the reification of 'ideas' prior to their analysis. *A priori* categorisation is an acknowledged problem with using network analysis methods,[43] and it may be helpful to think about the concurrent formation of ideas through the formation (or temporary stabilisation) of a network, both idea and connectivity in a dynamic process of becoming.

The scholars gathered here address what archaeological and documentary signatures we might find for religious and ideological innovation and change, how these might be tied into other technological changes or particular social networks, and how network methods—digital, methodological, and metaphorical—can help us in the search. Although there is a general focus on the ancient Mediterranean—in part because this is my own area of specialist knowledge and represents in part my own professional social network, other areas are represented, including South India and the United Kingdom; however, there are of course places, bodies of material, and approaches that are missing and through which useful further dialogue could be generated. The temporal range of the contributions stretch from the early Iron Age through to the fifteenth century CE; again, there are limitations in this timespan and interesting observations and comparisons could be made with material from before and after this period—however, the chapters do relate to each other usefully and the volume remains cohesive. It is my view that bringing together case studies from different regions and time periods introduces comparative aspects which helps to strengthen the usefulness of the volume to other sectors of scholarship. In addition, there is a mix presented here between case studies which make use of formal network models to demonstrate and illustrate their arguments, and others which do not. By blending these different approaches, we can highlight the usefulness of visualisations, but also the philosophical strength of the network heuristic.

The book is divided into three sections, each of which addresses a different element of the interface between social networks and the spread of new information or innovations, and the way we can find these in the archaeological and historical record: Part I: Sanctuaries, Part II: Storytelling, and

Part III: Systems. The chapters can all be read as standalone offerings, but by grouping them together in this way, the chapters all also speak to related themes. There is also a chronological flow here: we begin with the earliest case studies, in the Iron Age and Archaic period, and finish with chapters which address various different aspects of Christianity. This progression, although unintentional, is also useful to think through.

Sanctuaries: the places that bind

Part I comprises three chapters, all of which explore the power of sanctuaries to bring people together into temporary communities, and the power of these gatherings to generate strong tie connections even (especially) across great geographical space. The physical setting of the sanctuary in the ancient Greek world, with its neutral status and inviolability, enabled people from across the Mediterranean to participate in festivals—either at official, almost diplomatic levels as we see in the cases of *theoria* and *proxenia* in the Greek world, as a competitor in sacred games, or as a more ordinary general participant and observer of the festivities. The three chapters gathered here think through the role of sanctuaries as places of gathering, the exchange of ideas and of their wider dissemination: in the first, Megan Daniels examines the transmission of the image of standing female nudes across the Iron Age Aegean as a marker of participation in elite affiliation networks, and how this changes over time; in the second, John Mooring looks at the role that panhellenic sanctuaries played in the diffusion of the innovation of coinage in the Archaic Greek world and the relationship this has with elite social networks; and in the third, Sandra Blakely and Joanna Mundy explore the different social networks present at the cult of the Great Gods on Samothrace and their materialisation.

What is clear from all these chapters is that sanctuaries in the ancient Mediterranean were more than just places of worship. These locations were profoundly cross-cultural, and the participation of people from across Greece, Italy, and Asia Minor at events at these sanctuaries enabled the sharing and transmission of a number of new ideas. In her chapter, Megan Daniels examines the rise and fall in popularity of one particular common image, that of the nude female in a standing position, which formed part of the votive repertoire of sanctuaries across the east Mediterranean between c. 800 and 550 BCE. Traditionally, scholarship has seen this nude female imagery either as erotic, or as a symbol of fertility. Daniels demonstrates how both these interpretations are insufficient, and proposes instead that the nude female be seen through the lens of the shared religious and political world of the Iron Age Mediterranean, with a specific semiotic function relating to this interconnected setting. Sanctuaries offered the medium for this semiotic transfer. Daniels argues that sanctuaries, as places of elite meeting, exchange, and situated learning, were nodes which acted as 'wide bridges' for the transfer of ideas and materials that created and maintained elite social networks and identity of the Iron Age east Mediterranean. These nodes were connected by new forms of human mobility, which created strong ties within the Greek world, but also acted as channels for participants from the non-Greek world to engage and transfer new ideas.

Crucial to this argument is the role of dedicated objects within these sanctuaries. Many of these objects have long and complex biographies and far-flung points of origin, but they can also be seen as a demonstration of the multiple overlapping networks that existed in the Iron Age, formed through these multiple kinds of mobility, ranging from trade and diplomacy to military conquest, mercenary service, and marriage contracts. These objects embody and symbolise these points of origin and complex biographies, and so also come to represent these conceptual networks.

Although the image of the nude female had been widespread in the kingly courts of the Late Bronze Age Near East, it only begins to proliferate through the Greek world—as votives in internationally well-connected 'Orientalising' sanctuaries dedicated to female deities and also occasionally in graves—during the period of extremely high mobility and new encounter from around 800 BCE onwards. More than just an exotic import or a representation of a particular deity, these votives, which most often took the form of locally made terracotta figurines, indicate a new, popularised representation with particular meaning relating to ideologies of power and protection.

The nude female is shown by Daniels to represent an aspiration towards a connective kin relationship between elites and the divine manifest through the legitimising power of divine breastmilk. The proliferation of the image of the nude standing female across the cosmopolitan sanctuaries of the Iron Age Greek world demonstrates the power of shared cultic rituals and beliefs that were performed and learned in such places, and articulates the connective power of the sanctuary as a wide bridge: these are places which bind communities, however widely dispersed, into communities of practice, communities of behaviour, and communities of concept. As such, when the nude standing female disappears from the repertoire during the sixth century, this can be interpreted not simply as new artistic interests in other subjects, but rather as representative of changing attitudes towards human relationships with the divine, bound up in larger shifts in political and social life in Greece towards something more community-governed and *polis*-based.

Operating across a similar time period as that explored by Daniels, the transmission of the innovation of coinage within the context of panhellenic sanctuaries across the wider Archaic Greek world is the topic explored by John Mooring in his chapter. The invention of coinage represents a profound abstraction of exchange, and was a fundamental shift in the way transactions could be made, with far-reaching implications. The adoption of coinage was a complex process which involves characteristics of the innovation itself, such as the advantages it offers or the compatibility of the innovation regarding the social setting, but which also requires the consideration of the communication channels, the social ties through which such an innovation can progress. This is a two-step process, first involving the transfer of knowledge about the innovation, and second, the persuasion of people to take it up. Mooring argues that the Archaic Greek world, with homophilous connectivity provided by colonisation and mobility, language and cultural background, was a low-density network and an ideal situation in which weak ties could be powerful in spreading new ideas, as Granovetter has shown. These weak ties are represented by elites participating in festivals and gatherings at panhellenic sanctuaries.

It is the adoption of the innovation which is important though—the decision made in a number of different places to take on coinage and to use it, the persuasion that the innovation was useful, beneficial, and important. Persuasion requires two things: it requires the person bringing knowledge of the innovation to be an opinion leader with strong social network contacts, interpersonal channels through which they can persuade; and it requires trust. Trust in another person entails the recognition that we know how the other is likely to behave, that they are not going to cheat us—that the metal the coinage is made from is worth what they say it is, what the issuing city says it is. Trust is facilitated in small, close-knit and homophilous communities with shared norms and values, and in the small Greek communities of the Archaic period, which, despite being independent politically, shared common values and culture with each other, there was a comparatively high level of trust which enabled the persuasion needed for coinage to spread. Mooring argues that the places where this was possible were the international, panhellenic sanctuaries. He demonstrates through an analysis of official, elite social ties in the form of epigraphic records of *theorodokoi* and *proxenoi* to the major festivals of Hera at Argos, Asklepios at Epidauros, the Pythian games at Delphi, and the Nemean games, that 60% of the cities which engaged in these diplomatic activities were also minting coins in the Archaic period. He supports this with evidence provided by analysing the mints from which coins in Archaic hoards came: showing that in those hoards, 76% of the coins were from cities which also possessed an elite diplomatic presence at panhellenic festivals: the relationship between the spread of the innovation of coinage and participation of elites in international religious gatherings is abundantly clear. The power of these elites as local opinion leaders with strong ties in their local community enabled the innovation to be accepted and take hold.

Beyond the brave new world forged by migrating Greeks during the Iron Age and Archaic period in the east Mediterranean, and the key role that sanctuaries clearly played at this period in bringing these dispersed communities together and transferring new ideas, sanctuaries also acted as binders in other ways and throughout classical antiquity. In the final chapter of this first section, Sandra Blakely and Joanna Mundy focus in on the mysteries that operated at one particular sanctuary, that of the Great Gods on Samothrace. Becoming an initiate into the mysteries at the sanctuary at Samothrace—about which we know very little, as initiates were sworn to secrecy—afforded a number of benefits, both social and practical: the prestige of joining an elite group of *mystai* that included recent and mythical royalty; perhaps more pressing, given Samothrace's position, safety at sea; and the protection of a mutually non-aggressive network of city-states, where participation in the mysteries acted as a symbolic tie that underwrote social behaviour and cooperation.

They make use of the extensive epigraphic record for participation in the mysteries, largely dating between the second century BCE and the second century CE, although the mysteries themselves operated for far longer, at least from the Archaic period. Although the inscriptions contain little information about names, deities, narrative, or iconography, Blakely and Mundy demonstrate that the dynamics of the diffusion of knowledge about the mysteries (and associated political and social connections) can nevertheless be usefully illuminated using a social network analysis approach. Through the city-names that appear in inscriptions, the cities become the actors on the network and the individuals (mostly unknown) act as proxies for their home town. By assessing the strength of the relationships that the inscriptions record—initiations, the visits of *theoroi*, and grants of *proxenia*, which offer insights into strong ties of differing types, including contractual force, affective intensity, family lines, and the cultural *habitus* of secrecy—the authors are able to bring a focus on betweenness centrality: the nodes in the network that operated as bottlenecks or bridges for the flow of information. Places which demonstrate high betweenness are argued to have a significant role in the diffusion of the rites beyond the island, and to offer places where it is possible to see how the Samothracian rites were used for political and social ends. Such nodes offer insight into complex contagion, in which the diffusion of ideas is enabled by alignment with existing cultural traditions. Thera and Rome emerge from their analyses as bridges between other cities and Samothrace; Rhodes as particularly important as a bottleneck: demonstrating, through the presence of a number of different kinds of institutions on the island, the adaptation and integration of the cult of the Great Gods into the political and civic life of the cities and elites of Rhodes.

Storytelling: the narratives that bind

The second section of this book assesses the creative power of strong tie networks to bind people and places together through storytelling, orality, and the processes of sharing information. Storytelling and narrative in this case means a number of things and is expressed through a number of means—we see in Christina Williamson's chapter, which flows naturally on from the discussion of sanctuaries, an exploration of the transmission of common knowledge across Hellenistic festival networks; in Kevin Stoba's, the discussion is focused on the material and iconographic repertoire of the cult of Mithras in the Roman world, and what this can tell us about different regional practices; and in Nathanael Andrade's chapter, the analysis centres on texts and the role of social networks in the transformation of oral traditions about Christian saints across South India. Andrade's chapter also brings us towards the final section of the book, flowing into a discussion of systems that were operationalised through strong tie network connections.

As we have seen, the role of sanctuaries in the transfer of ideas is demonstrably crucial. Christina Williamson, whose work also acts as the bridging chapter between the section on sanctuaries and the section on storytelling, explores the strong social ties which were reiterated and renewed at sanctuaries and festivals, in particular the role of interstate diplomats and the inter-city bonding that occurred in these contexts. She argues that festivals—ritual gatherings—are the prime movers in the creation of communities, cities even, and are central in sustaining narratives of belonging. In particular, in the wake of Alexander's conquests, new or renewed festivals act as a fundamental marker of the interconnectivity of the Hellenistic period, but more than this, they offered a vital channel through which narratives of interlinkedness were performed. Her discussion centres on the creation of what she has termed 'portable communities' at festivals: the evidence for participation at Hellenistic festivals challenges the dominant ideas about transmission of information through weak ties, as these festival attendees were carefully invited or delegated, acting to reinforce an inter-city network already in place rather than filling any 'structural holes'.

In these festival spaces, Williamson argues, it was strong tie bonds which were reinforced, through the particular and unique quality of 'festival time' in these temporary gatherings: 'festival time' elided the present with the deep, heroic past, and acted to bring participants together through perceived kinship that was rooted in mythological ancestry. These narratives of community belonging were cemented by participation in collective ritual: music, sacrifice, feasting, and public awards suspend everyday life and create, in Jan Assman's words, 'islands of time'.[44] The powerful collective experiences and memories generated at festivals both enabled the spread of and also manifest the common knowledge and normative social values that were shared across the Hellenistic world, so demonstrating the globalised nature of that world.

The creation of community through collective experience and memory is also at the heart of Kevin Stoba's chapter exploring the subtle differences in the iconography of Mithras in the Roman world. On the surface, the bull-killing scene famous in Mithraism, the tauroctony, is universally understood as the key image for the cult, although debate remains about how it should be interpreted. However, there are iconographic variations in the tauroctony which are meaningful in terms of transmission of ideas, and which can be used to shed light on localised influences, principles, and social backgrounds of the believers. Stoba's approach arises from the fact that in formal network analysis, all evidence can be treated equally, meaning that there is no starting assumption of a standardised tauroctony or deviation from it; and as such, each piece of evidence that he compiles and details in the attributes database creates its own unique profile. Stoba's technical work reveals that the tendency for scholars to allocate tauroctonies to regional groups is not borne out by evidence of iconographic similarity.

He concludes this through two major analyses of his data. The first is based around homophily, the tendency in this case of a node to be tied with others that share a significant attribute on the tauroctony. He found that homophilous relationships were much less likely at a regional or provincial level, but significantly likely at a site or settlement level, indicative of the power of strong ties at a local level to foster conformity. This discovery also enables Stoba to suggest that certain groups of iconographic feature spread together, which might indicate different strains of Mithraism that can be extracted from the data, and which also have important connections to certain groups of people, ritual practices, and versions of sacred myths.

The second analysis used only the very strongest ties of similarity between tauroctonies, highlighting iconographic similarities especially across objects of the same medium (e.g. statues, all from Rome) in part because of the difficulties of expressing multiple iconographic elements in certain forms. However, the groupings are nevertheless revealing, with the similarities between the statues in Rome, for example, suggesting a high level of conservatism and conformity in this group of cult objects, some of which represent the earlier known images of Mithras. Stoba argues that the similarity between these representations and the continuing conservatism in Rome supports the proposition that the cult of Mithras originated in Rome and spread from there. The analysis identifies other strong site-based groupings at Sarmizegetusa, Carnuntum, and Alba Iulia, suggestive of localised cult innovations and cult narratives which bound these communities together, but which were able to spread across a strong-tie network of Mithraic adherents.

Geographical proximity can, of course, be crucial to the spread of ideas through strong social ties. However, the 'release from proximity' afforded by objects[45] which enabled community identities and ways of doing to be generated and to cohere even over long distances is also enabled by narratives. In the final chapter of this section, Nathanael Andrade uses divergent textual sources to reconstruct the social networks that enabled the transmission of narratives about St. Thomas across vast geographical distance, from Edessa in Osrhoene to South India.

The oral traditions surrounding the story of Thomas the Apostle and his journey through India are difficult to untangle in terms of dates and in terms of content, although it seems fairly likely that the original was composed in Edessa in the third century CE. Andrade contends that situating the *Acts of Thomas* narratives within the weak and strong ties which made up commercial, ecclesiastical, diplomatic and pilgrimage networks in operation between the Mediterranean and India allows us to consider afresh how the Thomas traditions were shaped across time and space. The *Acts of Thomas* is, in essence, a general imaginary of the apostle's journeys through and martyrdom in India; by contrast, oral traditions of Thomas Christians from South India are much more culturally specific. Andrade argues that the oral traditions of the Keralan coast are based on the general narrative of the *Acts*, but with deeper layers of regional and cultural specificity local to the area. But given their separation by 1300 years, Andrade uses networks to reflect on and identify the intermediary stages of the development and transmission of the narrative.

Key to this discussion is the narrative tradition that Thomas was, for a time, buried in India, but that his body was removed from this tomb—which remained empty—and transferred to Edessa, where the cult of his relics continued. This seems to have emerged in Thomas narratives only during the sixth century; prior to this, Edessa had been understood as the final resting place of Thomas, as seen in pilgrimage texts such as that of the fourth century pilgrim Egeria. Also during the sixth century, another crucial piece of information begins to be added to the narratives: that Thomas was martyred in a place in South India called Kalamene. Andrade contends that this is not coincidental nor a piece of historical fiction: rather, this new place-name indicates a shift in knowledge among Greek and Latin-speaking Christians in the Mediterranean about Thomas worship in South India. Andrade argues that this detail, as well as the augmentations of the Thomas narratives in the Mediterranean of an original burial place in India, and the Indian narrative of a final resting place for Thomas' relics at Edessa, illuminates knowledge transfer about sacred sites across new communication networks between Persian Christians, who had begun to settle in South India in the fifth century CE, and across commercial and pilgrimage networks which also began to operate between the Indian subcontinent and the Mediterranean.

Trust, to varying degrees, is key to the workings of these intertwined and co-dependent commercial, ecclesiastical, diplomatic, and pilgrimage networks.[46] Commercial networks in the ancient world were often based around regional origin, their participants bound into trust-based social relationships through direct personal contact, and Andrade demonstrates that these communities, of both Roman-Egyptian and Gujarati origin, were in operation across the Arabian Sea between India and the Red Sea. Ecclesiastical connectivity between Egypt, Ethiopia, Arabia, and Mesopotamia extended out across the Persian Gulf and India, crossing boundaries and exploiting these commercial connections. Diplomacy also used the social pathways laid by commercial connectivity; and on the basis of these intertwining and overlapping networks, pilgrimage routes followed—seen in both textual evidence, in coinage found in India and Sri Lanka, and in the transfer of knowledge as witnessed in the narratives about the body and burial of Thomas.

Systems: the structures that bind

In the final section of the book, we move from narratives that bound disparate communities together explicitly into the way that social systems and structures were generated through strong ties, and the ways in which these structures were used to control social behaviour. All three chapters explore the systems built by Christianity and the weak and strong ties that were used in the development and maintenance of the system—Rebecca Sweetman's discussion of how the architectural manifestation of Christianity spread across the Cyclades, Cyprus, and the coast of Asia Minor, situating this within topographical and social spaces for gathering; the examination, in Kilian Mallon's chapter, of the frameworks and boundaries of the Christian system as defined by and through Church councils and the construction of ecclesiastical law; and Esther Lewis' exploration of the transmission of the Lollard 'heresy' through the fifteenth century CE social networks in Bristol, where the use of networks destabilises what we thought was known.

That Christianity was beginning to find followers in the Mediterranean world of the Roman Empire in the first century CE is demonstrated by epigraphic and funerary evidence; but the places of gathering and worship of early Christians are largely lacking—make part partly for political and social reasons—until the Edict of Toleration and later adoption of Christianity by the Roman state in the third and fourth centuries CE. The choices made regarding the locating of new church buildings, and the processes by which Christianity spread and was physically established in these early years is the topic of Rebecca Sweetman's chapter. Sweetman argues that both horizontal and vertical transmission processes come into play in thinking about the spread of church building: the conversion and initial spread of churches operating simultaneously through horizontal peer-to-peer networks, and through vertical, generational ties to landscape, memory, and tradition. By examining the archaeological evidence for this ultimately physical process of conversion and monumentalisation in the east Mediterranean, Sweetman seeks to move beyond generalisations about how and why Christianity was so successful, or geographically deterministic suggestions that favour simple geographical proximity, towards a more topographically and temporally nuanced picture of the processes of church building—and so conversion—in particular cities.

Both weak and strong social ties are emphasised by Sweetman as playing an important role in the spread of new knowledge about Christianity and its monumentalised form. Using evidence for patronage, pilgrimage, and trade, she demonstrates that there were different network factors at play at different times during the process of Christianisation. In the earlier years of the spread of the new religion, mission, and the strong interpersonal social network bonds of tight-knit community were crucial to both the evangelists—who needed to find a place of social acceptance, that is, among family and friends, in order to evangelise—and to the evangelised: in order to convert to a new religion, the message needs to come from multiple, trusted sources. Churches, where they were built in these early years, were located in places with strong traditions of memory for the community, especially, for example, in cemeteries as at Corinth.

Once Christianity was adopted and endorsed at the level of the Roman state, the situation changed. Sweetman notes the strategically visible locations of churches at Edessa, Antioch, and Ephesus, situated either near points of entry and exit to the city, or near places with heavy footfall such as the baths or gymnasium, and argues that these city locations were known to the investors and were intended to bring in new worshippers. In other cities, such as Athens and Pergamon, Sweetman argues that the destruction of the civic core in the third century disrupted loci of memory and tradition and also caused breakdown in systems of patronage and elite investment in monument building that persisted until the fifth century, when direct imperial involvement seems to have resulted in new church building programmes. In all of these cases, Christianity began to be used to physically structure cities; but the sustaining of some places—in particular Antioch, Ephesus, and Salamis—through the regular visits by weak tie connections (pilgrims, traders, bankers, etc.) led to significantly more obvious monumentalisation processes in these locations.

With the establishment of the Christian system, the regulation of its boundaries—especially across the broad geographical swathes and political fractures of the Late Antique Mediterranean—became of paramount importance. Between the fourth and sixth centuries, bishops and Church officials drafted a complex series of laws designed to regulate the interactions between bishops, clergy, and the laity. Kilian Mallon takes this body of ecclesiastical law data and argues that it represents the maintenance and regulation of strong ties relating to the accessibility of work in churches, mobility, and the right to worship: the 'boundary work' that was necessary in order to maintain the structures of Christianity. Christian conversion is usually modelled using three elements: that of personal evangelism, that of secular privileges for Christians accorded by imperial benefactors, and that of familial, inter-generational habituation of children in Christian families. Mallon adds to this repertoire by positing that ecclesiastical regulations regarding strong ties of work and worship were fundamental in building the core of the trust-based institutional system that is the Christian Church. At this point in the development of the system, weak ties offer potentially dangerous methods for spreading doctrinal heresy and sedition—a point to which we shall return in Esther Lewis' chapter—and which needed heavy, top-down management.

Between the fourth and sixth centuries, the series of ecumenical councils and more regional Church synods met to discuss and clarify theological, doctrinal, and legal issues, in a period of time which saw the role of bishops evolving rapidly and radical changes in the social and political structures of the Roman world. The proceedings of these councils and synods were compiled, and alongside imperial documents such as the *Codex Theodosianus*, form the basis of Mallon's argument, and at this stage, are less coherent and consistent than they would later become. What is clear, however, is that the authors of these legal texts were attempting to shape the world as they wished it to be.

Canon law used the language of strong tie kin relationships to structure the responsibilities within the Church, with the bishop modelled on the *paterfamilias*. In addition to the spiritual duties of a bishop, financial management of the physical building of the church and its holdings was a key part of his role, as were judicial and arbitration responsibilities in the local society: all of these elements of the position would have required or developed a local network of strong social connections. However, because bishops were often chosen from the local property-owning elite, they continued to maintain weak tie social connections across geographical space, to other elites and through charitable acts, and with the military. These two factors, in effect, created both a close-knit, local clergy, and a broader network of socially and politically established elite church leaders. The close-knit ties of the local clergy formed the core of the moral boundary keeping of the Church and guarded against deviation from moral or theological values. In order to reduce the risk of new ideas entering Christian communities through weak tie contacts to other places, the bishops who created ecclesiastical law also placed strict controls on the mobility of their clergy, especially those of lower orders—the clerics, deacons, and monks. As well as defining 'orthodoxy', and by extension, 'heresy' (and so also sedition), the Church councils also expressly reduced possibilities for movement by these lower clerical orders; or if mobility were permitted, it was under the arbitration of the bishop. By controlling mobility, the highest officials of the Church also stifled the development and dissemination of new ideas that might damage their own positions. This self-serving approach had the effect of imposing a strong sense of localism in the late antique world, tying into secular legal developments which bound children to the professions and home towns of their fathers, restricting social mobility and creating a clergy that look very much like estate workers.

The restrictions on mobility and violation of ecclesiastical law were policed by the threat of excommunication, which severed a person's strong ties to Church and community, with profound social consequences, as well as physical violence and financial penalties. The bishops had structured the system of religious conformity through the production and control of state-sponsored fear. And ecclesiastical law rapidly extended its reach to become more heavily moralising, impinging with greater and greater severity on people's everyday lives: for example, charioteers had to give up their profession in order to become Christian, women were forbidden from spending time around men with long hair, and playing dice carried the threat of excommunication. But it was non-conformity of belief which caused the most anxiety among those who created the ecclesiastical law codes, and the excommunicated were still subject to the authority of the bishop: they were not permitted to leave the area, they could not attend another church, and a person's absence was of real social consequence. In the world of late antiquity, bishops made use of the strength of strong ties to control and oppress.

Structural oppression is also key to the subject of the final chapter of the book, in which Esther Lewis moves the discussion of the structures of control in Christianity into the medieval period, examining the turbulent political and social situation surrounding religious legitimacy and heresy in fifteenth-century Britain. Considering whether the city of Bristol was a 'hotbed of heresy', she uses social network analysis to approach the evidence anew: asking what SNA can add to the debate, whether there are differences between orthodox and heterodox networks of piety, and what the relationships were between heretics and their neighbours.

Lewis begins her investigation from the premise that heretics operated through informal and pre-existing social networks—through which they communicated new ideas, practised their faith, and discussed their beliefs—but that they also participated in mainstream orthodox religious behaviours such as attending church. The parish has generally been seen as the main defining feature of pious life in the medieval period; however, Lewis uses information from last wills and testaments to demonstrate an additional element of broader religious connectivity, through financial donations made beyond the parish to institutions in other parts of the city. These were largely guided by the topography of the town. In combination with the records of heresy trials, Lewis shows that there was a powerful correlation between new religious ideas and localised topography, suggesting that it is neighbourhood, associations, and participation in localised community that we should focus on as we try to understand the ways in which new or dissenting religious ideas or ideologies were able to grow and gain support.

Such geographically separate clusters or neighbourhoods need to be connected for the movement of new ideas to traverse longer distances and jump between disparate geographical areas. However, as both Andrade and Sweetman have shown, for these people—the 'weak ties' that are so well known in social network analysis—to be able to operate in these different communities across large and diverse geographical areas, they required solid bases and places where they would be welcome, where they would be met by trusted communities and known ideas. In other words, they needed to have strong tie connections into these diverse geographical communities. In addition to their power in helping to transmit new ideas, there is evidence that suggests that acting as a broker between groups is also socially advantageous. Burt[47] has outlined the theory of structural holes, which argues that being the 'bridge' between two clusters is a powerful position with regards to control of information and information transfer itself.

Lewis explores this powerful brokerage position, especially in relation to those at the edge of a minority religious network; in this case, the role of Christina More—between a heretical network of lollards and the wealthy members of the orthodox mainstream in fifteenth-century Bristol. More was indirectly involved in the Oldcastle Revolt; and she was tried for heresy with this group of dissenters in 1414. The lollard members of the Oldcastle Revolt, and their social connections, form the core of Lewis' case study. Close analysis of the social networks of these individuals demonstrates that far from being an isolated group of religious dissenters, these people were also members of burgess society at a number of different points, and so were also part of the religious mainstream and well-integrated into the communities around them. They possessed strong ties both into heretical clusters and into mainstream society. Furthermore, Lewis also shows clearly that there were several points of entry for religious dissent in Bristol, notably in the southern suburbs of the city, tying in with Centola and Macy's argument for multiple entry points for new ideas to take hold: South Bristol was in possession of such 'wide bridges' in the medieval period.

Human directions for networks in archaeology

What I hope the chapters gathered in this volume demonstrate is that there is a pressing need to reattach network analysis and network thinking to the lived realities of the past, the social relationships which generated and maintained communities, and the bonds of trust and reciprocity which constitute neighbourliness, friendship, and love. Network analysis in archaeology is in danger of becoming too much concerned with method, with models; we forget, perhaps, that our material is a proxy for real people's real-life interactions. Knappett has pointed out that much archaeological network analysis has been concerned with the regional and inter-regional scale, with the local scale ignored, perhaps because it is considered as consisting of 'largely predictable triadic connections'.[48] But it is these triadic connections which are trusted, which can enable the spread of new ideas about the divine or about how to interact with the divine, about new norms like coinage or elite relationships, about belief and its manifestations. Bringing strong social ties back into the discussion of network analysis in archaeology also ushers back in these human qualities too, the emotional connections and meaning contained within unfeeling nodes and edges: trust is a key part of the emotional timbre of the chapters collected here.

Just as we know that geography and topography make a difference to the way connectivity—or mobility—works, and networks work with and around such physical opportunities and limitations, likewise, emotional bonds of trust, friendship, respect, honour, and love also make a difference to how connectivity works. Identifying the different capacities of different kinds of strong tie in the past will also offer interesting and subtle ways forward, as well as thinking specifically about the places or occasions where such interactions take place, the narratives which are constructed and re-told which bind communities together, and the systemic use of human strong tie relationships for specific purposes. In all of these ways, and more, thinking about strong ties makes a difference to how we reconstruct and reimagine past lives, and bringing back strong ties into our networks offers us the chance to tell more human stories through our analyses.

Notes

1. Facebook has been fined £4 billion for violating user data privacy regulations and selling the data to Cambridge Analytica, a British consultancy firm with right-wing investors and founders including Steve Bannon and Robert Mercer. The company illegally used Facebook's data to psychologically profile individuals and target voters with material supporting Donald Trump's election campaign and the Leave campaign during the referendum on British membership of the EU: https://www.bbc.co.uk/news/world-us-canada-48972327, accessed 15th July 2019.
2. For example, Broodbank 2000; Knappett 2005; Malkin, Constantakopoulou and Panagopoulou 2006; Sindbæk 2007; Knappett 2011; Malkin 2011; Collar 2013; Knappett 2013; Collar et al. 2015; Brughmans, Collar & Coward 2016; Mills 2017; Donnellan 2020; Dawson and Iacono 2021.
3. Kuril: Gjesfjeld 2015; Pueblo: Peeples and Haas 2013; Crabtree 2015; Epipalaeolithic Near East: Coward 2013; Caribbean: Mol et al. 2015.
4. For useful recent introductions to network analysis in archaeology and history, see Collar et al. 2015; Mills 2017; Crabtree and Borck 2019.
5. Bentley and Maschner 2003.
6. Graham 2006; Crabtree 2015.
7. Van Oyen 2015.
8. Letters: Schor 2011; papyrus: Ruffini 2008.
9. For example, the work of Mills et al. in the US Southwest, and see discussion in Knappett 2018. Other important critiques levelled at network analysis in archaeology include the lack of a specific focus on the sociological, human element of these interconnections; the issue of equifinality (that is, the possibility that any number of scenarios might have led to the outcome that we observe on the ground); and the ontological problems that arise where the archaeologist makes decisions about categorising material prior to analysis. This negates the possibility to think about emergent connections via material but presupposes that the forms under study are categorical from the outset. See Van Oyen 2015 for full discussion.
10. Watts and Strogatz 2003; Sindbæk 2007; Malkin 2011.
11. Granovetter 1973.
12. The discussion by Knappett (2016) regarding strong ties and globalisation in the past is of relevance here.
13. Pitts and Versluys 2014; Knappett 2018.
14. Granovetter 1973.
15. Granovetter 1973; Easley and Kleinberg 2010 ch. 3; but see now Brashears and Quintane 2018 for a nuancing of the closed triad concept.
16. Granovetter 1973.
17. Duling 2013: 138.
18. Malkin 2011.
19. Buchanan, Hamilton & Kilby 2019.
20. Burt 1992.
21. Burt 1992: 28.
22. Burt 1992: 16; 47.
23. Aral and Van Alstyne 2011: 91.
24. White 2008; see further discussion in Williamson, this volume.

25. Rogers 1962: 282–283.
26. Woolf 2016: 52.
27. Knappett 2016: 31.
28. Roux, Bril and Karasik 2018.
29. Roux, Bril and Karasik 2018: 1025.
30. Bourdieu 1977.
31. Wildman 2017.
32. As argued by Czachesz 2011.
33. See argument outlined by Duling 2013: 143; for an alternative perspective, see discussion by Sweetman, this volume.
34. Granovetter 1973: 1361.
35. Brashears and Quintane 2018.
36. Woolf 2016: 46.
37. Brashears and Quintane 2018: 107. Their work goes on to identify other kinds of social tie with different capabilities which nuances the discussion of social networks and information transmission in important new ways.
38. Shi et al. 2006.
39. Centola 2018.
40. Centola and Macy 2007: 703.
41. Centola and Macy 2007.
42. Centola and Macy 2007: 707–708.
43. Van Oyen 2015.
44. Assman 1995: 129.
45. Gamble 1998; Knappett 2011; Feldman 2014.
46. See Whiting 2020 for a discussion of the ways in which commercial and pilgrimage networks intertwined both physically and metaphorically.
47. Burt 1992.
48. Knappett 2018: 975.

Bibliography

Aral, S., and Van Alstyne, M. 2011. 'The diversity-bandwidth trade-off'. *American Journal of Sociology* 117: 90–171.

Assman, J. 1995. 'Collective memory and cultural identity'. *New German Critique* 65: 125–133.

Bentley, R. A., and Maschner, H. D. G. 2003. *Complex Systems and Archaeology.* Salt Lake City, UT: University of Utah Press.

Bourdieu, P. 1977. *Outline of a Theory of Practice.* Cambridge: Cambridge University Press.

Brashears, M. E., and Quintane, E. 2018. 'The weakness of tie strength'. *Social Networks* 55: 104–115.

Broodbank, C. 2000. *An Island Archaeology of the Early Cyclades.* Cambridge: Cambridge University Press.

Brughmans, T. 2013. 'Thinking through networks: A review of formal network methods in archaeology'. *Journal of Archaeological Method and Theory.*

Brughmans, T., Collar, A., and Coward, F. 2016. *The Connected Past: Challenges to Network Studies in Archaeology and History.* Oxford: Oxford University Press.

Buchanan, B., Hamilton, M. J., and Kilby, J. D. 2019. 'The small world topology of Clovis lithic networks'. *Archaeological and Anthropological Sciences*, 11: 3537–3548.

Burt, R. S. 1992. *Structural Holes: The Social Structure of Competition*. Cambridge: Cambridge University Press.
Centola, D. 2018. *How Behavior Spreads: The Science of Complex Contagions*. Princeton: Princeton University Press.
Centola, D., and Macy, M. 2007. 'Complex contagions and the weakness of long ties'. *American Journal of Sociology* 113.3: 702–734.
Collar, A. 2013. *Religious Networks in the Roman Empire: The Spread of New Ideas*. Cambridge: Cambridge University Press.
Collar, A., Coward, F., Brughmans, T., and Mills, B. 2015. 'Networks in archaeology: Phenomena, abstraction, representation'. *Journal of Archaeological Method and Theory* 22.1: 1–32.
Crabtree, S. 2015. 'Inferring ancestral pueblo social networks from simulation in the Central Mesa Verde'. *Journal of Archaeological Method and Theory* 22.1:141–181.
Crabtree, S. A., and L. S. Borck. 2019. Social networks for archaeological research. In *Encyclopedia of Global Archaeology*, ed. C. Smith, 1–12. Cham: Springer International Publishing.
Coward, F. 2013. Grounding the net: Social networks, material culture and geography in the Epipalaeolithic and early Neolithic of the Near East (~21–6,000 cal BCE). In *Network Analysis in Archaeology*, ed. C. Knappett. Oxford: Oxford University Press.
Czachesz, I. 2011. 'Women, Charity and Mobility in Early Christianity. Weak Links and the Historical Transformation of Religions'. In *Changing Minds: Religion and Cognition through the Ages*, ed. I. Czachesz & T. Biró. Groningen Studies in Cultural Change 42, 129–154. Leuven, the Netherlands: Peeters.
Dawson, H., and Iacono, F. 2021. *Bridging Social and Geographical Space through Networks*. Leiden, the Netherlands: Sidestone Press.
Donnellan, L. 2020. *Archaeological Networks and Social Interaction*. London: Routledge.
Duling, D. C. 2013. 'Paul's Aegean network: The strength of strong ties'. *Biblical Theology Bulletin* 43.3: 135–154.
Easley, D., and Kleinberg, J. 2010. *Networks, Crowds and Markets: Reasoning about a Highly Connected World*. Cambridge: Cambridge University Press.
Feldman, M. 2014. *Communities of Style: Portable Luxury Arts, Identity, and Collective Memory in the Iron Age Levant*. Chicago: University of Chicago Press.
Gamble, C. 1998. 'Palaeolithic society and the release from proximity: A network approach to intimate relations'. *World Archaeology* 29.3: 426–449.
Gjesfjeld, E. 2015. 'Network analysis of archaeological data from hunter-gatherers: Methodological problems and potential solutions'. *Journal of Archaeological Method and Theory* 22: 182–205.
Graham, S. 2006. 'Networks, agent-based models and the Antonine itineraries: Implications for Roman archaeology'. *Journal of Mediterranean Archaeology* 19.1: 45–64.
Granovetter, M. 1973. 'The strength of weak ties'. *AJ S* 78.6: 1360–1380.
Granovetter, M. 1983. 'The strength of weak ties: A network theory revisited'. In *Sociological Theory*, 201–233. Philadelphia: Blackwell.
Jeffra, C. 2013. 'A re-examination of early wheel potting in Crete'. *Annual of the British School at Athens*, 108: 31–49.

Knappett, C. J. 2005. *Thinking through Material Culture: An Interdisciplinary Perspective. Archaeology, Culture, and Society*. Philadelphia: University of Pennsylvania Press.
Knappett, C. J. 2011. *An Archaeology of Interaction*. Oxford: Oxford University Press.
Knappett, C. J. (Ed.) 2013. *Network Analysis in Archaeology*. Oxford: Oxford University Press.
Knappett, C. J. 2016. 'Globalization, connectivities and networks: An archaeological perspective'. In *The Routledge Handbook of Archaeology and Globalization*, eds. T. Hodos, A. Guerds, P. Lane, I. Lilley, M. Pitts, G. Shelach, M. Stark, and M. J. Versluys. London: Routledge.
Knappett, C. J. 2018. 'From network connectivity to human mobility: Models for minoanization'. *Journal of Archaeological Method and Theory* 25: 974–995. https://doi.org/10.1007/s10816-018-9396-9
Knappett, C. J. 2020. Relational concepts and challenges to network analysis in social archaeology. In *Archaeological Networks and Social Interaction*, ed. L. Donnellan. London: Routledge.
Latour, B. 1987. *Science in Action: How to Follow Scientists and Engineers through Society*. Milton Keynes: Open University Press.
Latour, B. 1992. Where are the missing masses, sociology of a few mundane artefacts. In *Shaping Technology-Building Society. Studies in Sociotechnical Change*, eds. W. Bijker and J. Law, 225–259. Cambridge, MA: MIT Press.
Mills, B. 2017. 'Social Network Analysis in Archaeology'. *Annual Review of Anthropology* 46: 379–397.
Malkin, I. 2003. 'Networks and the emergence of Greek identity'. *Mediterranean Historical Review* 18: 56–74.
Malkin, I. 2011. *A Small Greek World*. Cambridge: Cambridge University Press.
Malkin, I., Constantakopoulou, C., and Panagopoulou, K. 2009. *Greek and Roman Networks in the Mediterranean*. Abingdon: Routledge.
Mol, A. A. A., Hoogland, M. L. P., and Hofman, C. L. 2015. 'Remotely local: Ego-networks of late pre-colonial (AD 1000–1450) Saba, north-eastern Caribbean'. *Journal of Archaeological Method and Theory* 22.1: 275–305.
Peeples and Haas 2013. 'Brokerage and social capital in the prehispanic U.S. Southwest'. *American Anthropologist* 115.2, 232–247.
Pitts, M., and Versluys, M-J. 2014. *Globalisation and the Roman World: World History, Connectivity and Material Culture*. Cambridge: Cambridge University Press.
Rogers, E. M. 1995. *Diffusion of Innovations*. 4th edition. New York: Free Press.
Roux, V., Bril, B., and Karasik, A. 2018. 'Weak ties and expertise: Crossing technological boundaries'. *Journal of Archaeological Method and Theory* 25: 1024–1050.
Ruffini, G. 2008. *Social networks in Byzantine Egypt*. Cambridge: Cambridge University Press.
Schor, A. 2011. *Theodoret's People. Social Networks and Religious Conflict in Late Roman Syria*. Berkeley University Press.
Shi, X., Adamic, L., and Strauss, M. 2006. 'Networks of strong ties'. *arXiv:arch-ive.condmat/0605279v1*, http://arxiv.org/abs/cond-mat/0605279
Sindbæk, S. 2007. 'The small world of the Vikings: Networks in early medieval communication and exchange'. *Norwegian Archaeological Review* 40.1: 59–74.
Van Oyen, A. 2015. 'Actor-network theory's take on archaeological types: Becoming, material agency, and historical explanation'. *Cambridge Archaeological Journal* 25: 63–78.

Van der Leeuw, S. E., and Torrence, R. 1989. *What's New?: A Closer Look at the Process of Innovation*. One World Archaeology. London: Routledge.
Watts, D., and Strogatz, S. H. 2003. 'Collective dynamics of "small world" networks. *Nature* 393:440–442.
White, L. M. 1992a. 'Finding the ties that bind'. In *Semeia 56, Social Networks in the Early Christian Environment: Issues and Methods for Social History*, 3–22. Atlanta, GA: Scholar's Press.
White, L. M. 1992b. Social networks: Theoretical orientation and historical applications. In *Semeia 56, Social Networks in the Early Christian Environment: Issues and Methods for Social History*, 23–36. Atlanta, GA: Scholar's Press.
White, H. C. 2008. *Identity and Control. How Social Formations Emerge*, Princeton: Princeton University Press.
Whiting, M. 2020. Braided networks: Pilgrimage and the economics of travel infrastructure in the late antique holy land. In *Pilgrimage and Economy in the Ancient Mediterranean*, eds. A. Collar and T. M. Kristensen. Leiden: Brill.
Wildman, W. J., 2017. 'Modelling religion and the integration of the sciences and the humanities in the bio-cultural study of religion'. The Religious Studies Project (Podcast Transcript). 9 October 2017.
Woolf, G. 2016. 'Only connect? Network analysis and religious change in the Roman World'. *Hélade* 2.2: 43–58.

Part I

Sanctuaries: the places that bind

2 'Orientalising' networks and the nude standing female

Synchronic and diachronic dimensions of ideology transfer

Megan Daniels

Introduction

The nude standing female figurine, appearing in Greek religious contexts between ca. 800 and 550 BCE, has often been chalked up to an artistic borrowing from the Near East in the so-called 'Orientalising' period (ca. 750–600 BCE),[1] or else is used to pin down an identity of the god(s) being worshipped.[2] Yet, when considering the long-standing socio-economic interconnections linking Greece and the Near East over the Late Bronze and Iron Ages, deeper clues surrounding the meaning of this imagery and its associated ideologies emerge. When placed in its synchronic context—namely, large cross-cultural sanctuaries emerging in the Iron Age—and its diachronic context—namely, its long history of use stemming from early Bronze Age Mesopotamia—the symbolism of the nude female has the power to illuminate influential ideologies centred on human sovereignty and relationships with the divine that were shared between numerous cultural groups.

My aim in this chapter is to push the dialogue surrounding the nude standing female beyond discussion of artistic transmission or identification of deities and into the realm of shared religious and political ideologies between Greece and western Asia via their social contexts—namely, cross-cultural sanctuaries. The rise and decline of the nude female in the votive repertoire of numerous Greek sanctuaries between ca. 800 and 550 BCE captures these changing ideologies through the placement of this imagery's religious symbolism within long-standing social and commercial networks that united communities over long distances in the Bronze and Iron Ages. Numerous theoretical and methodological advances have infiltrated the study of these interconnections, from macro-scale histories that blur and transcend strict cultural boundaries[3] to micro-scale examinations of objects as agents structuring human interaction over long distances.[4] The move towards studying and modelling connections through social networks offers further opportunity to move beyond old paradigms.[5] In particular, the study of how these connections bolster—and are bolstered by—strong

DOI: 10.4324/9780429429217-3

ties between people based on trust, collective memory, and shared beliefs, allows us to stand on new ground when analysing the semiotic function of votive imagery spread over long distances.

This argument will proceed on several levels: (a) a contextual-synchronic approach that examines sanctuaries as representative of the 'wide bridges' of elite social networks and identity formation in the Iron Age; (b) a close reading of the religious and political symbolism represented by the nude standing female found within these sanctuaries; and (c) a long-term (diachronic) analysis of elite associations with powerful goddesses represented through the nude female. This chapter will thus start by examining the evidence for Iron Age sanctuaries as emergent nodes in long-distance social and commercial networks, whose edges were powered by various types of human mobility, creating the necessary strong ties within the Greek world for the spread of ideologies, as well as wide bridges to regions beyond the Greek world where the nude standing female imagery first emerged. With the other two levels, expressed in (b) and (c), I examine how symbolism associated with nude female imagery was embedded not only in synchronic networks reflected by the sanctuaries, but in diachronic ones as well. In particular, I examine the long history of nude female imagery across western Asia, coming to focus on her immediate meanings within Iron Age sanctuaries. Long seen by scholars as a symbol of eroticism and fertility, I argue instead that the nude female symbolised aspirations towards kinship between worshippers and the deity. Her disappearance in the sixth century represents, rather than changing artistic preferences, shifting attitudes towards human relationships with the divine, bound up in larger political and social changes in Greece. Before delving into this argument, I will frame it through considering theoretical and methodological advancements made in recent years with Mediterranean connectivity and networks, beginning with the concept of 'Orientalising,' which has coloured our view of the nude standing female.

Reorienting 'orientalising:' synchronic and diachronic pathways of exchange and interaction

Two dominant paradigms have affected how the imagery of the nude female in the Iron Age/Archaic Greek world is understood. An examination of these paradigms can help frame how the concept of 'wide bridges' offers better inroads to understanding religious and ideological connections between Greece and the peoples of Egypt and western Asia. The first paradigm is the 'Orientalising' perspective, which saw the Greeks in a particular period (between ca. 750 and 600 BCE) intentionally adopting and adapting literary and artistic traits from the more 'venerable' cultural groups to the east. The second is overarching views on what nudity symbolised in the ancient world. In the case of the nude standing female, these views centred on all-encompassing concepts such as fertility and sexuality, sometimes attempting to link this imagery to particular deities who represented these concepts.

Both paradigms have already undergone refinements, but bringing the role of wide bridges and strong ties into the equation can refine them further. I will concentrate on the first one here, the 'Orientalising' paradigm, which informs how the Iron Age sanctuaries in which the nude female imagery appears are envisioned, and turn back to the question of nudity below.

The most traditional definition of the Orientalising Period comes from Walter Burkert's seminal book on this subject: 'that decisive epoch in which, under the influence of the Semitic East, Greek culture began its unique flowering, soon to assume cultural hegemony in the Mediterranean.'[6] The interest in Greek–Near Eastern interactions and their role in formulating a distinctively 'Greek' culture emerged in the post-World War II era[7] and picked up pace with seminal works in the 1980s and 1990s.[8] As Sarah Morris has outlined, the study of Greek–Near Eastern interactions has undergone several shifts, from narratives that polarised and reified 'Greece' and the 'Near East' to much more critical self-reflections about cultural boundaries and transfers.[9] These self-reflections include recognition of the much deeper ancestries of these interactions and exchanges going back to the Bronze Age.[10]

The focus on longer-term histories of interaction and exchange aligns with new approaches to Mediterranean history in general, at both the macro- and micro-scales. These approaches offer ways to capture more accurately the realities of this Mediterranean–Near Eastern socio-cultural continuum in the Iron Age and its longer-term roots. At the macro-scale are the big history approaches to Mediterranean history starting with Braudel's 1949 opus. More recently, these approaches have centred on paradigms of mobility and interconnectivity in attempts to overcome the 'crisply bounded cellular entities' that characterised scholarly views of Mediterranean cultures in the 1960s and 1970s.[11] Such approaches seek more encompassing histories that see movement and exchange as constant features of the Mediterranean world and its bordering regions. These approaches are geographically as well as temporally extensive, and allow the questioning of the epistemological divide drawn between the Bronze and Iron Ages.[12]

Added to these developments are the network approaches to conceptualising and modelling human connectivity.[13] These approaches have helped to move beyond earlier critiques of interconnectivity's 'static' and 'timeless' nature[14] and envision more dynamic, multidirectional processes of interaction over time. Along with the interest in networks and the diffusion of innovations they allow through human actors,[15] a renewed focus on objects has emerged. In particular, scholars have stressed the agency of objects in structuring human interaction over long distances into 'communities of practice,' a concept adopted from Jean Lave and Etienne Wenger's 1991 publication to describe and analyse processes by which humans become full participants in sociocultural praxes.[16] The role of objects and their semiotic meanings in joining human groups into recognised communities who share ideologies, norms, and identities along synchronic and diachronic networks are central to the arguments in this chapter, and I will come back to this concept at the end.

Finally, as outlined in the introduction to this volume, the most recent research on social networks and their role in communication and diffusion has overturned the previous focus on 'weak ties'[17] and 'small worlds'[18] for the spread of innovations. Instead, scholars like Damon Centola emphasise the need for *strong* ties through multiple, overlapping relationships in order to transmit ideas and practices. The well-trodden analogy of viral diffusion, whereby germs spread over wide distances via simple node-to-node links, or 'weak ties,' between individuals, seems to fall apart when it comes to studying the diffusion of what Centola calls 'complex contagions,' such as ideologies, ideas, technologies, and norms. These complex contagions, which can spread within a cluster of strongly tied nodes, need multiple sources of exposure and reinforcement to travel between clusters and take root. Centola describes such a scenario as a 'wide bridge.'[19] Yet, it is not simply multiple connected nodes that guarantee the spread and adoption of complex contagions: research into recruitment and engagement activities on social media platforms, for instance, found that it was not so much influence from multiple neighbours full stop that predicted the likelihood of recruitment and engagement to a platform, but multiple neighbours *from distinct social contexts* of an individual's life that raised the likelihood of adoption.[20] More specifically to religion, research has found that the more religious institutions interacted with one another through shared organisations, the more likely they were to behave like one another.[21] This research has important implications for understanding both the imagery of the naked female and its contexts of use beyond 'Orientalising,' namely, the strong ties emerging within Iron Age sanctuaries, to which I now turn.

Iron Age sanctuaries and wide bridges

One of the hallmarks of the period between 800 and 550 BCE is the proliferation of monumental sanctuaries around the eastern Mediterranean world, many of which also exhibited a startling quantity of dedicated objects of both local and foreign nature.[22] These sanctuaries include the panhellenic sanctuaries at Delphi and Olympia, the temples to Hera at Perachora and Samos, the temples to Athena from Rhodes, Smyrna, and Miletos, the temple to Aphrodite at Miletos, the temples to Apollo at Delos, Knidos, and Eretria, the temples to Artemis at Ephesos and Thasos, the sanctuary to Orthia at Sparta, and several sanctuaries on Crete. The foreign dedications included metalwork in the form of sculpture, weapons, equestrian equipment, vessels, cauldrons, carved ivory containers and furniture fittings, and terracotta sculptures in Phoenician, Cypriot, Syrian, Phrygian, and Egyptian styles, alongside specimens from farther-flung production centres such as Urartu.[23] These exotica, excavated in the late nineteenth and twentieth centuries, joined other astonishing finds, such as the 'orientalia' from Etruscan and Greek tombs and decorated metalwork and ivories from the Assyrian palaces at Nimrud ascribed to Phoenician, North Syrian, and Assyrian schools. All of these assemblages, alongside locally made works showing Egyptian or Near Eastern stylistic influences, helped to define the period and style termed 'Orientalising.'[24]

As Gunter has pointed out, more recent scholarship takes the notion of Orientalising not as a style or period but as a process, and attempts to understand 'the circumstances under which new images and technologies were selectively adopted and why—instead of assuming the inevitability of cultural change or a simple and unidirectional notion of influence.'[25] These 'Orientalising' sanctuary assemblages, however, have proved elusive in clarifying either circumstances or directions of influence, adoption, and adaptation.[26] At the heart of this difficulty lies the issue of equifinality, which overburdens to a certain extent most votive assemblages in ancient sanctuaries: essentially, there remain multiple ways and means by which objects end up as dedications, and very rarely is the donor's identity (nor the circumstances by which they acquired this object) made explicit. Scholars have put forth numerous arguments over the years concerning the individuals behind these dedications in Iron Age sanctuaries: itinerant artisans, eastern kings as described in Herodotus, Greek mercenaries returning with souvenirs, foreign women and/or brides, tomb-robbers, and other types of foreigners such as merchants and travelling priests.[27]

Yet, only in rare cases can the identities of foreigners via these objects be demonstrated, which makes reconstruction of the social and geographical networks represented by objects in these sanctuaries difficult.[28] As Jan-Paul Crielaard notes, these votives likely changed hands multiple times (e.g. the Sidonian silver krater in the *Iliad* 23.740–749), as many came from lands quite remote from the Aegean and circulated with complex biographies in very specific contexts of elite consumption and identity.[29] Yet, stepping back and releasing the need to pinpoint origins and individuals behind these objects, it is possible to see much broader patterns emerge that position these sanctuaries as loci of multiple overlapping networks imbued with long-standing ideologies wrought by all kinds of activity: mercenary service, diplomacy, migration, trade, and conquest. The breakdown in origins of objects from a single sanctuary alone, such as the Heraion on Samos, shows the range of locales and cultural groups that shaped, to varying extents, religious activity at this sanctuary (see Figure 2.1).[30] Furthermore, the few instances in which there is direct information about the origin and/or dedicator—for instance, the Ionian soldier Pedon, who served under an Egyptian pharaoh and dedicated an Egyptian block statue in Priene—also suggest a range of cross-cultural human actors, traditions, and objects structuring religious activity.[31] Recent studies of cosmopolitan graves from around the Iron Age Mediterranean also suggest the existence not only of hybridised objects, but also hybrid individuals whose tastes, actions, and livelihoods fused a number of cultural and ethnic identities, and who possibly acted as mediating individuals or 'brokers' between different communities.[32]

Examining broader groups of objects reveals different, yet related patterns. The high amounts of Cypriot figurines documented from sanctuaries

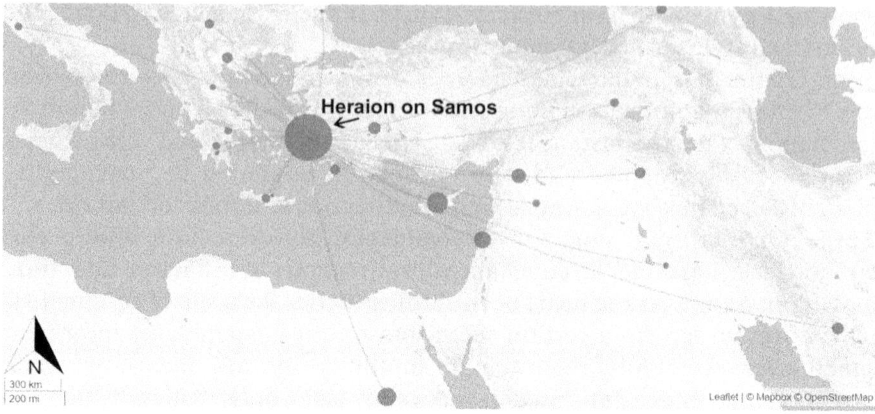

Figure 2.1 Visual representation created in Palladio of percentages of foreign dedications amongst the votives from the Heraion on Samos, eighth-seventh centuries BCE.

Source: Percentages adapted from Kilian-Dirlmeier 1985, 237, Abb. 18.

at Naukratis, Knidos, Miletus, and Samos,[33] Phrygian bronze objects from sanctuaries across Asia Minor,[34] the large amounts of 'aegyptiaca' from numerous sanctuaries,[35] or bronze bowls and cauldrons[36] allow the pinpointing not of individual dedicators, but well-established networks along which these objects travelled, which formed overlapping Venn Diagrams with one another.[37] Particular classes of imagery, for instance seals and bowls showing lyre-players and other musicians, seemed to follow specific routes of transmission, and were also linked into much longer-term religious and political ideologies as John Franklin has recently demonstrated.[38] These networks were not merely physical but conceptual: foreign votives appearing in these sanctuaries were imbued with prestigious biographies and symbolic valence.[39] They were subject to conscious choice within local cultural contexts as well: scholars studying these 'Orientalising' assemblages have noted distinct patterning in foreign object types at different sanctuaries. For instance, the sanctuary to Athena and the sanctuary to Aphrodite at Miletus, while producing assemblages of foreign objects from the Levant, Cyprus, and Egypt, displayed dissimilarity in their object types, suggesting different roles, meanings, and perhaps statuses ascribed to different categories of imports.[40]

Yet from all of these foreign locales—Phrygia, Lydia, Cyprus, Egypt, and Mesopotamia—we find the evidence of nude standing female imagery, which emerges in 'Orientalising sanctuaries' dedicated to female deities in the Greek world between 800 and 550 BCE. As discussed previously, multiple sources of exposure from different social contexts are seen to be vital in the adoption of complex contagions, and with these ideas, we can start to paint a picture of the mechanisms by which the nude standing female became popular in multiple Greek sanctuaries to goddesses, as well as wealthy graves. In their influential 2007 paper, Centola and Macy found that 'the weaker the ties to acquaintances compared to friends, the wider must be the bridges connecting otherwise distant neighborhoods.'[41] Their findings suggest that complex ideas and symbolisms could spread via weaker links (e.g. a weak link being two nodes alongside a trade route as opposed to two nodes linked by close, familial connections), provided there are multiple weak ties between clusters. Furthermore, these complex contagions are likely to take root within a cluster that has strong ties *within* it—in other words, ties that are reinforced by strategic complementarity (the more people adopt an innovation, the more its value increases); credibility; legitimacy; and emotional contagion.[42]

The networks within the Greek world during the period between 800 and 500 BCE have been outlined most recently by Irad Malkin in his 2011 monograph, although Malkin relies more on small-world models developed by Watts and Strogatz in their 1998 article. Nonetheless, Malkin demonstrates within this book the mechanisms by which a unified Greek identity converged in a period when many Greeks were in fact moving apart, a process that was largely facilitated through various diacritical markers like shared cultic traditions employed in the contexts of overseas settlements. Likewise, Mooring (this volume) characterises the Archaic Greek world as made up of multiple weak ties around the Mediterranean and Black Sea, with diffusions spreading via the interaction of elites at panhellenic sanctuaries, who acted as early adopters and spread new practices to their respective *poleis* via strong ties within their communities. Returning to the 'Orientalising' sanctuaries, while their assemblages are most often studied via their foreign connections, we should not ignore the ties they suggest within the Greek world as well. Sanctuaries close by one another could indeed display subtle differences in their object types, as discussed previously with Miletus. Yet, the repetition of a common assemblage of imageries at these sanctuaries in the period between 800 and 550 BCE suggests a high degree of interaction and common identity-building within the Greek world; the multiple mechanisms of interaction (the wide bridges) with the Near East and Egypt necessarily affected these identities.

The nude female imagery can thus be seen as an outcome of these wide bridges connecting Greek communities (and their sanctuaries) to multiple far flung parts of the Mediterranean and Middle East, whose meaning and imagery took root in burgeoning networks within emerging Greek communities, ties which were fostered by many means including shared cultic rituals and beliefs performed at these sanctuaries. Moreover, whereas the 'Orientalising' assemblages from these sites might be chalked up as mere trinkets, or *athyrmata*, with no real meaning attached to them, once again, a broader examination of the evidence suggests otherwise. As Villing notes, for instance, the connotations inherent in Egyptian objects in sanctuaries appear to have been shared widely, since the rather limited selection of *aegyptiaca* and their use appears fairly consistent across the Mediterranean world.[43] Furthermore, the nude standing female imagery in Greek sanctuaries appears most often as locally made terracotta figurines, indicating adoption, reproduction, and adaptation of this imagery and its symbolic value at the local scale. In the following pages, I will outline the long-standing ideology of the imagery across the Bronze and Iron Ages and consider evidence for the Greek engagement with its meaning, stressing in particular the *diachronic* importance of networks for fostering vital means of reinforcing strong ties—namely through credibility, legitimacy, and emotional contagion.

Yet, remaining for the moment within a synchronic perspective, it is important to stress that the very contexts in which this imagery travelled—the networks themselves—should also be understood as imbued with symbolisms and ideologies that structured how these objects were received and put to use—a situation anthropologists refer to as a 'social field.'[44] In the Iron Age, these networks were wrought, largely, by the elite exchange systems created by an expanding Neo-Assyrian Empire, although they certainly mapped on to much earlier routes of exchange as well.[45] These elite Iron Age 'social fields' can also relate to the concepts touched on above of 'situated learning' and 'communities of practice,' whereby the contexts and relations in which actors—including objects—interact create shared repertoires of doing and being (mutuality) as well as shared histories of learning.[46] I thus suggest releasing the need to identify specific individuals or groups dedicating at these sanctuaries and instead to see sanctuaries and their variegated assemblages as facilitating the 'wide bridges'—or contexts of situated learning and experience—that allowed communities to create and transmit ideologies and symbolisms in the Iron Age via these overlapping networks. It is within this context that I now turn to a specific type of imagery appearing within these locales: the nude standing female, and the diachronic significance of this imagery within these networks.

The nude female and 'orientalising' networks

The image of the nude female has a long history across the Near East and Egypt. It appears at Iron Age Greek sanctuaries dedicated to female deities amidst larger 'Orientalising' assemblages in the ninth-sixth centuries BCE, most commonly as terracotta mould-made plaques and figurines (see Figures 2.2a and 2.2b). Importantly, while the nude female replicated Near Eastern imagery[47] and likely derived from moulds from Cyprus or else directly from the Levant, its production, with few exceptions, was local.[48] Given the symbolism of this imagery and its transmission over long periods across numerous cultural groups, the nude female should be seen as more than just an exotic import. This imagery, rather, represented particular ways of relating to a deity bound up with ideologies of power and protection shared between groups around the Near East and Mediterranean.

The most comprehensive study on the nude female in the Iron Age Greek world comes from Stephanie Böhm's 1990 publication. Böhm catalogued the

Figure 2.2 (a) Cypriot mould-made figurine from the Heraion on Samos, before 570/560 BCE. (b) Terracotta figurine from Praisos, Crete, seventh century BCE in the Metropolitan Museum of Art (53.5.23).

Source: (a) Schmidt 1968, 16, 19, and Pl. 29, T 2395; image and permission courtesy of the Deutsches Archäologisches Institut. (b) © Metropolitan Museum of Art, Public Domain.

appearance of the nude female across various media in Greece, Crete, and Asia Minor (see Figure 2.3). Given the age of publication, more specimens would have to be added to this catalogue, in particular very large numbers from some Cretan sanctuaries.[49] From the map-graph in Figure 2.3, however, it is clear that a large range of locales across Ionia, Crete, and mainland Greece produced this image, primarily at sanctuaries to goddesses.

Furthermore, when the data is arranged according to the dating of individual specimens, certain trends stand out. When the date range is limited to the Iron Age and early Archaic period (ca. 1000–700 BCE), regions generally

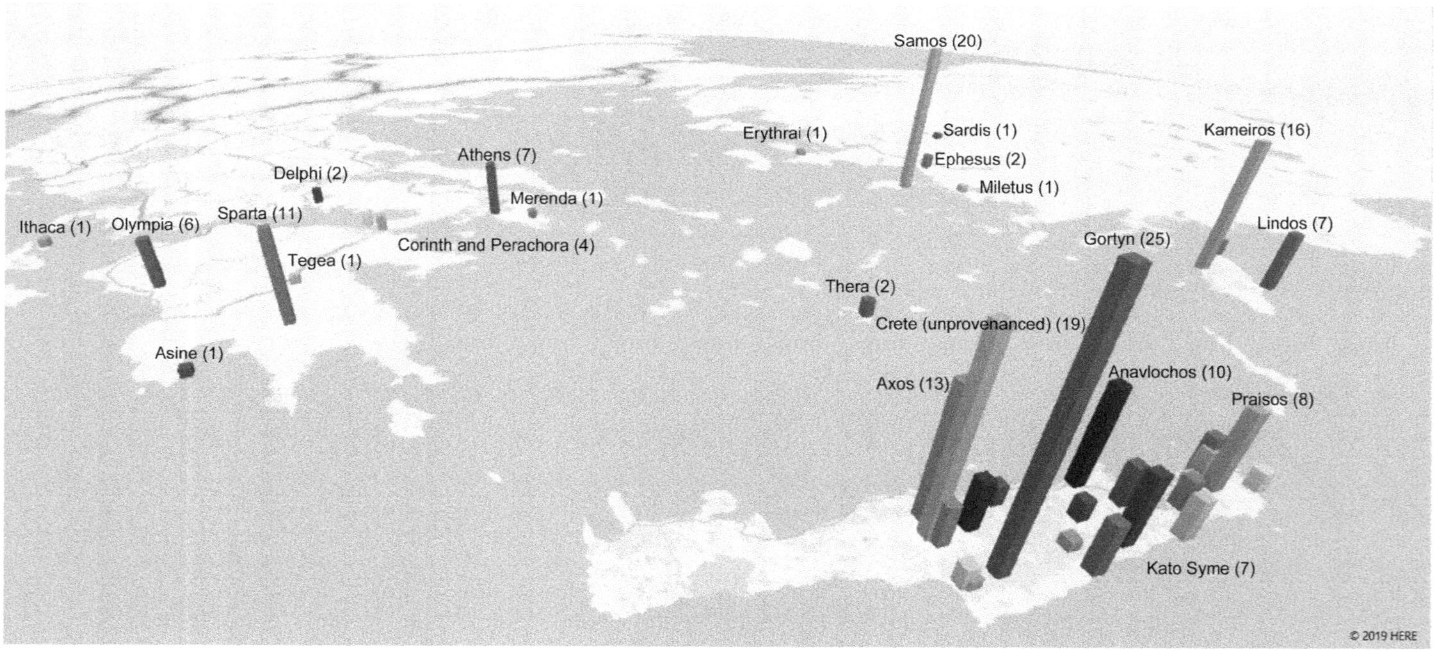

Figure 2.3 Counts and locations of nude female figurines from the Greek mainland and islands, eighth-seventh centuries BCE.

Source: Adapted from Böhm 1990.

regarded as 'early adopters' of eastern motifs and media, including Crete, Rhodes, Samos, Athens, and Olympia, are represented (see Figure 2.4). Eastern Crete in particular is a forerunner in adopting Phoenician and Syrian styles, including the nude standing female.[50]

Furthermore, the Iron Age/early Archaic specimens tended to be made in higher-value materials like bronze and ivory, alongside terracotta, whereas in the later Archaic period, nude female figurines tended to be made primarily as mould-made terracotta statuettes (Figures 2.5a and 2.5b). These trends suggest that a large number of early figurines were, in fact, imports or at least singular luxury items,[51] which eventually became more widely established as mould-made terracotta figurines dedicated primarily at sanctuaries in mainland Greece, Ionia, and east Crete.[52]

Böhm grouped these figures and plaques via the various poses they took: with arms straight by her sides (Type A); with both hands supporting the breasts (Type B); with both hands against the lower abdominal/genital area (Type C); and with one hand against the lower abdominal/genital region and one hand against the breast (Type D). Occasionally, the nude female in the Greek world could also appear holding attributes in both hands and standing atop animals, particularly lions.[53] All of these image types correspond to earlier precedents stemming from the Bronze Age Near East, and it is vital to consider the long-term developments of the symbolism and style of this imagery, as well as its contexts of use, to assess its transmission to the Greek world.

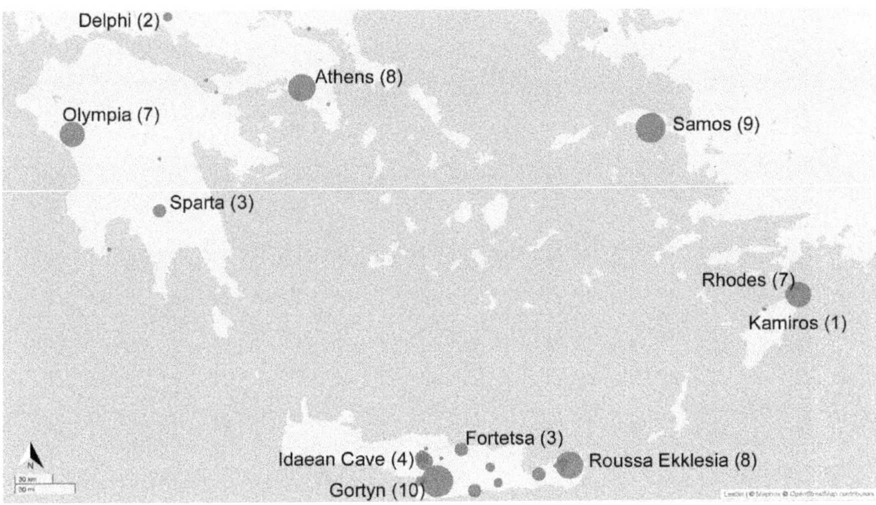

Figure 2.4 Areas of the Greek world with nude standing females, 1000–700 BCE, created in Palladio.

Source: Adapted from Böhm 1990.

The iconography of the nude female is first encountered on storage jar handles from Cemetery A by the temple of Inanna at Kish dated to ca. 2500 BCE.[54] Throughout the second half of the third millennium BCE, the nude female spread and developed into three-dimensional form across several Sumerian city-states and also further afield in Syria. Figurines of a bird-faced female with hands grasping or supporting the breasts or with arms out to the sides appeared at Alalakh, Ebla, and elsewhere in Syria

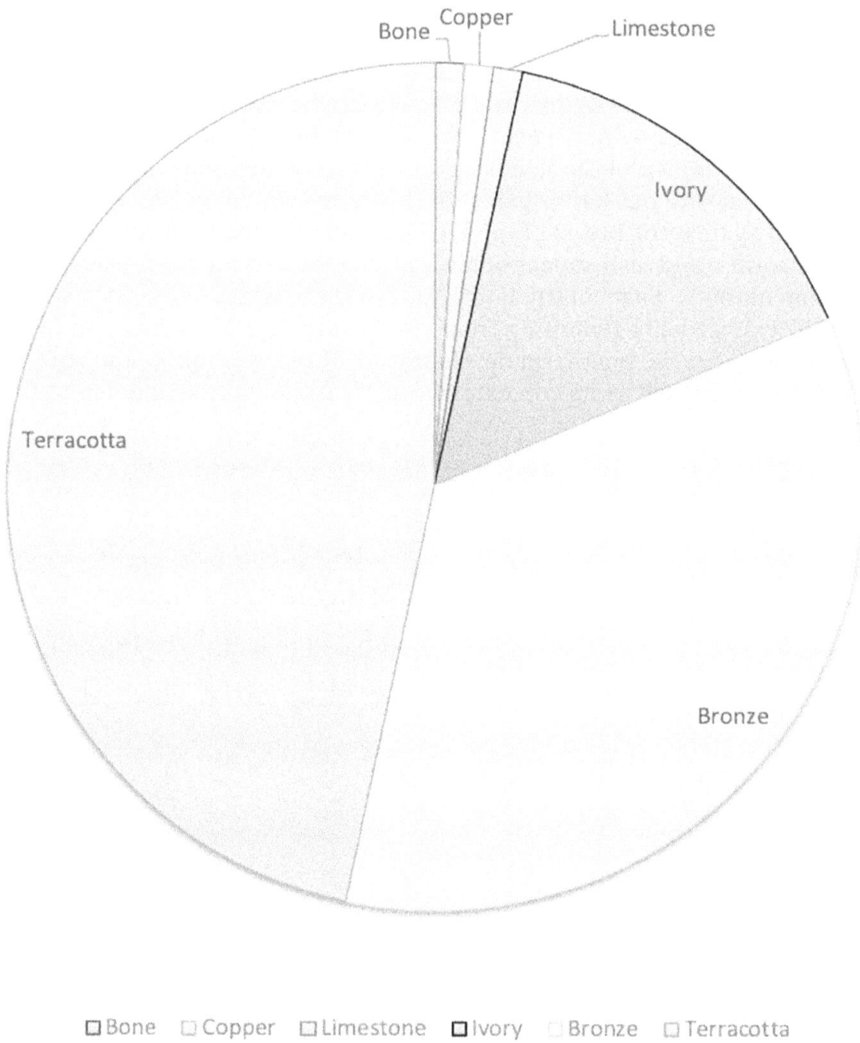

Figure 2.5 (a) Proportion of materials used 1000–700 BCE ($n = 92$).

Source: From Böhm 1990.

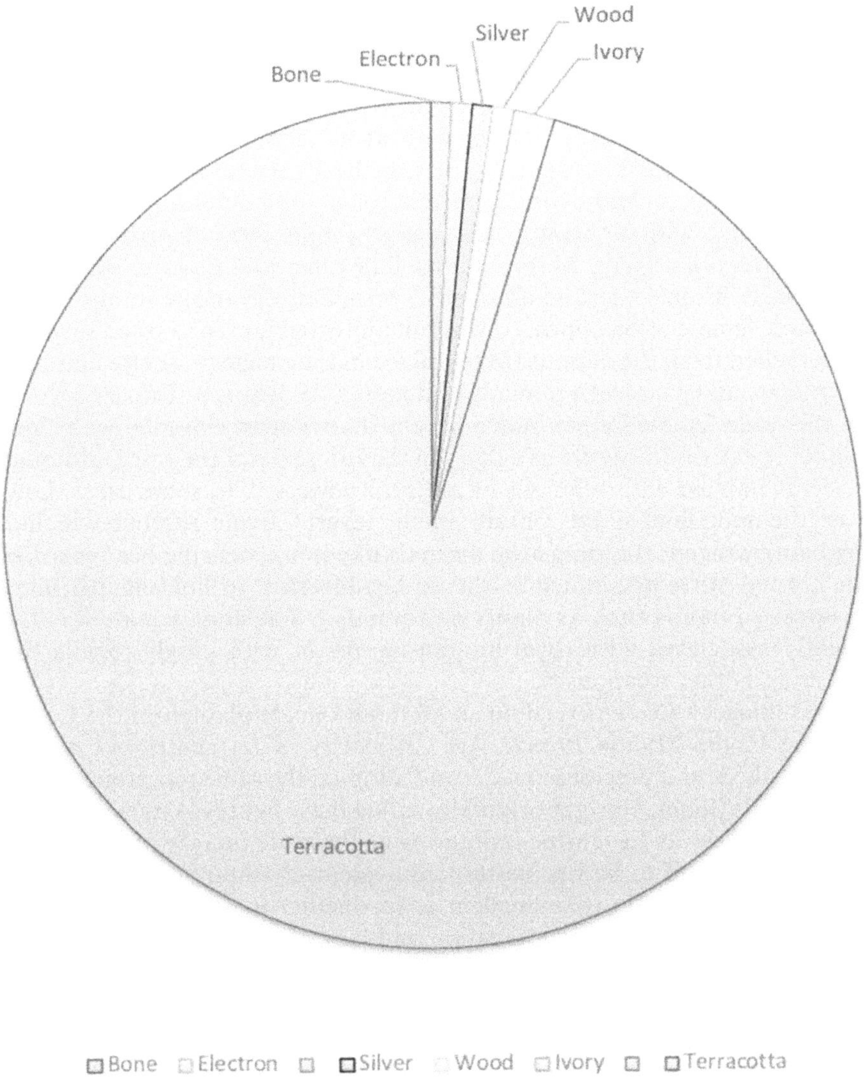

Figure 2.5 (b) Proportion of materials used 700–500 BCE (n = 129).

in the early second millennium.[55] The bird-faced figurines were replaced by 'normal-face' mould-made figurines and plaques by the Late Bronze Age.[56] Both Alalakh and Ebla had Ishtar as their patron deity by the Old Syrian period, whose worship was associated with nude female imagery. In some cases, particularly in instances when the nude female was winged, this imagery may have depicted Ishtar herself.[57] As we will see below, there seems to emerge two general types of nude female: ones that contain divine attributes and may represent an actual goddess like Ishtar, and others without divine attributes, that seem imbued with magical or religious properties, but seem to act more as intermediaries between humans and gods.

The diagnostic features of this female imagery from the start included nudity and frontal-facing design, with arms in positions similar to the later Greek examples: straight against the body, stretched outwards, supporting breasts, or indicating genitalia (or a combination of these positions). Along with figurines, this imagery appeared on cylinder seals across Anatolia, Syria, Mesopotamia Palestine, and Cyprus starting in the Middle Bronze Age and on later Cypriot and Levantine stamp seals.[58] The nude female often appeared without any attributes, and often seems to be detached from the actions taking place in the imagery, or else acting as an intermediary between humans and gods.[59] In Beatrice Teissier's words on the nude female's appearance on Old Babylonian cylinder seals, 'one cannot speak of this figure as a deity in the full sense of the word, although it seems imbued with religious or magical powers.'[60] In some cases, however, the nude female did contain one or several divine attributes including being winged, standing atop animals like lions, wearing headgear like the horned mitre or plumed/feathered headdresses, or holding attributes in upraised hands such as plants or animals.[61] The nude female was frequently associated with royal human figures or with kingly regalia like maces and other weapons.[62]

This imagery also emerged on small mould-made plaques in the Levant starting in the Middle Bronze Age, primarily in terracotta but also in faience, glass, and precious metals, and adopted the same repertoire of arm positions. William Albright originally called these figures 'Astarte plaques,' as he saw them as Levantine imitations of the nude images of Ishtar, and Astarte appeared to be the western equivalent of Ishtar.[63] This name has stuck, but scholarship is ambivalent as to whether these females represent goddesses, given the simple execution and lack of divine attributes.[64] Most of these plaques in the Late Bronze Age and Iron I and II periods were found in living quarters, with a few coming from graves and cult sites, prompting many to see them as apotropaic charms, possibly used in female-oriented rituals.[65] We might assume a function similar to the plain, nude female on cylinder seals as described by Teissier: as an intermediary between humans and gods.

Like some cylinder seal imagery, however, examples have emerged in the Late Bronze Age Levant, Syria, Anatolia, and Mesopotamia of a nude, frontal-facing female with the divine attributes mentioned above. One example includes the gold foil pendant of a female atop a lion holding animals in either hand with hair in the style of the Egyptian great goddess Hathor, attributes that scholars take to symbolise divinity (see Figure 2.6).[66] Yet, another medium of this iconography is a set of metal piriform pendants from the Middle-Late Bronze Age showing a schematically rendered nude female with Hathoric hairstyle, sometimes associated with a branch motif in her navel area.[67] To the Levantine corpus can also be added the Iron Age Judaean Pillar Figurines, which show women with bell-shaped bottoms supporting their breasts, relating to Cretan and Cypriot styles.[68]

Cyprus and Egypt also yielded nude female imagery from the Early Bronze Age onwards, which eventually converged in style and form with the

Figure 2.6 Gold foil pendant from Ugarit showing naked female mounted on a lion and framed by serpents, controlling two goats; fourteenth–thirteenth centuries BCE. 6.5 cm × 2.84 cm.

Source: AO 14.714, © RMN-Grand Palais/Art Resource, NY.

Levantine imagery by the Late Bronze Age—Egypt in the form of 'paddle dolls' and 'potency figurines' in the Middle Kingdom and Cyprus in the form of Early and Middle Cypriot plank figurines.[69] In the Late Cypriot period, Cyprus adopted the Syrian and Mesopotamian bird-faced terracotta figurines (LCII) and then figurines in bronze (LCIII). These specimens were largely confined to domestic and funerary contexts, with some examples at urban sanctuaries (e.g. Kition, Paphos, and Enkomi). Egypt, in the New Kingdom, began producing small figurines of nude females appearing to lie on a bed, as well as steles showing a female, nude and frontal with Hathoric hairstyle, often standing on a lion and holding attributes in either hand, a type directly influenced by Levantine iconography.[70] Many of these latter types of images name Levantine goddesses like Anat, Astarte, and the mysterious Qedeshet/Qudshu, whose personas and symbolism intermixed with major Egyptian goddesses, in particular Hathor, through her distinctive hairstyle and long-standing links to the Levant and Cyprus.[71]

Thus, across Mesopotamia, Anatolia, the Levant, Cyprus, and Egypt, there is a fascinating convergence of traditions (and deities) by the Late Bronze Age, whereby all regions adopt the imagery of the nude standing female, although this imagery is expressed in various media: figurines, plaques, cylinder seals, pendants, steles, and, in rare cases, statues.[72] This imagery also emerges in various contexts: domestic, funerary, cultic, and palatial. While Egyptian sources frequently name this female as a divinity (Anat, Astarte, or Qedeshet), in most cases the identity is ambiguous, although the most basic feature of this imagery—the nudity—is generally not associated with mortal women.[73]

Conversely, the nude female is rare in the Bronze Age Aegean. There are a few examples: the gold foil pendants of a nude female supporting her breasts and associated with doves from Grave Circle A at Mycenae, a seated terracotta female from Knossos with arms under her breasts and Type A figurines from the Cult Centre at Mycenae, which support their breasts but do not seem to be nude.[74] Otherwise, the nude female emerges in Iron Age Greece notably with five ivory figurines from a Late Geometric grave near the Dipylon Gate in Athens with arms at their sides wearing a decorated polos.[75] This imagery appears primarily, however, in sanctuaries, as discussed above, in both terracotta and in higher value media. The Greek types hold their arms straight at their sides, or else hold their breasts or indicate genitalia, more akin to the 'Astarte' figurines than the elaborate Levantine/Egyptian 'Qedeshet/Qudshu' imagery. These Greek types primarily come from sanctuaries to Hera, Athena, Artemis, and Aphrodite. Thus, they cannot be identified as any specific goddess, but their relation to all of these goddesses is significant. Given their place within these 'international' sanctuaries, which brought together cosmopolitan assemblages by a variety of human actors along ideologically charged networks, these nude females represent more than just artistic influence or a particular goddess. Rather, I suggest seeing them as ritual and symbolic means of interacting with deities that developed amidst these 'Orientalising' networks. I turn now to further considerations of what the nude female symbolised religiously and ideologically.

The meaning of nudity

The Orientalising paradigm has coloured our notions of the nude female in the Greek world, namely positioning this imagery as an artistic expression of Greece's 'courtship' with the Near East in the Iron Age.[76] The second overarching influence involves scholarly interpretations of nudity and its symbolism, which have coloured much more than just ancient Greek cultural paradigms. Nudity—in particular female nudity—has long been weighed down by misogynistic attitudes in earlier scholarship of the Greek and Near Eastern worlds.[77] As Cornelius notes, 'Representations of naked women seemed to have somewhat bothered earlier scholars working in the ancient Near East.'[78] The nude female was often describe in terms of her 'sensuousness and self-indulgence,'[79] associated with Canaanite 'orgiastic nature worship.'[80]

Nudity was also taken as a catch-all symbol of generic 'fertility.' For instance, in his seminal work on female iconography in 1983, Urs Winter wrote of nude female imagery: 'Dabei darf man voraussetzen, dass das Bild einer nackten Frau immer irgendwie in Beziehung zur Sexualität und – im Kontext des Alten Orients – zur weiblichen 'Fruchtbarkeit' steht…'[81]—'One can assume that the image of a naked woman is always somehow related to sexuality and—in the context of the ancient Orient—to female fertility'. Jo Ann Hackett attacked this paradigm back in 1989, arguing that it was a one-word explanation reserved solely for female Canaanite deities, whereas male deities always received much more nuanced analyses (despite certain male gods being also heavily associated with fertility).[82] Stephanie Budin noted this same tendency for scholars who studied this imagery on Cyprus. In fact, during the twentieth century, the fertility paradigm/mother goddess interpretation has coloured most readings of female imagery from the Neolithic onwards.[83]

More recent readings have deconstructed the notion of female fecundity in the ancient world.[84] As Budin suggests, female imagery did not symbolise a generic 'fertility,' but could be tied to particular concepts, in particular a female's 'physical eroticism, her ability to stimulate desire, and her eventual ability to rear and nourish an infant provide by a male.'[85] Other interpretations have also emphasised roles for this imagery beyond symbolising fertility, for instance Marinatos' reading of the nude female in her 2000 monograph as signifying potency through her sexuality for the warrior male, as a dangerous character whose power had to be harnessed for purposes of protection, whereas other interpretations of nude females focus on their use in rituals of procreation and apotropaism.[86]

While scholarship has moved beyond the fertility fetish, however, it continues to emphasise the eroticism of the nude standing female.[87] This interpretation is understandable from a modern perspective, in particular given the tendency of this imagery to highlight sexualised characteristics such as breasts and genitalia. Inanna-Ishtar, and to a certain extent, Astarte, were associated with sexuality in various Near Eastern traditions (and, at least for Inanna-Ishtar, this sexuality combined with notions of fertility as well).[88] Yet, while viewing an image of a nude female cupping her breasts in the twenty-first century might automatically bring up associations with sex, did this imagery hold the same meaning for ancient Mediterranean and Near Eastern cultural groups? What did an emphasis on the nude female in general and specific attributes like the breasts and genitalia symbolise? I suggest looking more closely at the ancient literature to better interpret the imagery of the nude female within these long-standing elite social networks that span the Mediterranean and Near East.

Emphasis on the breast in Near Eastern literature in some cases is bound up with sexuality and fertility. In the Sumerian text known as 'The Courtship of Inanna and Dumuzi,' Dumuzi sings: 'Inana, let your breasts be your fields, your wide fields which pour forth flax, your wide fields which pour forth grain! Make water flow from them!'[89] Yet, more often, the breasts in literature are associated with nurturing, rejuvenation, and legitimisation through kinship, particularly in divine and royal spheres. The Babylonian god Marduk in the *Enuma Elish* was emphasised to have 'sucked the breasts of goddesses,'[90] while Gilgamesh, son of a human father, suckled his mother, 'the exalted cow Wild-Cow Ninsun.'[91] Gilgamesh was declared to be two-thirds divine. Likewise, two Ugaritic gods, Dawn and Dusk (Shahar and Shalim), were born from El and two human women, and were nursed at the breast of the goddess Athirat to become the 'Gracious Gods.'[92]

Fully mortal humans also gained status and power through suckling goddesses: the Ugaritic epic of King Kirta, *KTU* 1.15, contains a prophecy that Kirta's son will 'drink the milk of A[thi]rat; he will drain the breast of Virgin…; the suckling of [goddesses.'[93] Most scholars identify the 'Virgin' as Anat, an adolescent warrior goddess and sometimes counterpart to Astarte, since Anat was consistently called 'Virgin' in Ugaritic literature as well as in Nineteenth-Dynasty Egyptian sources.[94] An ivory plaque from Ugarit, dated to ca. 1250 BCE and thought to be one of several furniture attachments for a bed, shows a winged goddess suckling two infants and may symbolise Anat, who is the only goddess portrayed with wings in Ugaritic literature (see Figure 2.7).[95] The Ugaritic hymn to Anat calls her the 'Breast of the nations.'[96]

It is important to emphasise here the overall trope being expressed in these passages and in contemporary art, namely the suckling of gods, heroes, and kings by goddesses, who in turn convey divinity, authority, and legitimacy through their milk, as well as rejuvenation and prosperity in the afterlife.[97] This trope bears a direct link to Egypt, notably through goddesses like Hathor, who was depicted suckling the king in both human and bovine form. Hathor and other goddesses acted as the 'divine wet nurse,'

Figure 2.7 Ivory plaque with winged goddess suckling youths from Ugarit, drawn by Paul Butler.

Source: Damascus Museum 3599, © Paul Butler 2014.

with the divine suckling symbolising the potency and legitimacy of kingship.[98] In the Nineteenth Dynasty, the goddess increasingly took on the image of the Queen Mother, while the imagery also appeared in non-royal tombs.[99] We see these same traditions infiltrating Neo-Assyrian literature in the Iron Age: in prophetical texts, Ishtar of Arbela is called the 'good wet nurse of king Ashurbanipal.' In one text, Ishtar is said to have four breasts to feed the king.[100]

The claim to have nursed at the breasts of a goddess was significant in the ancient world, particularly for kings whose legitimacy needed to be enforced like Ashurbanipal.[101] The act of nursing from a goddess did not necessarily mean a king was thus born from the goddess' womb, but it did symbolise kinship with the gods via the breastmilk, a substance seen in the ancient world to confer power and status on an individual. Once again departing from modern ideas, breastmilk appeared to have greater kinship-forging properties in the ancient world than blood. Shared beliefs across cultural groups concerning breastmilk and its power in bestowing kinship and legitimacy can be observed in various literary sources. For instance, Hebrew texts feature stories of Israelite or Judaean women of royal and priestly status breastfeeding and thus legitimising male eponymous ancestors of priests and kings. Exilic and post-exilic texts position Jerusalem as a child nursing 'the milk of nations' and the 'breasts of kings' (*Isa.* 60:16), a tradition Cynthia Chapman links to the Judaean Pillar Figurines who, like many of the nude standing females, cup their breasts with their hands.[102]

The emphasis on the breasts in nude standing females can arguably indicate this sense of kinship with deities. More elaborate examples, such as the ivory plaque from Ugarit, certainly indicate the 'divine wet nurse' nature of goddesses.[103] Yet, images of females nursing children, often referred to as *kourotrophos* imagery, have evoked debate and ambivalence amongst scholars as to whether they depict deities—akin to the ambivalence over the 'Astarte plaques,' the nude females without divine attributes discussed above. Like the 'Astarte plaques,' *kourotrophos* imagery occurs in all types of contexts. There are examples where it is definitely divine, such as the Egyptian examples mentioned above, but also more ambiguous cases, particularly New Kingdom plaques and figurines showing a female lying on a bed with a child without any divine attributes. While these more mundane objects were often associated with goddesses like Hathor and Mut, their meaning is debated.[104] Like the 'Astarte plaques,' scholars have linked them to rituals involving fertility, procreation, and healing,[105] and Budin suggests seeing these Egyptian figurines as ones used in rituals to increase the potency of an individual or spell. Once again, a similar interpretation has been offered for the nude standing female imagery.[106]

In some cases, nude female and *kourotrophos* imagery intermixed, particularly in the southern Levant.[107] Budin, in a 2015 article, however, has attempted to clarify differing meanings between, on the one hand, nursing imagery, and, on the other hand, nude standing female imagery. The nude standing female, according to Budin, derived from the imagery of Inanna-Ishtar in Mesopotamia and symbolised a goddess in the Levant, namely Astarte. This nude female imagery from Mesopotamia was erotic and not maternal—and it was strictly differentiated from nursing maternal goddesses and other nursing females: 'The Nude Goddess is not maternal; the maternal goddess is not nude.'[108] Budin sees the examples from the southern Levant of nude and nursing females as simply a mixing of artistic traditions.

Yet, the literary tropes discussed above open the suggestion that much of this imagery, divine (and clothed) wet nurses like Hathor to more generic nude standing females, can be seen as symbolising kinship with deities and, consequently, protection and even some cases legitimacy stemming from their powers. The nude standing female imagery derives from the more erotic images of Inanna-Ishtar, as Budin argues, yet Inanna-Ishtar is also one of the quintessential goddesses who confers legitimacy on a king, including through nursing. In some cases, this relationship was simultaneously sexual and maternal.[109] These cross-associations must therefore leave open the possibility that the nude standing female, often with hands cupping the breasts, represented the potency of the life-giving substance from these goddesses that was so often evoked as conferring heroism, divinity, and legitimacy on individuals in the ancient Near East. The emphasis on the genitalia as well can be linked to the transference of vitality and power.[110] The entire process—from conception, to birth, to breastfeeding, and weaning was seen in the Hebrew Bible as forming a fully complete human.[111] The various gestures taken by nude female figurines across the ancient world—from emphasising the breasts, to sometimes having one or both hands on the abdomen or genital area—may allude to this the widespread ideology of conferring ruling power and legitimacy on an individual through their kinship with the gods.

Furthermore, while the literary sources stem from royal praise poetry and epics, the widespread distribution of this imagery, particularly in Levantine, Egyptian, and Cypriot domestic contexts, also suggests that, over time, a larger array of social classes made use of this trope in ritual activity, possibly in relation to domestic concerns such as protection of children, pregnancy, and lactation,[112] or else funerary concerns revolving around notions of rebirth and vitality in the afterlife. A reading of the nude standing female as symbolising, above all, intimacy and kinship with a goddess thus also leaves the door open to related readings concerned with divine protection in matters pertaining to daily life and the afterlife. Beyond the 'Orientalising' sanctuaries, this imagery itself can be taken as an outcome of the spread of complex contagions across clusters given that it was used, accepted, and understood across cultural groups for multiple, overlapping purposes. In particular, Centola and Macy's notions of *credibility* and *legitimacy*: innovations spread more easily when neighbours adopt innovations and can be witnessed employing these innovations. The association of the nude female with powerful ruling structures in Mesopotamian and Syrian city-states in the Early and Middle Bronze Ages may have facilitated the necessary legitimacy and credibility of this complex contagion; the fact that this imagery explodes across western Asia and the eastern Mediterranean in the Late Bronze Age period, the so-called age of the 'Brotherhood of Kings,' further strengthens the role of elites in imbuing this imagery with powerful meanings that endured for centuries.

Goddesses and kinship in Iron Age Greece

Bringing the argument back to the emergence of the nude standing female imagery along the multiple networks converging within Iron Age sanctuaries, one must ask whether this interpretation of the nude goddess is convincing for the Greek world. Nudity had many meanings across different genres and genders, which could be both positive and negative.[113] The instances where mortal women bare their breasts in Greek literature generally appear in negative circumstances, symbolising moments of great danger, weakness, and vulnerability.[114] Examples include Hecuba when she bares her breast to her son Hector when beseeching him not to fight Achilles (*Iliad* 22.77–92) and Clytemnestra before Orestes when he is about to kill her (Aeschylus, *Choephori*, 885–930). Goddesses like Aphrodite also express vulnerability when they are nude—for instance, in *Homeric Hymn 5*.[115] Yet, despite the negative connotations that stem from these scenes, both Hecuba and Clytemnestra are clearly displaying their breast in an attempt to invoke pity in their sons through emphasising kinship, while Aphrodite's nudity leads to the begetting of Aeneas. The scene with Hecuba underscores this link: 'And for her part his mother in her turn wailed and shed tears, loosening the folds of her robe, while with the other hand she held out her breast, and shedding tears she spoke to him winged words: "Hector, my child, respect this and pity me, if ever I gave you the breast to lull your pain."'[116]

With goddesses, Hera is the prime example of a deity who confers legitimacy on gods and heroes—in particular with Herakles, via her breastmilk.[117] The nursing of Herakles by Hera is absent from Greek art, but occurred several times in Italic and Etruscan art, in two instances of Red-Figure vase painting and on four mirrors from the fourth century BCE.[118] In one instance, Hercle (the Etruscan Herakles), interestingly, is depicted as an adult nursing at Uni's breast, suggesting Uni's breastmilk bestowed immortality on him upon his death and thus allowed for his apotheosis. Anthony Corbeill, in fact, notes several episodes in Etrurian and Italic cultures where breastmilk is associated with funerary ritual as a means to offer new life (immortality) to the deceased.[119]

Larissa Bonfante has stressed the differences between Italic art, which shows a greater acceptance of the act of breastfeeding to confer legitimacy, and Greek art, which tended to avoid the depiction of Hera breastfeeding Herakles.[120] Tom Rasmussen, however, has raised the possibility that this act of breastfeeding to signify adoption was not unknown in Greek tradition: late accounts have Hera breastfeed Dionysus (Nonnus 35.319–335) and Hermes (Hyginus, *Astronomica* 2.43; Martianus Capella 1.34).[121] Other goddesses as well, namely Demeter and Athena, are linked to the nursing of heroes, although the explicit mention of breastfeeding is avoided, a point to which I will return.[122] Bonfante suggests that nudity in general in Classical Greek art was 'something to be avoided. The nursing image in particular was a sign of non-civilisation, characterising the animal nature of such creatures as centaurs and barbarians.'[123] Yet, given the link between breastfeeding and kinship from the Homeric epics onwards, are different attitudes at play in Iron Age worship, in particular via the 'Orientalising' networks as discussed above?

Imagery of the popular Egyptian motif of Isis nursing Horus can be found at places along these networks. Four figurines from the cave sanctuary of Eilithiya at Inatos on Crete replicate Egyptian images of Isis nursing Horus, for instance, and other examples have appeared in the form of figurines and amulets at Amnisos and Rhodes.[124] As well, faience figurines of Isis nursing Horus were found at the Heraion on Samos.[125] Several faience examples of this imagery also come from Rhodes, along with numerous *kourotrophos* figurines from deposits at the sanctuary of Athena Lindia, including a standing female in a polos nursing a child.[126] Certainly, this other class of figurines, the *kourotrophos*, formed part of the cult assemblage of female goddesses into later periods. Once again, this imagery was rare in the Bronze Age, but proliferated in the Archaic Period in both sanctuary and funerary domains.[127] The *kourotrophos* in the Greek world most often appears clothed, and Böhm and others have taken this imagery to be distinct from the nude female, associated with nursing and motherhood.[128] As discussed above, I suggest that this distinction—between nude female and nursing female—was not so clear-cut.

We can, for instance, detect a mixing of nude female and *kourotrophos* imagery from the mainland in an eighth-century figurine in the Benaki Museum of a nude female seated sideways on a saddle atop a horse nursing a young male. Papageorgiou's assessment of the stylistic characteristics of the horse suggests an Argive workshop (the piece comes from a private collection).[129] A comparable bronze statuette is known from the altar at the Heraion on Samos, dated to the first half of the eighth to the seventh century BCE, which bears stylistic similarities to specimens from the Colchis region. Of particular interest for the Benaki example is the fact that the child is a youth, not an infant, suggesting the nude female atop a horse played a role in rearing aristocratic males through the intimate and protective act of breastfeeding, which has obvious parallels to mythical traditions from the Near East and Mediterranean worlds, and ties into earlier interpretations of the ideological meanings of the nude female.[130] The warrior aspect—connected to the youth and the horse—is also within the purview of the nude female in myth, who frequently appears in Indo-European traditions as a source of vitality for wearied warriors.[131] Furthermore, the female seated sideways on a throne atop a horse has fascinating parallels with examples from Cyprus and the Levant, as well as from imagery in Egypt; in both Cyprus and Egypt, the female seated atop a horse is generally interpreted as Astarte, whose role in legitimising and protecting the king—and also a warrior herself—is clear.[132]

The wide variety of emotional experiences associated with this imagery—vulnerability towards danger, the conveyance of power and vitality to humans via gods, and the guiding of youths through important milestones in life—also suggests another social mechanism by which complex contagions spread and take root, namely emotional contagion. Centola describes this mechanism as 'expressive and symbolic impulses in human behavior that can translate into emotional, rather than deliberative, forms of social contagion.'[133] This mechanism admittedly must be more suggested than clearly demonstrated for the appearance of the nude female in the Greek world; nonetheless, the raw power associated with the gestures often assumed by this imagery needs emphasising.[134] Overall, however, nude female imagery is short-lived in the Greek world, and nude-nursing female imagery is extremely rare. Furthermore, it is possible to detect a certain circumspection in art and literature concerning goddesses who bestow their potency on worshippers through breastmilk, although the theme of goddesses nourishing a hero as a key component of their quasi-divinity continues.[135] The meaning of nude standing female appearing in the 'Orientalising' sanctuaries needs further discussion.

The nude standing female and 'communities of practice'

Scholars have indeed emphasised the 'non-Homeric' character of the nude standing female,[136] and this imagery certainly did not fit into the repertoire of attributes that we traditionally associate with Greek goddesses based on Homer, Hesiod, and later authors. While analyses by scholars like Böhm and Marinatos have helped clarify the function of this imagery in cult situations, the ideological meanings of this imagery can be further explained through placing it in the contexts of the converging synchronic and diachronic social networks discussed above. Specifically, its emergence in the 'Orientalising' sanctuaries situated along numerous overlapping cultural and commercial routes linking Greece to regions with a very long and ideologically charged usage of this imagery is highly significant.

In this sense, focusing on the importance of objects and their agency in uniting far flung communities into similar practices and belief systems—termed 'communities of practice'—can help elucidate the meanings of the nude female in Greek sanctuary contexts as well as the mechanisms of strong ties. As discussed above, the concept of 'communities of practice' was originally outlined by Lave and Wenger in their 1991 work as a process by which humans adopt certain sociocultural praxes through situated learning. As Carl Knappett argues, 'communities of practice' must be conceived of as both human and non-human, the non-human referring to artefacts and the situations in which humans and artefacts participate.[137] In particular, objects serve to 'scaffold' lived experiences within these webs of relations by catalysing shared identities and experiences.[138] In a later publication, Wenger also emphasised the role of form or 'thingness' in learning through the duality of participation-reification. In Wenger's words, reification refers to 'the process of giving form to our experience by producing objects that congeal this experience into 'thingness'. In so doing we create points of focus around which the negotiation of meaning becomes organised.'[139] Beyond Lave and Wenger, Knappett and others have also stressed the ability of objects to enable a 'release from proximity,' allowing group identities and praxes to take form over long distances via objects and their symbolic associations.[140] Still other research has considered the power of objects—specifically 'Orientalising' objects in sanctuaries—as powerful media of knowledge transfer that preceded and set in motion new systems of meaning.[141]

All of these phenomena—humans, objects, and spaces—thus form multipartite networks of situated learning that exist through the constant creation of relationships and knowledge by human and non-human actors.[142] The concept of strong ties can certainly gel with the model of communities of practice, whereby cross-cultural norms and behaviours are constantly reaffirmed amongst practitioners through practice, the creation of meaning, and the reification of these practices and meanings into object form. In fact, a network approach can help to alleviate one of the main concerns of Wenger and others: namely, issues of scale and the stretching of the concept of communities of practice beyond clear localities, and thus 'beyond recognition or usefulness.'[143] Drawing network analyses into realms that look beyond the weak links, as Centola and others have done, promises to adapt the study of situated learning to larger scales and 'constellations of practice,'[144] which are certainly representative of the long-term interconnectivity of the eastern Mediterranean. The idea that some objects can act as boundary objects—things that 'can bridge, mediate, or form liminal spaces between communities,' be adapted to 'distinct perspectives, meanings, and values,' and are wrapped up in complex power relations[145]—can also be nested within the mechanisms of complex contagions such as legitimacy, credibility, and emotional contagion. All of these concepts further enhance our understanding of the nude standing female.

The ideologies bound up in the nude standing female appear to have held special significance in emergent sanctuaries to poliadic goddesses like Hera, Athena, Artemis, and Aphrodite. The associations of all of these goddesses with the Near East—in particular, Aphrodite and Ephesian Artemis—have been explored in previous literature. These relations, combined with the exotic repertoires in sanctuaries, thus position these places as apt arenas where non-material ideas and ideologies could be practised and shared. The long history of use of this cross-cultural symbolism and its current cachet as part of the culture repertoire of the Neo-Assyrian Empire[146] lent the nude female imagery the necessary legitimacy and credibility for its expression within elite assemblages in Greek Iron Age sanctuaries alongside broader 'Orientalising' imagery. Of course, in line with the concept of boundary objects discussed above, this imagery might have been repurposed in local contexts over time: for instance, scholarly interpretations have examined how the nude female imagery eventually evolved to celebrate maturation rites on Crete.[147] This reality is nicely captured by research on situated learning and communities of practice, which stresses knowledge as being in a constant process of negotiation and adaptation through use, and also recognises that actors can belong to multiple and often overlapping communities of practice that affect these processes of negotiation and adaptation.[148]

Moreover, the exotic and high-value nature of many of the items occurring alongside nude female imagery, which often displayed the imagery of elite activities such as hunting and warfare, suggests that the ideology of kinship with goddesses held special relevance for elite males in the Iron Age, a suggestion already explored by Marinatos in her 2000 monograph. While many examples of the nude female occurred in simple terracotta form, such lower-value media do not necessarily imply that their donors were poor, as Gina Salapata has recently argued.[149] Rather, I suggest that the appearance of the nude female in 'Orientalising' sanctuaries dedicated to powerful goddesses was bound up in broader ideological expression of power and legitimacy expressed by elite males in Iron Age societies through special kinship with these goddesses and with other elites across cultural boundaries. These practices replicated earlier (Bronze Age Near Eastern) use of this imagery as a form of legitimisation, but they also mapped onto heroic ideals expressed in Archaic Greek literature—for instance, the numerous examples of female deities protecting heroes in the Homeric epics.

Finally, the disappearance of this imagery in the sixth century was not simply due to Greeks shifting their artistic preferences or becoming embarrassed by nudity, but represents, rather, a shift in ideas of how humans related to the gods, and specifically how elites could claim intimate connections with goddesses as a basis of their social power, an argument I explore in a forthcoming monograph. Such a change emerged alongside moves towards egalitarianism in various parts of the Greek world, reflected in shifts away from personal and aristocratic dedications at sanctuaries and expensive votive materials like bronze, not to mention a broader disappearance of 'Orientalising' imagery.[150] I suggest that these ideological shifts influenced the later chariness towards explicit breast-feeding (specifically by goddesses) and general familial relationships to goddesses that we witness in literature.[151] Lastly, the recognition of these shifts aligns with calls to account for processes of discontinuity and transformation in the study of communities of practice, which—as with social network analyses—tends to over-emphasise connectivity and continuity.[152]

Conclusion

Nicholas Purcell offered three 'givens' in order to understand the so-called Orientalising Period: (1) the Mediterranean experienced dramatic social, economic, and cultural change in this period; (2) it exhibited an uptake in 'large-scale, long-distance interaction'; and (3) it is impossible to understand this period (and these phenomena), without encompassing Anatolia, western Asia, Egypt, and the Mesopotamian heartland.[153] Purcell also advocated modelling these complex interrelations 'without the dead weight of monolithic constructs such as the Orient.'[154] Emphasising the strong social ties inherent in the arenas where these interconnections are traced in tandem with their long-term routes can thus help refocus questions towards the processes of sustained interaction over long periods and distances. Of course, while sanctuaries and their associated cultic activities are one way to characterise these strong ties, there were undeniably other overlapping commercial and even familial long-distance connections emerging from the Bronze to Iron Age transition.[155]

I have emphasised especially in this chapter the two overarching types of networks that formed sanctuaries into effective wide bridges for the transmission of the imagery and meaning of the nude standing female: synchronic ones, represented through the multicultural assemblages of numerous 'Orientalising' sanctuaries, and diachronic ones, represented through the imagery's long-standing associated ideologies particularly as read through literary paradigms. These two types of networks converged within sanctuaries where worshippers emphasised specific ways of relating to the deity through votive symbolism and deposition. While, in the context of these sanctuaries, I have suggested male elite uses of this ideology, it undoubtedly surfaced in other contexts as well, suggesting that multiple levels of society understood and used this symbolism to communicate with the divine. Its disappearance from the Greek repertoire is thus fascinating not only for artistic purposes but also—and especially—for religious and ideological ones.

An obvious next step for this research would be to nuance the emergence and spread of the nude female imagery in the Greek world and to further explain why certain areas adopt this motif earlier and more intensively than others—for instance, eastern Crete emerges as a clear anomaly on the map in Figure 2.3, with very high concentrations of nude female figurines at several sites. Such further research would aid in elucidating the 'small world' cluster of Greece in terms of the spread of ideologies and symbolisms from Egypt and western Asia. Overall, this chapter's main aim has been to emphasise that we cannot understand the Iron Age merely in terms of traded or emulated objects and styles. These objects and styles, however, when placed in the context of wide bridges that stretched across social, cultural, and temporal boundaries, can illuminate the immaterial heritage of long-term connectivities sustained by these 'Orientalising' networks.

Abbreviations

ETCSL Electronic Text Corpus of Sumerian Literature
KTU Keilalphabetischen texte aus Ugarit
LP Lobel and Page = Lobel, E. and D. Page. 1955. *Poetarum Lesbiorum Fragmenta*. Oxford: Oxford University Press (corrected ed. 1963).

Ancient Sources

Homer. *Iliad*. Vol. 2: Books 13–24. Translated by A. T. Murray and W. F. Wyatt. 1999. Cambridge: Harvard University Press.

Notes

1. E.g., Boardman 1999, 62–63.
2. E.g., Miller Ammerman 1991.
3. E.g., Horden and Purcell 2000; Broodbank 2013.
4. E.g., Knappett 2011.
5. Collar 2013; Brughmans, Collar, and Coward 2016; Knappett and Leidwanger 2018.
6. Burkert 1992, 6.

7. See López-Ruiz 2014 for a lengthier review and bibliography on works from this period.
8. To name a few: Burkert 1992 [1984]; Bernal 1987, 1991, 2006; Morris 1992; West 1997.
9. Morris 2006, 66–67. See, in general, the edited volume by Riva and Vella (2006) in which Morris' chapter appears. These shifts of course largely went hand-in-hand with the rise of postcolonialism post-World War II and especially notions of orientalism that emerged in the late 1970s (Said 1978), but were acknowledged to have a much longer genealogy (Said 1978, 55–57). See also Schweizer 2005, 355–360; Gunter 2014.
10. In terms of Greece's social and cultural relationships with the Near East over the Bronze to Iron Age transition, Sarah Morris' 1992 book, *Daidalos and the Origins of Greek Art*, was ground-breaking. More and more works are bridging the divide between these periods and illuminating the social, cultural, political, and economic transitions over the period between ca. 1100 and 700 BCE, most recently Murray 2017 (and see Murray 2017, 3, n. 2 for a list of works since 1990 dealing with the Bronze-Iron Age transition).
11. Horden and Purcell 2000, 74; Morris 2003; Broodbank 2013.
12. E.g., Morris 1992; Papadopoulos 2014; Murray 2017; Knappett and Leidwanger 2018, 9–11.
13. Malkin 2011; Knappett 2011; 2013; Tartaron 2013; Collar 2013; Brughmans, Collar, and Coward 2016; Knappett and Leidwanger 2018.
14. Morris 2003; Concannon and Mazurek 2016, 8–9; Manning 2018, 76.
15. Collar 2013.
16. E.g., Knappett 2011; Feldman 2014; Roddick and Stahl 2016. Lave and Wenger's 1991 model was later modified by Wenger (1998).
17. Granovetter 1973.
18. Watts and Strogatz 1998.
19. Centola 2018. Centola and Macy, in their influential 2007 article in the *American Journal of Sociology*, outline the means of reinforcement for the spread of complex contagions: strategic complementarity (the more people adopt an innovation, the more its value increases); credibility; legitimacy; and emotional contagion (2007, 707–708).
20. Ugander et al. 2012.
21. Everton 2018, building on Chaves 1997.
22. de Polignac 2009, 427; Crielaard 2016, 351.
23. Curtis 1994; Braun-Holzinger and Rehm 2005; Aruz et al. (eds.) 2014.
24. Gunter 2014, 79–80. Contemporary literature, in the form of epic and lyric poetry and cosmogonies, have also been studied through the Orientalising lens—e.g., Martin West: 'Greek literature is a Near Eastern literature' (1966, 3). See, more recently, López-Ruiz 2010; Louden 2011; Haubold 2013; Whitley 2013; Metcalf 2015 (the latter for a more critical view on the extent of borrowing between Greek and Near Eastern literature).
25. Gunter 2014, 84–85; cf. Osborne 1993, 233; Morris 2006; Whitley 2013, 410; Villing 2017, 564.
26. I will continue to use the term 'Orientalising' sanctuaries throughout for the sake of brevity.
27. See Crielaard 2016, 351–352 in general. On kings in Herodotus: Midas and Gyges: 1.14; Croesus: 1.50–52; Necho: 2.159; Amasis: 182; Datis: 6.97, 118; Xerxes: 7.42. On itinerant professionals: Burkert 1992; Schweizer 2006, 189–193. On dedications by foreign kings in Greek sanctuaries: Kaplan 2006. On the possibility of foreigners other than kings dedicating in Greek sanctuaries: Kyrieleis 1979; Bumke 2007; 2012 (among many others). On the role of mercenaries returning from abroad in depositing foreign votives: Morris 2006,

67–68; Ebbinghaus 2006, 213; Villing 2017, 570 (cf. Alkaios fr. 48 and 350 *LP*). On women dedicating in these sanctuaries: Kilian-Dirlmeier 1985, 228–230; Morris 2006, 68; Bumke 2007; Weiss 2012, 510. On the possibility of tomb-robbing: Sherratt 2012; On the more general role of trade and reciprocity in generating multicultural assemblages: Klebinder-Gauß and Pülz 2008, 202 (cf. Herodotus 4.152 on the Samian traders and the Heraion of Samos; Schweizer 2007, 307–309).

28. Mylonopoulos 2006, 377; Klebinder-Gauß and Pülz 2008, 201; Crielaard 2016, 352. Other classes of finds, for instance, animal bones used in sacrifices, have in some instances been used to demonstrate the presence of certain cultural groups (e.g., Bammer 1985 on donkey bones at the Artemision at Ephesus linked to Phoenician cult activity).
29. Crielaard 2016, 354.
30. Figure 2.1 is adapted from Imma Kilian-Dirlmeier's chart in a 1985 publication. As with all old data, these counts should be updated and refined, but this network visualisation nonetheless still gives a good example of the range of foreign dedications at the Samian Heraion.
31. Kilian-Dirlmeier 1985, 235–243; Brize 1997; Mylonopoulos 2006, 362–363, 376; Morris 2006, 67–68; Moyer 2011, 57–58. Ian Moyer notes the combination of practices on this piece: Greek writing in boustrophedon style, but placed on the statue in a way that conforms to Egyptian practices; a biographical text deposited in a Greek sanctuary that celebrates honours from a king (the latter a very common Egyptian practice). 'These mixed features of Pedon's statue and its use suggest that Pedon brought back with him to Ionia not only a statue, but also some cultural knowledge surrounding its function in its original context' (Moyer 2011, 58).
32. Crielaard 2018; Wenger 1998, 109–110.
33. Schmidt 1968, 113–119; Fourrier 2001; Henke 2017.
34. Klebinder-Gauß and Pülz 2008, 202–203.
35. Skon-Jedele 1994; Hölbl 1999; Ebbinghaus 2006; Villing 2017.
36. Strøm 1992, 49–57.
37. See Villing 2017 for more bibliography on *aegyptiaca*.
38. Franklin 2015; Fletcher 2004.
39. Duplouy 2006; Crielaard 2016, 355. On object biographies in this period see also Whitley 2002; Ebbinghaus 2006: 212; Duplouy 2006, 174–177; Gunter 2009; Feldman 2014; Villing 2017, 569.
40. Saint-Pierre 2009.
41. Centola and Macy 2007, 726.
42. Centola and Macy 2007, 707–708.
43. Villing 2017, 565. cf. Morris 2006, 72; Skon-Jedele 1994, 1779–1780; Apostola 2015, 110; Fletcher 2004.
44. Postill 2015.
45. Gunter 2009, 16; Moyer 2006. See also Guralnick 2004 and Aruz 2014 on the role of the Neo-Assyrian Empire in the circulation of objects, people, and ideas.
46. Wenger 1998; Roddick and Stahl 2016, 5.
47. Contemporary examples of nude females emerge in the Near East in this period—for instance, as part of the carved ivory repertoires from Assyrian palaces like Nimrud (Gubel 2005, 129–133).
48. Miller Ammerman 1991, 221; Marinatos 2000, 29–30; Erickson 2009, 364–365; Pilz 2009, 99. On the nude female transmitted directly from the Near East (to Crete): Böhm 1990, 107–116. Cyprus received new mould technology from the Levant in the Iron Age: terracotta figurines and relief plaques of the nude standing female have been found in tombs and sanctuaries from a number of sites including Arsos, Kition, Salamis and Amathus. The use of the mould is attested in the seventh

century BCE at Samos, Ephesus, Corinth, Argos, Laconia, Attica and Western Greece (Prent 2005, 406, n. 1045; Böhm 1990, 77–87, 103, 141; Higgins 1954, 11), and possibly as early as the eighth century BCE on Crete (Pilz 2009, 99).

49. Many finds from Crete remain unpublished (Pilz and Krumme 2011, 323). Recent finds from Roussa Ekklesia have added a number of nude standing female plaques, made from one or two moulds, to this repertoire (Erickson 2009, 357–365). As well, examples of seated nursing female figurines (*kourotrophos*), along with Daedalic-style plaques showing standing clothed females wearing a high headdress (*polos*) and nude standing females, have come to light at Anavlochos (Gaignerot-Driessen 2018; Pilz and Krumme 2011). See also Pilz 2009; Prent 2005, 404–411; Marinatos 2000. New finds are also emerging from Miletus: von Graeve 2013; Senff 2003; Hölbl 1999 and Naukratis. Böhm's catalogue also excludes western Mediterranean sites (see Miller Ammerman 1991).
50. Whitley 2013; Hoffman 1997.
51. For instance: the bronze horse frontlet from the Heraion at Samos bearing an inscription identifying it as a gift to King Hazael of Aram-Damascus from the ninth century BCE (but deposited in the Heraion in the seventh century) (Aruz et al. (eds.) 2014, 296–297, no. 156; Jantzen 1972, Taf. 52). Another well-known example is the bronze bowl from Olympia with Aramaic inscription showing various figures, including a nude standing female and female musicians (Aruz et al. (eds.) 2014, 311, no. 183).
52. Once again, when large numbers of eight-century mould made terracotta statuettes from sites like Roussa Ekklesia are added in (Erickson 2009), the numbers would certainly shift towards terracotta; nonetheless, earlier periods still show higher numbers of high-value materials like ivory and bronze used to make nude female imagery.
53. Böhm 1990, 59–69.
54. Budin 2015, 315; 2019.
55. Woolley 1955, 244–247, Pls. LIV, LV; Marchetti 2007; Matthiae 2013; Marchetti and Nigro 1997, 22; Ramazzotti 2014; Yener 2015, 208, Figs. 6a, 6b, 8a, 8b.
56. See Budin 2015, 317, Figs. 3 and 4; Yener 2015.
57. E.g., Matthiae 1993, 655–661; 2013, 5; see also Dolce 2008.
58. Frankfort 1939, nos. 270–271; Winter 1983; Budin 2003, 199–241; Keel and Uehlinger 1998, 26–29; Matthiae 2013, 3–4; Tsukimoto 2014, 23; Pizzimenti 2014.
59. For instance, BM 132358, an Old Babylonian hematite cylinder seal, shows a goddess facing a male (a king?) with a small nude standing female between them, with hands under the breasts.
60. Teissier 1984: 26.
61. Some scholars identify the winged nude female with the Hurrian goddess Shaushka, who was equated with Ishtar in the early-mid second millennium BCE (Beyer 2001, 95; Ornan 2005, 31–32; Tsukimoto 2014, 23). Lipiński (1986, 90) instead argues that the nude winged female represents Ishara (see Budin 2003, 217–218). Old Babylonian terracotta plaques also depict a winged, naked, frontal-facing female in horned headdress—e.g., the famous 'Burney Relief,' sometimes identified as Ishtar (Collon 2005, 6, Fig. 1), but whose authenticity is in doubt (Albenda 2005). See Yener (2015, 211) on various headdresses associated with Anatolian, Hurrian, Syrian and Levantine goddesses.
62. See, for instance, Matthiae (2013, 4–7) on cylinder seals from Middle Bronze Age Ebla, which associate the nude standing female with the king. This imagery is repeated in contemporary Old Syrian and Old Assyrian cylinder seals as far north as Anatolia (Kultepe/Kanesh).
63. Albright 1939; 1946, 76. For instance, tablet RS 20.24 from Ugarit is a bilingual text that lists the names of the gods of Ugarit and their Mesopotamian equivalents: Athtart is named as the western equivalent of Ishtar (*'ttrt* = ^{d}ISTAR.istar)

(see Bonnet 1996, 144–147). On the figurines see also Pritchard 1943; Riis 1949; 1960–1961; Nishiyama and Yoshizawa 1997; Keel and Uehlinger 1998: 97–108; Cornelius 2004; Jackson 2006; Budin 2015.

64. Pritchard 1943, 194–195; cf. Riis 1949, 81–83; Patai 1990, 58–60; Dever 2005, 185–189. Miriam Tadmor (1982) reanalysed much of this imagery and suggested nude female figures represented instead 'concubines of the dead,' akin to Late Bronze Age Egyptian figurines (cf. Lipiński 1986, 89). Othmar Keel and Christoph Uehlinger suggest that this figure does represent a divinity, but not a specific divinity (1998, 105; cf. Marinatos 2000, 1; Winter 1983, 194–199). Likewise, Böhm also disputed the identification of the nude female in the Greek world with particular goddesses, and argued that they are more divine figures associated with a female divinity (1990, 23, 86–87, 125–126, 134–141). Meike Prent largely agrees, and suggested they might be neutral images who 'would assume a more specific meaning in the context of specific cults' (2005, 411).
65. Keel and Uehlinger 1998: 100–105.
66. Although Pritchard (1943, 84) also had his doubts about the female holding attributes, and interpreted this type, alongside the simple nude females, as 'divine courtesans or sacred harlots.' For more examples of Levantine nude female figurines see, in general, Winter 1983; Keel and Uehlinger 1998; Cornelius 2004, esp. pls. 5.19–62. The Hathoric iconography has earlier precedents in Syria-Palestine—e.g. a large basin from Temple P2 at Ebla showing a nude winged female with Hathoric hair (Matthiae 1993, 655).
67. Hestrin 1987; Keel and Uehlinger 1998; Ziffer 2010, 413–415; Budin 2016. Budin (2016) suggests these images evoke the Egyptian goddess Hathor, although other scholars have drawn connections with Ugaritic Athirat/Asherah, notably through the tree and female genitalia (see Hestrin 1987 on the Lachish ewer and Ziffer 2010). Palestinian scarab seals that show up beginning in the MBA IIB period also show nude goddess imagery with Egyptian branch motifs (Keel and Uehlinger 1998, 26–27).
68. Darby 2014; Budin 2005; Prent 2005, 400–403; Kletter 1996; Patai 1990, 58–60.
69. Rotté 2012, 14. Budin suggests that these earlier EC III-MC plank figurines likely represented mortal women in the context of family and lineage, based on their findspots. It was only with the emergence of the new imagery ca. 1450–1200 BCE that female figurines became associated with divinity, an outgrowth of both internal developments on Cyprus in this period as well as close contact with Levantine centres like Alalakh (Budin 2014, 198–200; 2002, 317; cf. Bolger 1996).
70. A number of these steles were likely from the village at Deir el-Medina (Cornelius 2004, 68–70). Nude female figures showing Syrian and Anatolian influences also come from sanctuaries (Waraksa 2009; Budin 2011, 121).
71. New Kingdom textual sources betray religious interchange between these two regions in the Eighteenth and Nineteenth Dynasties, including the syncretising of goddesses like Astarte, Anat, Hathor, Sekhmet, Isis and others (Tazawa 2014; Wilson-Wright 2016, 27–71). For religious interrelations with Cyprus, see Ulbrich 2008; Carbillet 2011.
72. This statue comes from the Ishtar temple at Nineveh, dating to the second quarter of the eleventh century BCE (Ornan 2005, 31).
73. Bloch-Smith 2014, 193. As Izak Cornelius (2004, 5) notes, there is only one stele with the name of Anat and one cylinder seal with the name of Astarte known from Syria-Palestine.
74. Budin 2003, 47–57; Böhm 1990, 14–15, 145–146; Lupack 2012, 265–266.
75. Nicholas Coldstream linked these ivories to North Syrian carvers, and specifically to the centre at Hama, whose influence also reached Rhodes and Crete. He also draws parallels with the Loftus Group, a collection of ivories including the standing nude female from the southeast palace at Nimrud 'whither

they were probably carried off by the Assyrians who destroyed the kingdom of Hama in 720 B.C.' (Coldstream 2004, 130). The acropolis at Athens also yielded two bronze figurines of a nude female from about the same date as the grave figurines (Böhm 1990, 151, Taf 10a—B1 and 20e—B2; see also Miller Ammerman 1991, 95 and n. 73 for bibliography).

76. E.g., Marinatos 2000, 27.
77. See Assante 2006, 180–191.
78. Cornelius 2004, 11.
79. Edwards 1955, 51.
80. Albright 1957, 281.
81. Winter 1983, 95.
82. Hackett 1989, 74; cf. Sherwood 2017, 498–499.
83. For Cyprus: Budin 2002, 315; also Budin 2014; Budin 2011, 11–12; Morris (C.) 2006.
84. E.g., Assante 2003; 2006.
85. Budin 2011, 35.
86. Keel and Uehlinger 1998: 100–105. E.g. Darby 2014, 331 on the Judaean Pillar Figurines; Waraksa 2009 on Egyptian nude figurines from the Precinct of Mut; Bloch-Smith 2014 on Astarte figurines found alongside figurines of pregnant females at Sarepta.
87. E.g., Bahrani 1996, 10; Budin 2002; 2014; Pizzimenti 2014, 140–141; cf. Assante 2006.
88. Many hymns from the late third and second millennia associate Inanna-Ishtar with fertility. A cult song, for example, celebrates Inanna's marriage to Dumuzi amidst scenes of growth and fecundity: 'As she arises from the king's embrace, the flax rises up with her, the barley rises up with her. With her, the desert is filled with a glorious garden' (*ETCSL* 4.08.16)
89. *ETCSL* 4.07.7. Likewise, Inanna's vulva is implicated in sexual activity and notions of fecundity (e.g., *ETCSL* 4.08.16; Pryke 2017: 35). Images of the pubic triangle regularly occur on imagery across the Levant and Aegean world in the Late Bronze and Early Iron Ages, often associated with animals and sacred trees, and also appear as part of the nude female's anatomy (Lightbody 2013; Ziffer 2010; Hestrin 1987). For the associations between Inanna, trees/gardens, kingship, and sexuality, see Anagnostou-Laoutides 2017, 66–72.
90. Lines 83–88; Lamber 2008, 39.
91. Lines 35–36; George 2003, 540–541.
92. *KTU* 1.23.24, 59, 61; Chapman 2012, 11.
93. *KTU* 1.15.ii.27–29, as cited in Wyatt 2002, 209.
94. E.g., Allen 2015. Nicholas Wyatt disputes the identification of 'Virgin' (*btlt*) as Anat. Egyptian sources, however, especially under Ramesses II, call Anat a wet nurse and cow. Ramesses II refers to himself as *mhr 'nt*, which can be translated as either 'suckling of Anat' or 'warrior of Anat' (Bowman 1978, 225–234; Ward 1969, 229). Note also that Edward Greenstein reads 'Athirat' as 'Attart' (Astarte) (1998, 25, 45, n. 66).
95. But see Budin (2011, 154–159) for discussion. The goddess is winged and nurses two young boys. The hair is in Hathoric style, with horns flanking a solar disk in 'Hittite style,' although the dress is in 'Syrian fashion' (Ward 1969, 225; Barrelet 1955). Ward (1969, 232–234) suggests that the two infants represent a Syrian adaptation of the Egyptian custom of portraying the king with his spirit (the *ka*)(Budin 2011, 158–159).
96. *KTU* 1.13.19–22 (as cited in Becking 1999, 177, following de Moor's reading—cf. Wyatt 2002, 172, n. 27). See Peggy Day (1992) on Anat's role as wet nurse.
97. Budin 2011, 39.

98. See Budin 2011, 38ff. For instance, Mentuhotep II portrayed himself in his funerary chapel as suckling the breast of Hathor. Above the image is the passage: 'Utterance of Hathor, Mistress of Dendera: 'I protect you, my arms unite you with life. I rejuvenate you with milk so that your enemies may fall under you" (as cited in Budin 2016, 170–171). Hatshepsut records that she was nursed by the Seven Hathors in her funerary chapel at Deir el-Bahari, while Seti I depicted himself suckling the breasts of Hathor at Abydos (Lesko 1999, 106–107; Budin 2016, 171). Many other goddesses took on the role of 'divine wet nurse,' including Isis and Nephthys.
99. Budin 2011, 85.
100. Becking 1999, 177 This imagery extends back to at least the Ur III Period, in a text where Inanna-Ishtar promises king Shulgi protection in battle: 'You are worthy to delight yourself on my holy breast like a pure calf' (*ETCSL* 2.4.2.24; Pryke 2017, 124). The text on the earlier Stele of the Vultures (ca. 2550 BCE) describes Inanna as taking Eannatum, declaring him worthy, and then giving him to the goddess Ninhursaga to nurse (Frankfort 1948, 301).
101. Chapman 2016, 130–131.
102. Chapman 2012, 14–17. Also Sugimoto 2014: 146, n. 17. See also *Isa.* 49:23, where the exiles are prophesied to have kings as their foster fathers and queens as their nursing mothers, and *Isa.* 66:10–13, where the nation is said to nurse at the comforting breasts of Jerusalem (Powell 2018, 74; Chapman 2012, 12). See Ackerman (1993, 398–401) on the role of the Hebrew Queen Mother as a hypostasis of the goddess Asherah, an identification that was tied up in matters of succession and legitimation of kings. See Budin (2011, 61–62) on similar tendencies in New Kingdom Egypt, linking queens with the great goddess of Thebes, Mut.
103. Another example from the Iron Age comes from the eighth-century North Gate at Karatepe displaying a female suckling a child under a date palm, which relates to Egyptian imagery (Ziffer 2010, 416, Fig. 8).
104. Pinch 1993; Waraksa 2009.
105. Budin 2011, 117–135.
106. E.g., Marinatos 2000, 30–31, who suggests that the nude female acts as a source of vigour and luck for the male in the Greek world. For instance, several images from Crete, Olympia, Samos and Sparta show a male between two nude females, or else a nude female flanked by two men who grasp the female's arms (Prent 2005, 430; Böhm 1990, Taf. 35, 36). There are other examples of scenes with anywhere between two and five nude females (e.g., the horse frontlet from the Samian Heraion), an arrangement which Blome suggests was a way of emphasising and augmenting the divine presence (1982, 77; cf. Prent 2005, 410).
107. A striking example is the Late Bronze Age naked female figurine from Revadim shown suckling two small infants, with tree-and-caprid tattoos on the thighs. The female wears a moon pendant along with a Hathoric hairstyle and holds open her labia. Three more fragmentary figures from Tel Harasim, Tel Burna and Aphek appear to be made from the same mould (Ziffer 2010, 415–416, Fig. 7; Budin 2015, 321, Fig. 10; 2011, 167, Fig. 16; Sharp et al. 2015, 64, Fig. 3). Beth Shean also yielded figurines, including a nude female nursing a child from Late Bronze Age levels, another nude female nursing a child from Iron I levels, and an amulet showing an Egyptian goddess with child upon her lap (Budin 2011, 161–163). In some cases, more 'traditional' nude standing female imagery also appears nursing infants (Ben-Arieh 1983; Sharp et al. 2015). Bird-face figurines from Cyprus also frequently held infants to their breasts (Bolger 1996). See Winter 1983, Abb. 57–61.
108. Budin 2015, 319.

109. Thus Ishme-Dagan's hymn to Inanna: 'May my spouse [Inanna], a ewe cherishing its lamb, be praised with sweet admiration!' (*ETCSL* 2.5.4.10; Pryke 2017, 125; see also Pongratz-Leisten 2008 on the sacred marriage ritual as a representation of knowledge transfer between divinity and king).
110. See, for instance, the tale of Hathor flashing her genitalia at Re to cheer him up in a text known as *The Contendings of Horus and Seth*, and a similar trope in Greek myth, where Demeter's maid, Baubo, cheers her up with the same action in an Orphic fragment preserved in later sources (Bonfante 2008, n. 5).
111. See Chapman 2012, 24–25.
112. Böhm 1990, 136–137.
113. E.g., Bonfante 1989a; Assante 2006, 178–180 Hurwit 2007; Pizzimenti 2014, 136.
114. Bonfante 1989a, 560; 1997, 175.
115. Kenaan 2010, 40–43.
116. *Iliad* 22.79–81 (trans. Murray). See Philochorus *FGrH* 328 F35 and Aristotle, *Politics* 1.2.1252b18 for mentions of the *homogalaktes* in Athenian custom, which Aristotle connects with members of a village who are like a colony of a household—'children and children's children'; Philochorus links them to the concept of *gennetai* (Lambert 1998, 60, cf. Derks 1995).
117. Pedrucci 2017, 311; Pirenne-Delforge and Pironti 2016, 270–275. In some versions, Hera unknowingly suckled Herakles (e.g., Pausanias 9.25.2; Diodorus Siculus 4.9.6–7), and, in other versions she gave Herakles her daughter Hebe following his apotheosis (Apollodorus 2.7.7). Pseudo-Eratosthenes and Hyginus recount how Herakles tugged too hard on Hera's breast, causing Hera to pull away and her milk to spill out, forming the Milky Way (see Pedrucci 2017, 314, n. 15 for bibliography; Rasmussen 2005, 36). Vinciane Pirenne-Delforge (2010, 694) suggests that both episodes represent the cycle of upbringing, from nursing (represented by Hera suckling Herakles) to adolescence (represented by Hera giving Herakles Hebe in marriage) (also Pirenne-Delforge and Pironti 2016, 276). Diodorus also recounts an adoption scene where Hera mimicked giving birth to Herakles (4.39.2).
118. Bonfante 1989a, 567–568; 1989b; 1997, 180; Rasmussen 2005, 33–34.
119. Corbeill 2004, 100–106. cf. Rasmussen 2005, 30. See Budin 2011 for similar arguments concerning Egyptian funerary art and Chapman (2012, 16) for similar arguments concerning the Judaean Pillar Figurines.
120. Bonfante 1989a; 1989b; 1997, 180.
121. Rasmussen 2005.
122. Demeter nurses Demophon, who was said to grow up like a *daimôn*, not drinking the milk of his mother (οὐ θησάμενος [γάλα μητρὸς), but was held at the breast of Demeter (ἐν κόλποισιν ἔχουσα). The goddess attempted to make him immortal by immersing him in the fire, but was interrupted by Demophon's mother, who spied on Demeter. As a result of the interruption, Demophon was given over to his sisters, 'inferior nursemaids' (χειρότεραι ... τιθῆναι), but would remain a hero and enjoy undying honour because he was held by Demeter (*Homeric Hymn to Demeter* 213–291; Pirenne-Delforge 2010, 687–691). Likewise, Athena is said to have nourished Athens' mythical king, Erechtheus in *Iliad* 2.547–549, and this nourishing seems directly related to the next lines (550–551), which states that the Athenian youths sacrifice to Erechtheus to win his favour (see Frame 2009, Chapter 9 and n. 268 on the linkages between these two episodes of Demeter and Athena).
123. Bonfante 2000, 280. Bonfante and others have also noted that the Greek word for nurse, τιθήνη, is never used to describe a goddess, unlike the Latin *nutrix* (Bonfante 1997, 186–187; Laskaris 2008, 481).

124. Isis and Horus at Inatos: Skon-Jedele 1994, 1790–1792; Amnisos and Rhodes: Skon-Jedele 1994: 1692–1694; 2013–2016. Amnisos was home to a large range of exotic votive objects in the eighth–sixth centuries BCE, possibly deposited by seafarers and women who invoked the protection of the birth-goddess (Marinatos 1996).
125. Laskaris 2008, 461; Webb 2016, 127–130. Samos was also connected to a later cult of Kourotrophos, known from the *Vita Herodotea, Hom. Vitae* 399 (fourth century BCE), possibly connected to Hera (Hadzisteliou-Price 1978, 152). A figurine of Isis was also found at the sanctuary of Aphrodite at Zeytintepe at Miletus, although Isis figurines are comparatively rare in the Aegean in Archaic Period (Hölbl 1999: 356).
126. Hadzisteliou-Price 1978, 154–156.
127. See Hadzisteliou-Price 1978; Prent 2005, 413–414 and also n. 1089 for examples of nursing females in the Bronze Age. For instance, examples of kourotrophos figurines occur in Geometric-Archaic graves from Boeotia at Rhitsona, Akraiphia, and Tanagra (Higgins 1954, ns. 846–51, 860, pls. 118, 122).
128. Böhm 1990, 136–137; Budin 2015.
129. See Papageorgiou 2005: Taf. 1 (Benaki Museum Inv. 31475), and also Taf. 3.4 for a bronze standing nude kourotrophos. While increasingly popular in the Archaic period, Geometric examples of females on horses are rare, and include examples from the sanctuary of Artemis Hemera at Lusoi, the sanctuary at Olympia, the sanctuary of Athena Alea at Tegea, the sanctuary of Hera on Samos, one in a private collection (Goulandris collection), and from two sites in Colchis (Papageorgiou 2005, 10, 19). Bronze Age examples are rare, but include examples of goddesses seated on thrones atop lions and dragon-like creatures on seals, and a terracotta example of a Mycenaean figurine with arms raised atop a horse with a similar saddle to the Benaki bronze (Papageorgiou 2005, 14–15). A bronze pendant of a standing kourotrophos dating to between 1150 and 1000 BCE from the G. Ortiz collection suggests some amount of continuity of this imagery from Bronze to Iron Age (Papageorgiou 2005, 17).
130. Marinatos 2000; Papageorgiou 2005.
131. Woodard 2019.
132. Leclant 1960. On similar imagery from Cyprus and the Levant: Karageorghis 2006; from Egypt: Hoffmeier and Kitchen 2007.
133. Centola 2018, 39–40.
134. See note above on the motifs of Hathor and Baubo flashing their genitalia to evoke emotional states.
135. Pirenne-Delforge 2010.
136. Prent 2005, 408; Böhm 1990, 125–126.
137. Knappett 2011, 102–105.
138. See also Feldman 2014.
139. Wenger 1998, 58; see also Roddick and Stahl 2016.
140. E.g., Gamble 1998; Malkin 2011; Feldman 2014; Bonnet and Bricault 2016. Marian Feldman, in her analysis of the spread of artistic styles in the Iron Age Levant (2014), has argued against tying cultural practices to place, and has instead emphasised the 'affective properties' of style and symbolism in activating collective identities and cultural memories along wide-ranging interconnected networks.
141. Kyrieleis 1996; Schweizer 2005, 363.
142. In this sense, such approaches draw, ontologically, on Actor-Network Theory (ANT), a set of methodologies grounded in material-semiotic analyses of networks of human and non-human interactions (see Latour 2005; Baron and Gomez 2016). In Bruno Latour's words, ANT 'has been made by the fusion of three hitherto unrelated strands of preoccupations: a semiotic definition of

entity building; a methodological framework to record the heterogeneity of such a building; an ontological claim on the 'networky' character of actants themselves' (Latour 1996, 373).

143. Wenger 1998, 123.
144. Roddick and Stahl 2016.
145. Roddick and Stahl 2016, 10.
146. Gubel 2005.
147. Pilz 2009. Crete certainly appears as a special case, since several sites (e.g., Roussa Ekklesia), show high numbers of mass-produced nude female figurines. See Prent 2005, 480–481; Pilz 2009; Erickson 2009; Marinatos 2000. While Pilz disputes Marinatos' association of the nude female with male maturation rites, versions of nude female imagery do appear at Kato Syme, a site long associated with male rites of passage (Lebessi 2009).
148. Roddick and Stahl 2016, 9.
149. Salapata 2018.
150. Snodgrass 1989–1990; Hodkinson 1998.
151. See Eisenfeld 2015, on the *Homeric Hymn 5* to Aphrodite and its relationship to earlier Mesopotamian royal tropes surrounding Ishtar. Eisenfeld suggests that this hymn interfaces with these tropes through, for instance, Aphrodite begetting a future king, but also subtly criticises and denounces them. See also Metcalf 2015, 185.
152. Roddick and Stahl 2016, 20.
153. Purcell 2006, 21. See also Burkert 2004.
154. Purcell 2006, 28.
155. For instance, see Valdés 2005 on the on Athenian Salaminian clan and its connections to Cyprus through cultic and commercial activity possibly dating back to ca. 1200 BCE.

Bibliography

Ackerman, S. (1993). "The Queen Mother and the Cult in Ancient Israel." *JBL* 112 (3): 385–401.

Anagnostou-Laoutides, E. (2017). *In the Garden of the Gods: Models of Kingship from the Sumerians to the Seleucids.* London and New York: Routledge.

Albenda, P. (2005). "The 'Queen of the Night' Plaque: A Revisit." *JAOS* 125 (2): 171–190.

Albright, W. F. (1939). "Astarte Plaques and Figurines from Tell Beit Mirsim." In *Mélanges syriens offert à Monsieur René Dussaud, Vol. I*, Académie des inscriptions et belles-lettres, 107–120. Paris: P. Geuthner.

Albright, W. F. (1946). *Archaeology and the Religion of Israel*, 2nd Ed. Baltimore: The Johns Hopkins University Press.

Albright, W. (1957). *From the Stone Age to Christianity: Monotheism and the Historical Process.* Reprint of 1940 edition. Garden City: Doubleday.

Allen, S. (2015). *The Splintered Divine: A Study of Ištar, Baal, and Yahweh Divine Names and Divine Multiplicity in the Ancient Near East.* Boston: Walter de Gruyter.

Apostola, E. (2015). "Cross-Cultural Relations between Egypt and Greece During the Early Iron Age; Representations of Egyptian Lion-Headed Deities in the Aegean." In *Current Research in Egyptology 2014: Proceedings from the Fifteenth Annual Symposium University College London & King's College London 2014*, edited by M. Pinarello, J. Yoo, J. Lundock, and C. Walsh, 100–112. Oxford: Oxbow.

Aruz, J. (2014). "Art and Networks of Interaction Across the Mediterranean." In *Assyria to Iberia at the Dawn of the Classical Age*, edited by J. Aruz, S. B. Graff, and Y. Rakic, 112–124. New York: Metropolitan Museum of Art.

Aruz, J., S. B. Graff, and Y. Rakic, eds. (2014). *Assyria to Iberia at the Dawn of the Classical Age*. New York: Metropolitan Museum of Art.

Assante, J. (2003). "From Whores to Hierodules: The Historiographic Invention of Mesopotamian Female Sex Professionals." In *Ancient Art and Its Historiography*, edited by A. A. Donohue and M. D. Fullerton, 13–47. Cambridge: Cambridge University Press.

Assante, J. (2006). "Undressing the Nude: Problems in Analyzing Nudity in Ancient Art, with an Old Babylonian Cast Study." In *Images and Gender: Contributions to the Hermeneutics of Reading Ancient Art. Orbis Biblicus et Orientalis 220*, edited by S. Schroer, 177–207. Fribourg: Academic Press and Göttingen: Vandenhoeck & Ruprecht.

Bahrani, Z. (1996). "The Hellenization of Ishtar: Nudity, Fetishism, and the Production of Cultural Differentiation in Ancient Art." *Oxford Art Journal* 19 (2): 3–16.

Bammer, A. (1985). "Spuren der Phöniker im Artemision von Ephesos." *Anatolian Studies* 35: 103–108.

Baron, L. F. and R. Gomez. (2016). "The Associations between Technologies and Societies: The Utility of Actor-Network Theory." *Science, Technology & Society* 21 (2): 1–20.

Barrelet, M.-T. (1955). "Les déesses armées et ailées." *Syria* 32 (2–4): 222–260.

Barrett, C. (2007). "Was Dust Their Food and Clay Their Bread? Grave Goods, the Mesopotamian Afterlife, and the Liminal Role of Inana/Ishtar." *Journal of Ancient Near Eastern Religions* 7: 7–65.

Becking, B. (1999). "Breasts and Womb." In *Dictionary of Deities and Demons in the Bible*, 2nd Ed., edited by K. van der Toorn, B. Becking, and P. W. van der Horst, 177–178. Leiden: Brill.

Ben-Arieh, S. (1983). "A Mould for a Goddess Plaque." *Israel Exploration Society* 33 (1/2): 72–77.

Bernal, M. (1987, 1991, 2006). *Black Athena: The Afroasiatic Roots of Classical Civilization, 3 Vols.* New Brunswick: Rutgers University Press.

Beyer, D. (2001). *Emar IV: Les sceaux. Orbis Biblicus et Orientalis 20.* Fribourg: Academic Press and Göttingen: Vandenhoeck & Ruprecht.

Blake, E. (2005). "The Material Expression of Cult, Ritual, and Feasting." In *The Archaeology of Mediterranean Prehistory*, edited by E. Blake and A. B. Knapp, 102–129. Malden: Blackwell.

Bloch-Smith, E. (2014). "Archaeological and Inscriptional Evidence for Phoenician Astarte." In *Transformation of a Goddess: Ishtar-Astarte-Aphrodite*, edited by D. T. Sugimoto, 167–194. Fribourg: Academic Press and Göttingen: Vandenhoeck & Ruprecht.

Blome, P. (1982). *Die figürliche Bildwelt Kretas in der geometrischen und früharchaïschen periode.* Mainz am Rhein: Philipp von Zabern.

Boardman, J. (1999). *The Greeks Overseas: Their Early Colonies and Trade*, 4th Ed. New York: Thames and Hudson.

Böhm, S. (1990). *Die >nackte Göttin<: Zur Ikonographie und Deutung unbekleideter weiblicher Figuren in der frühgriechischen Kunst.* Mainz am Rhein: Philipp von Zabern.

Bonfante, L. (1989a). "Nudity as a Costume in Classical Art." *AJA* 93 (4): 543–570.

Bonfante, L. (1989b). "L'iconografia della madre nell'arte dell'Italia antica." In *Le donne in Etruria*, edited by A. Rallo, 85–106. Rome: "L'Erma" di Bretschneider.
Bonfante, L. (1997). "Nursing Mothers in Classical Art." In *Naked Truths: Women, Sexuality, and Gender in Classical Art and Archaeology*, edited by A. O. Koloski-Ostrow and C. L. Lyons, 174–196. London and New York: Routledge.
Bonfante, L. (2000). "Classical Nudity in Italy and Greece." In *Ancient Italy in Its Mediterranean Setting: Studies in Honour of Ellen Macnamara. Accordia Specialist Studies on the Mediterranean* 4, edited by D. Ridgway, F. Serra-Ridgway, M. Pearce, E. Herring, R. D. Whitehouse, and J. B. Wilkins, 271–293. London: Accordia Research Institute, University of London.
Bonfante, L. (2008). "Freud and the Psychoanalytical Meaning of the Baubo Gesture in Ancient Art." *Source: Notes in the History of Art* 27 (2/3): 2–9.
Bonnet, C. (1996). *Astarté: Dossier documentaire et perspectives historiques.* Rome: Consiglio nazionale delle ricerche.
Bonnet, C. and L. Bricault. (2016). *Quand les dieux voyagent: cultes et mythes en mouvement dans l'espace méditerranéen antique.* Histoire des religions. Genève: Labor et Fides.
Bolger, D. (1996). "Figurines, Fertility, and the Emergence of Complex Society on Cyprus." *Current Anthropology* 37 (2): 365–373.
Bowman, C. H. (1978). *The Goddess 'Anatu in the Ancient Near East.* Unpublished Ph.D. Dissertation. Berkeley: University of California.
Braudel, F. (1949). *La Méditerranée et le monde méditerranéen à l'époque de Philippe II.* Paris: Armand Colin.
Braun-Holzinger, E. A. and E. Rehm. (2005). *Orientalischer Import in Griechenland im frühen 1. Jahrtausend v. Chr.* AOAT 328. Münster: Ugarit-Verlag.
Brize, P. (1997). "Offrandes de l'époque géométrique et archaïque à l'Héraion de Samos." In *Héra: Images, espaces, cultes*, edited by J. de la Genière, 123–139. Naples: Centre Jean Bérard.
Broodbank, C. (2013). *The Making of the Middle Sea: A History of the Mediterranean from the Beginning to the Emergence of the Classical World.* Oxford: Oxford University Press.
Brughmans, T., A. Collar, and F. Coward, eds. (2016). *The Connected Past: Challenges to Network Studies in Archaeology and History.* Oxford: Oxford University Press.
Budin, S. L. (2002). "Creating a Goddess of Sex." In *Engendering Society: Women and Society in Ancient Cyprus. CAARI Monographs Volume 3*, edited by D. Bolger and N. Serwint, 315–324. Boston: American Schools of Oriental Research.
Budin, S. L. (2003). *The Origin of Aphrodite.* Maryland: CDL Press.
Budin, S. (2005). "Minoan Asherah?" *Archaeological Perspectives on the Transmission and Transformation of Culture in the Eastern Mediterranean*, edited by J. Clarke, 188–197. Oxford: Oxbow.
Budin, S. L. (2011). *Images of Woman and Child from the Bronze Age: Reconsidering Fertility, Maternity, and Gender in the Ancient World.* Cambridge: Cambridge University Press.
Budin, S. L. (2014). "Before Kypris was Aphrodite." In *Transformation of a Goddess: Ishtar-Astarte-Aphrodite*, edited by D. T. Sugimoto, 195–215. Fribourg: Academic Press and Göttingen: Vandenhoeck & Ruprecht.
Budin, S. L. (2015). "The Nude Female in the Southern Levant: a Mixing of Syro-Mesopotamian and Egyptian Iconographies." *Cult and Ritual on the Levantine Coast, and its impact on the Eastern Mediterranean Realm.* Baal Hors-Series X: 315–336.

Budin, S. L. (2016). "Reduced to Her Bare Essentials: Bronze Age Piriform Pendants in the Levant." *Near Eastern Archaeology* 79 (3): 166–173.

Budin, S. L. (2019). "Jar Handles, Nudity, and the Female." In *Gender and Methodology in the Ancient Near East: Approaches from Assyriology and Beyond.* Barcino monographica orientalia 10, edited by S. L. Budin, M. Cifarelli, A. Garcia-Ventura, and A. Millet Albà, 179–198. Barcelona: Edicions de la Universitat de Barcelona.

Bumke, H. (2007). "Fremde Weihungen für griechische Götter: Überlegungen zu den Bronzestatuetten Ägyptischer Götter und Priester im Heraion von Samos." In *Kult und Kommunikation: Medien in Heiligtümern der Antike.* ZAKMIRA 4, edited by C. Frevel and H. von Hesberg, 349–380. Wiesbaden: Reichert Verlag.

Bumke, H. (2012). "Fremde Votive Oder Fremde Dedikanten? Ägyptische Weihgaben in Ionischen Heiligtümern und ihr Zeugniswert für Kulturtransfer." In *Tryphe und Kultritual im archaischen Kleinasien. – ex oriente luxuria?*, edited by L.-M. Günther, 11–31. Wiesbaden: Harrassowitz Verlag.

Burkert, W. (1992). *The Orientalizing Revolution: Near Eastern Influence on Greek Culture in the Early Archaic Age.* Cambridge: Harvard University Press.

Burkert, W. (2004). *Babylon, Memphis, Persepolis: Eastern Contexts of Greek Culture.* Cambridge, MA and London: Harvard University Press.

Carbillet, A. (2011). *La figure hathorique à Chypre (IIe-Ier mill. av. J.-C.). Alter Orient und Altes Testament 388.* Münster: Ugarit-Verlag.

Centola, D. (2018). *How Behavior Spreads: The Science of Complex Contagions.* Princeton: Princeton University Press.

Centola, D. and M. Macy. (2007). "Complex Contagions and the Weakness of Long Ties." *American Journal of Sociology* 113 (3): 702–734.

Chapman, C. (2012). "'Oh that you were like a brother to me, one who had nursed at my mother's breasts.' Breast Milk as a Kinship-Forging Substance." *Journal of Hebrew Scriptures* 12 (7): DOI:10.5508/jhs.2012.v12.a7

Chapman, C. (2016). *The House of the Mother: The Social Roles of Maternal Kin in Biblical Hebrew Narrative and Poetry.* New Haven: Yale University Press.

Chaves, M. (1997). *Ordaining Women: Culture and Conflict in Religious Organizations.* Cambridge: Harvard University Press.

Coldstream, N. (2004). *Geometric Greece 900–700 BCE*, 2nd Ed. London and New York: Routledge.

Collar, A. (2013). *Religious Networks in the Roman Empire: The Spread of New Ideas.* Cambridge: Cambridge University Press.

Collon, D. (2005). *The Queen of the Night.* London: British Museum Press.

Concannon, C. and L. Mazurek. (2016). "Introduction: A New Connectivity for the Twenty-first Century." In *Across the Corrupting Sea: Post-Braudelian Approaches to the Ancient Eastern Mediterranean*, edited by C. Concannon and L. Mazurek, 1–16. London: Routledge.

Corbeill, A. (2004). *Nature Embodied: Gesture in Ancient Rome.* Princeton: Princeton University Press.

Cornelius, I. (2008). *The Many Faces of the Goddess: The Iconography of the Syro-Palestinian Goddesses Anat, Astarte, Qedeshet, and Asherah C. 1500–1000 BCE. Orbis Biblicus et Orientalis 204.* Fribourg: Academic Press and Göttingen: Vandenhoeck & Ruprecht.

Crielaard, J.-P. (2016). "Powerful Things in Motion: A Biographical Approach to Eastern Elite Goods in Greek Sanctuaries." In *Sanctuaries and the Power of Consumption: Networking and the Formation of Elites in the Archaic Western*

Mediterranean World. Proceedings of the International Conference in Innsbruck, 20th–23rd March 2012, edited by E. Kistler, B. Öhlinger, M. Mohr, and M. Hoernes, 351–372. Wiesbaden: Harrassowitz Verlag.

Crielaard, J.-P. (2018). "Hybrid Go-Betweens: The Role of Individuals with Multiple Identities in Cross-Cultural Contacts in the Late Bronze Age and Iron Age Central and Eastern Mediterranean." In *Change, Continuity, and Connectivity: North-Eastern Mediterranean at the Turn of the Bronze Age and in the Early Iron Age*, edited by Ł. Niesiołowski-Spanò and M. Węcowski, 196–220. Wiesbaden: Harrassowitz Verlag.

Curtis, J. (1994). "Mesopotamian Bronzes from Greek Sites: The Workshops of Origin." *Iraq* 56: 1–25.

Darby, E. (2014). *Interpreting Judean Pillar Figurines: Gender and Empire in Judean Apotropaic Ritual. Forschungen zum Alten Testament 2. Reihe 69*. Tübingen: Mohr Siebeck.

Day, P. (1992). "Anat: Ugarit's 'Mistress of Animals'." *JNES* 51 (3): 181–190.

de Polignac, F. (2009). "Sanctuaries and Festivals." *In A Companion to Archaic Greece*, edited by K. A. Raaflaub and H. van Wees, 427–443. Malden: Blackwell.

Derks H. (1995). "A Note on Ὀμογαλακτες: In Aristotle's Πολιτικα." *Dialogues d'histoire ancienne* 21 (2): 27–40.

Dever, W. (2005). *Did God Have a Wife? Archaeology and Folk Religion in Ancient Israel*. Grand Rapids and Cambridge: W. B. Eerdmans.

Dolce, R. (2008). "Human Beings and Gods at Ebla in the Early and Old Syrian Periods: Some Suggestsions." *Orientalia, Nova Series* 77 (2): 145–172.

Duplouy, A. (2006). *Le prestige des Élites: Recherches sur les modes de reconnaissance sociale en Grèce entre les Xe et Ve siècles*. Paris: Les Belles Lettres.

Ebbinghaus, S. (2006). "Begegnungen mit Ägypten und Vorderasien im archaischen Heraheiligtum von Samos." In *Stranieri e non cittadini nei santuari Greci: Atti del convegno internazionale, Udine 20 al 22 novembre 2003*, edited by A. Naso, 187–229. Florence: Le Monnier.

Edwards, I. E. S. (1955). "A Relief of Qudshu-Astarte-Anath in the Winchester College Collection." *JNES* 14: 49–51.

Eisenfeld, H. (2015). "Ishtar Rejected: Reading a Mesopotamian Goddess in the *Homeric Hymn to Aphrodite*." *Archiv für Religionsgeschichte* 16: 133–162.

Erickson, B. (2009). "Roussa Ekklesia, Part 1: Religion and Politics in East Crete." *AJA* 113 (3): 353–404.

Everton, S. F. (2018). *Networks and Religion: Ties that Bind, Loose, Build-up, and Tear Down*. Cambridge: Cambridge University Press.

Feldman, M. (2014). *Communities of Style: Portable Luxury Arts, Identity, and Collective Memory in the Iron Age Levant*. Chicago: University of Chicago Press.

Fletcher, R. (2004). "Sidonians, Tyrians, and Greeks in the Mediterranean: The Evidence from Egyptianising Amulets." *Ancient West and East* 3: 51–77.

Fourrier, S. (2001). "Naucratis, Chypre et la Grèce de l'Est: le commerce des sculptures 'chypro-ioniennes'." In *Naukratis: Die Beziehungen zu Ostgriechenland, Ägypten und Zypern in archaischer Zeit. Akten der Table Ronde in Mainz, 25–27 November 1999*, edited by U. Höckmann and D. Kreikenbom, 39–54. Möhnesee: Bibliopolis.

Frame, D. (2009). *Hippota Nestor*. Cambridge: Harvard University Press.

Frankfort, H. (1939). *Cylinder Seals: A Documentary Essay on the Art and Religion of the Ancient Year East*. London: Macmillan and Co.

Frankfort, H. (1948). *Kingship and the Gods: A Study of Ancient Near Eastern Religion as the Integration of Society and Nature*. Chicago: University of Chicago Press.

Franklin, J. C. (2015). *Kinyras the Divine Lyre*. Washington, DC: Center for Hellenic Studies.

Gaignerot-Driessen, F. (2018). "Terracotta Figures, Figurines, and Plaques from the Anavlochos, Crete." *Les Carnets de l'ACoSt* 17: 1–9.

Gamble, C. (1998). "Palaeolithic Society and the Release from Proximity: A Network Approach to Intimate Relations." *World Archaeology* 29 (3): 426–449.

George, A., trans. (2003). *The Epic of Gilgamesh*. London: Penguin.

Greenstein, E. I. (1998). "New Readings in the Kirat Epic." *Israel Oriental Studies* 18: 105–123

Granovetter, M. S. (1973). "The Strength of Weak Ties." *American Journal of Sociology* 78 (6): 1360–1380.

Gubel, E. (2005). "Phoenician and Aramaean Bridle-harness Decoration: Examples of Cultural Contact and Innovation in the Eastern Mediterranean." In *Crafts and Images in Contact: Studies on Eastern Mediterranean Art of the First Millennium BCE. Orbis Biblicus et Orientalis 210*, edited by C. E. Suter and C. Uehlinger, 111–147. Fribourg: Academic Press and Göttingen: Vandenhoeck & Ruprecht.

Gunter, A. C. (2009). *Greek Art and the Orient*. Cambridge: Cambridge University Press.

Gunter, A. C. (2014). "Orientalism and Orientalization in the Iron Age Mediterranean." In *Critical Approaches to Ancient Near Eastern Art*, edited by B. A. Brown and M. H. Feldman, 79–108. Boston: de Gruyter.

Guralnick, E. (2004). "A Group of Near Eastern Bronzes from Olympia." *AJA* 108 (2): 187–222.

Hackett, J. (1989). "Can a Sexist Model Liberate Us? Ancient Near Eastern 'Fertility' Goddesses." *Journal of Feminist Studies in Religion* 5: 65–76.

Hadzisteliou-Price, T. (1978). *Kourotrophos: Cults and Representations of the Greek Nursing Deities*. Leiden: Brill.

Haubold, J. (2013). *Greece and Mesopotamia: Dialogues in Literature*. Cambridge: Cambridge University Press.

Henke, J.-M. (2017). *Die zyprischen Terrakotten aus Milet. Neue Überlegungen zur Einordnung der archaischen zyprischen Terrakotten aus Ostägäischen Fundkontexten und Ihrer Werkstattspezifischen Zueweisung*. Milesische Forschungen, Band 7. Berlin: De Gruyter.

Hestrin, R. (1987). "The Lachish Ewer and the 'Asherah'." *Israel Exploration Journal* 37 (4): 212–223.

Higgins, R. A. (1954). *Catalogue of the Terracottas in the Department of Greek and Roman Antiquities, British Museum. Greek: 730–330 B.C.* 2 Vols. London: British Museum.

Hodkinson, S. (1998). "Patterns of Bronze Dedications at Spartan Sanctuaries, c. 650–350 BC: Towards a Quantified Database of Material and Religious Investment." In *Sparta and Laconia: From Prehistory to Pre-Modern*, edited by W. G. Cavanagh, C. Gallou, and M. Georgiadis, 55–63. *BSA Studies* 16. London: British School at Athens.

Hoffman, G. (1997). *Imports and Immigrants: Near Eastern Contacts with Iron Age Crete*. Ann Arbor: University of Michigan Press.

Hoffmeier, J. K. and K. A. Kitchen. (2007). "Reshep and Astarte in Norht Sinai: A Recently Discovered Stela from Tell el-Borg." *Egypt and the Levant* XVII: 127–136.

Hölbl, G. (1999). "Aegyptiaka von Aphrotidetempel auf dem Zeytintepe." *Archäologischer Anzeiger* 114: 345–371.

Horden, P. and N. Purcell. (2000). *The Corrupting Sea: A Study of Mediterranean History*. Oxford and Malden: Blackwell.

Hurwit, J. (2007). "The Problem with Dexileos: Heroic and Other Nudities in Greek Art." *AJA* 111 (1): 35–60.

Jackson, H. (2006). *Jebel Khalid on the Euphrates II: The Terracotta Figurines*. Sydney: Sydney University Press.

Jantzen, U., (1972). *Ägyptische und Orientalische Bronzen aus dem Heraion von Samos. Samos VIII*. Bonn: R. Habelt.

Kaplan, P. (2006). "Dedications to Greek Sanctuaries by Foreign Kings in the Eighth through Sixth Centuries BCE." *Historia* 55 (2): 129–152.

Karageorghis, V. (2006). "Aphrodite/Aštarte on Horseback." In *Confronting the Past: Archaeological and Historical Essays on Ancient Israel in Honor of William G. Dever*, edited by S. Gitin, J. E. Wright, and J. P. Dessel, 75–79. Winona Lake: Eisenbrauns.

Keel, O. and C. Uehlinger. (1998). *Gods, Goddesses, and Images of God in Ancient Israel*. Trans. T. H. Trapp. Minneapolis: Fortress Press.

Kenaan, V. L. (2010). "Aphrodite: The Goddess of Appearances." In *Brill's Companion to Aphrodite*, edited by S. Pickup and A. C. Smith, 27–49. Leiden: Brill.

Kletter, R. (1996). *The Judaean Pillar-Figurines and the Archaeology of Asherah*. BAR International Series 636. Oxford: Tempus Reparatum.

Knappett, C. (2011). *An Archaeology of Interaction: Network Perspectives on Material Culture and Society*. Oxford: Oxford University Press.

Knappett, C., ed. (2013). *Network Analysis in Archaeology: New Approaches to Regional Interaction*. Oxford: Oxford University Press.

Knappett, C and Leidwanger, J. (2018). "Maritime Networks, Connectivity, and Mobility in the Ancient Mediterranean." In *Maritime Networks in the Ancient Mediterranean World*, edited by C. Knappett and J. Leidwanger, 1–21. Cambridge: Cambridge University Press.

Kilian-Dirlmeier, I. (1985). "Fremde Weihungen in griechischen Heiligtümern vom 8. bis zum beginn des 7. Jahrhunderts v. Chr." *JRGZM* 32: 215–254.

Klebinder-Gauß, G. and A. M. Pülz. (2008). "'Fremdes' in der materiellen Kultur im Artemision von Ephesos." In *Die Archäologie der ephesischen Artemis: Gestalt und Ritual eines Heiligtums*, edited by U. Muss, 201–208. Wien: Phoibos Verlag.

Kyrieleis, H. (1979). "Babylonische Bronzen im Heraion von Samos." *Jahrbuch des Deutschen Archäologischen Instituts* 94: 32–48.

Kyrieleis, H. (1996). *Der große Kuros von Samos. Samos X*. Bonn: R. Habelt.

Lamber, W. G. (2008). "Mesopotamian Creation Stories." In *Imagining Creation. IJS Studies in Judaica, Volume: 5*, edited by M. Geller and M. Schipper, 15–59. Leiden: Brill.

Lambert, S. D. (1998). *The Phratries of Attica*. 2nd Ed. Ann Arbor: University of Michigan Press.

Laskaris, J. (2008). "Nursing Mothers in Greek and Roman Medicine." *AJA* 112 (3): 459–464.

Latour, B. (1996). "On Actor-Network Theory: A Few Clarifications." *Soziale Welt* 47 (4): 369–381.

Latour, B. (2005). *Reassembling the Social: An Introduction to Actor-Network Theory*. Oxford: Oxford University Press.

Lave, J. and E. Wenger. (1991). *Situated Learning: Legitimate Peripheral Participation*. Cambridge: Cambridge University Press.

Lebessi, A. (2009). "The Erotic Goddess of the Syme Sanctuary, Crete." *AJA* 113 (4): 521–545.

Leclant, J. (1960). "Astarté à cheval d'après les représantations Égyptiennes." *Syria* 37: 1–67.

Lesko, B. S. (1999). *The Great Goddesses of Egypt*. Norman: University of Oklahoma Press.

Lévêque, P. and A. Verbanck-Piérard. (1992). "Héraclès héros ou dieu?" In *Héraclès: D'une rive à l'autre de la Méditerranée. Bilan et perspectives*, edited by C. Bonnet and C. Jourdain-Annequin, 43–65. Bruxelles, Brussels, Rome: Institut Historique Belge de Rome.

Lev Kenaan, V. (2010). "Aphrodite: The Goddess of Appearances." In *Brill's Companion to Aphrodite*, edited by A. C. Smith and S. Pickup, 29–50. Leiden: Brill.

Lightbody, D. I. (2013). *The Hybridising Tree of Life: A Postcolonial Archaeology of the Cypriot Iron Age City Kingdoms*. Unpublished PhD Thesis. Glasgow: University of Glasgow.

Lipiński, E. (1986). "The Syro-Palestinian Iconography of Woman and Goddess (Review Article)." *Israel Exploration Journal* 36 (1/2): 87–96.

López-Ruiz, C. (2010). *When the Gods Were Born: Greek Cosmogonies and the Near East*. Cambridge: Harvard University Press.

López-Ruiz, C. (2014). "Greek and Near Eastern Mythologies: A Story of Mediterranean Encounters." In *Approaches to Greek Myth*, 2nd Ed., edited by L. Edmunds. 154–199. Baltimore: Johns Hopkins University Press.

Louden, B. (2011). *Homer's Odyssey and the Near East*. Cambridge: Cambridge University Press.

Lupack, S. (2012). "Mycenaean Religion." In *The Oxford Handbook of the Bronze Age Aegean (ca. 3000–1000 BC)*, edited by E. H. Cline, 263–276. Oxford: Oxford University Press.

Malkin, I. (2011). *A Small Greek World: Networks in the Ancient Mediterranean*. Oxford: Oxford University Press.

Manning, J. G. (2018). *The Open Sea: The Economic Life of the Ancient Mediterranean World from the Iron Age to the Rise of Rome*. Princeton: Princeton University Press.

Marchetti, N. (2007). "Chronology and Stratification of Middle Bronze Age Clay Figurines in Syria and Northern Palestine." In *From Relative Chronology to Absolute Chronology: The Second Millennium BC in Syria-Palestine. Proceedings of the International Colloquium* (Rome, 29th November–1st December 2001), edited by P. Matthiae, P. Pinnock, L. Nigro, and L. Peyronel, 247–283. Rome: Bardi.

Marchetti, N. and L. Nigro. (1997). "Cultic Activities in the Sacred Area of Ishtar at Ebla during the Old Syrian Period: The Favissae F.5327 and F.5328." *Journal of Cuneiform Studies* 49: 1–44.

Marinatos, N. (1996). "Cult by the Seashore: What Happened at Amnisos?" In *The Role of Religion in the Early Greek Polis. Proceedings of the Third International Seminar on Ancient Greek Cult*, Athens 16–18 October 1992, edited by R. Hägg, 135–139. Stockholm: Swedish Institute at Athens.

Marinatos, N. (2000). *The Goddess and the Warrior: The Naked Goddess and the Mistress of Animals in Early Greek Religion*. London and New York: Routledge.

Matthiae, P. (1993). "L'aire sacrée d'Ishtar à Ebla: résultats des fouilles de 1990–1992." *Comptes rendus des séances de l'Académie des Inscriptions et Belles-Lettres* 137 (3): 613–662.

Matthiae, P. (2013). "La déesse nue et le dieu au panache. Aux origins de l'iconographie de l'Ishtar d'Ébla." In *Ritual, Religion and Reason: Studies in the Ancient World in Honour of Paolo Xella*, edited by O. Loretz, S. Ribichini, W. G. E. Watson, and J. A. Zamora, 1–24. Münster: Ugarit-Verlag.

Metcalf, C. (2015). *The Gods Rich in Praise: Early Greek and Mesopotamian Religious Poetry*. Oxford: Oxford University Press.

Miller Ammerman, R. (1991). "The Naked Standing Goddess: A Group of Archaic Terracotta Figurines from Paestum." *AJA* 95 (2): 203–230.

Morris, C. (2006). "From Ideologies of Motherhood to 'Collecting Mother Goddesses.'" *Creta Antica* 7: 69–78.

Morris, I. (2003). "Mediterraneanization." *Mediterranean Historical Review* 18 (2): 30–55.

Morris, S. P. (1992). *Daidalos and the Origins of Greek Art*. Princeton: Princeton University Press.

Morris, S. P. (2006). "The View from East Greece: Miletus, Samos and Ephesus." In *Debating Orientalization: Multidisciplinary Approaches to Processes of Change in the Ancient Mediterranean*, edited by C. Riva and N. C. Vella, 66–84. London: Equinox.

Moyer, I. (2006). "Golden Fetters and Economies of Cultural Exchange." *JANER* 6: 225–256.

Moyer, I. (2011). *Egypt and the Limits of Hellenism*. Cambridge: Cambridge University Press.

Murray, S. C. (2017). *The Collapse of the Mycenaean Economy: Imports, Trade, and Institutions 1300–700 BCE*. Cambridge: Cambridge University Press.

Mylonopoulos, J. (2006). "'Fremde' Weihungen in Heiligtümern der Ostägäis im 7. Und 6. Jh. V.Chr." In *Austausch von Gütern, Ideen und Technologien in der Ägäis und im östlichen Mittelmeer: Von der prähistorischen bis zu der Archaischen Zeit*, edited by A. Kyriatsoulis, 363–383. Weilheim: Verein zur Förderung der Aufarbeitung der Hellenischen Geschichte.

Nishiyama, S. and S. Yoshizawa. (1997). "Who Worshipped the Clay Goddess? The Late First Millennium BC Terracotta Figurines from Tell Mastuma, Northwest Syria." *Bulletin of the Ancient Orient Museum* 18: 73–79.

Ornan, T. (2005). *The Triumph of the Symbol: Pictorial Representation of Deities in Mesopotamia and the Biblical Image Ban. Orbis Biblicus et Orientalis 213*. Fribourg: Academic Press and Göttingen: Vandenhoeck & Ruprecht.

Osborne, R. (1993). "'À la grecque' (Review discussion of Burkert's Orientalizing Revolution and Morris's Daidalos)." *Journal of Mediterranean Archaeology* 6: 231–237.

Papadopoulos, J. (2014). "Greece in the Early Iron Age: Mobility, Commodities, Polities, and Literacy." In *The Cambridge Prehistory of the Bronze and Iron Age Mediterranean*, edited by A. B. Knapp and P. van Dommelen, 178–195. Cambridge: Cambridge University Press.

Papageorgiou, I. (2005). "Eine reitende Kourotrophos-Göttin geometrischer Zeit im Benaki-Museum." *Mitteilungen des Deutschen Archäologischen Instituts* 120: 1–34.

Patai, R. (1990). *The Hebrew Goddess*. Detroit: Wayne State University Press.

Pedrucci, G. (2017). "Motherhood, Breastfeeding and Adoption: The Case of Hera Suckling Heracles." *Acta Antiqua Academiae Scientiarum Hungaricae* 57: 311–322.

Pilz, O. (2009). "Some Remarks on Meaning and Function of Moldmade Terracotta Relief Plaques Depicting Naked and Dressed Female Figures." In *La donateur, l'offrande et la déesse: Systèmes votifs dans les sanctuaires de déesses du monde grec*. Kernos Supplement 23, edited by C. Prêtre, 97–110. Liège: Centre International d'Étude de la Religion Grecque Antique.

Pilz, O. and M. Krumme. (2011). "Il deposito votivo di Κακό Πλάι sull'Anavlochos: risulti preliminari dello studio dei materiali." In *Identità Culturale, Etnicità, Processi Di Trasformazione a Creta fra Dark Age a Arcaismo: Per i Cento Anni Dello Scavo di Priniàs 1906–2006: Convegno Di Studi, Atene, 9–12 Novembre 2006*, edited by G. Rizza, 323–332. Catania: Consiglio nazionale delle ricerche I.B.A.M. and Università di Catania, Centro di archeologia cretese.

Pinch, G. (1993). *Votive Offerings to Hathor.* Oxford: Griffith Institute, Ashmolean Museum.

Pirenne-Delforge, V. (2010). "Nourricières d'immortalité: Déméter, Héra et autres déesses en pays grec." *Paedagogica Historica* 46 (6): 685–697.

Pirenne-Delforge, P. and G. Pironti. (2016). *L'Héra de Zeus: Ennemie Intime, Épouse Definitive.* Paris: Les Belles Lettres.

Pizzimenti, S. (2014). "The Kassite 'Naked Goddess': Analysis and Interpretation." In *Šime ummiānka Studi in onore di Paolo Matthiae in occasione del suo 75° Compleanno Offerti Dall'ultima Generazione di Allievi*, edited by S. Pizzimenti and L. Romano, 135–150. Rome: Sapienza.

Pongratz-Leisten, B. (2008). "Sacred Marriage and the Transfer of Divine Knowledge: Alliances between the Gods and the King in Ancient Mesopotamia." In *Sacred Marriages: The Divine-Human Sexual Metaphor from Sumer to Early Christianity*, edited by M. Nissinen and R. Uro. Winona Lake: Eisenbrauns.

Postill, J. (2015). "Fields: dynamic configurations of practices, games, and socialities." In *Thinking Through Sociality: An Anthropological Interrogation*, edited by V. Amit, 47–68. Oxford: Berghahn

Powell, S. D. (2018). *Narrative Desire the Book of Ruth.* London: Bloomsbury.

Prent, M. (2005). *Cretan Sanctuaries and Cults: Continuity and Change from Late Minoan IIIC to the Archaic Period.* Leiden: Brill.

Pritchard, J. (1943). *Palestinian Figurines in Relation to Certain Goddesses Known through Literature.* New Haven: Yale University Press.

Pryke, L. (2017). *Ishtar.* London and New York: Routledge.

Purcell, N. (2006). "Orientalizing: Five Historical Questions." In *Debating Orientalization: Multidisciplinary Approaches to Processes of Change in the Ancient Mediterranean*, edited by C. Riva and N. C. Vella, 21–30. London: Equinox.

Ramazzotti, M. (2014). "The Mimesis of a World: The Early and Middle Bronze Clay Figurines from Ebla-Tell Mardikh." In *Figuring Out the Figurines of the Ancient Near East*, edited by S. M. Langin-Hooper. Occasional Papers in Coroplastic Studies 1, 39–64. Association for Coroplastic Studies.

Rasmussen, T. (2005). "Herakles' Apotheosis in Etruria and Greece." *Antike Kunst* 48: 30–39.

Riis, P. J. (1949). "The Syrian Astarte Plaques and their Western Connections." *Berytus* 9 (2): 36–90.

Riis, P. J. (1960–1961). "Plaquettes syriennes d'Astarté dans des milieux grecs." *Mélanges de l'Université Saint-Joseph* 37: 194–198.

Riva, C. and N. C. Vella, eds. (2006). *Debating Orientalization: Multidisciplinary Approaches to Processes of Change in the Ancient Mediterranean.* London: Equinox.

Roddick, A. P. and A. B. Stahl. (2016) "Introduction: Knowledge in Motion." In *Knowledge in Motion: Constellations of Learning across Time and Place*, edited by A. P. Roddick and A. B. Stahl, 3–35. Tucson: The University of Arizona Press.

Rotté, E. (2012). "Egyptian Plaque Terracottas of the Standing Nude Women from the Late Period: Egyptian Heritage or Foreign Influences." *Newsletter of the Coroplastic Studies Interest Group* 7: 13–16.
Said, E. (1978). *Orientalism*. New York: Pantheon.
Saint-Pierre, C. (2009). "Offrir l'Orient aux déesses grecques." In *Le donateur, l'offrande et la déesse: Systèmes votifs des sanctuaires de déesses dans le monde grec*, edited by C. Prêtre, 111–122. Liège: Presse universitaires de Liège.
Salapata, G. (2018). "Tokens of Piety: Inexpensive Dedications as Functional and Symbolic Objects." *Opuscula* 11: 97–109.
Schmidt, C. (1968). *Kyprische Bildwerke aus dem Heraion von Samos. Samos VII*. Bonn: R. Habelt.
Schweizer, B. (2005). "Fremde Bilder – andere Inhalte und Formen des Wissens. Olympia in der "orientalisierenden" Epoche des 8. und 7. Jh. v. Chr." In *Die Dinge als Zeichen: Kulturelles Wissen und materielle Kultur. Internationale Fachtagung an der Johann Wolfgang Goethe Universität Frankfurt am Main 3..–5. April 2003*, edited by T. L. Kienlin, 355–382. Bonn: R. Habelt.
Schweizer, B. (2006). *Griechen und Phöniker am Tyrrhenischen Meer. Repräsentation kultureller Interaktion im 8. Und 7. Jr. v. Chr. In Etrurien, Latium und Kampanien*. Charybdis 16. Münster: Lit-Verlag.
Schweizer, B. (2007). "Zwischen Naukratis und Gravisca: Händler im Mittelmeerraum des 7. und 6. Jhs. v. Chr. empórion – Port of Trade – extraurbanes Heiligtum: von der xenía zur emporia." In *Das Heilige und die Ware Zum Spannungsfeld von Religion und Ökonomie*, edited by M. Fitzenreiter, 307–324. London: Golden House Publications.
Senff, R. (2003). "Das Aphroditeheiligtum von Milet." In *Neue Forschungen zur Religionsgeschichte Kleinasiens: Elmar Schwertheim zum 60. Geburtstag gewidmet. Asia Minor Studien 49*, edited by G. Heedemann and E. Winter, 11–25. Bonn: Habelt.
Sharp, C., C. McKinny, and I. Shai. (2015). "The Late Bronze Age Figurines from Tel Burna." *Strata: Bulletin of the Anglo-Israel Archaeological Society* 33: 61–75.
Sherratt, S. (2012). "The Intercultural Transformative Capacities of Irregularly Appropriated Goods." In *Materiality and Social Practice: Transformative Capacities of Intercultural Encounters*, edited by J. Maran and P. W. Stockhammer, 152–172. Oxford and Oakville: Oxbow.
Sherwood, Y. (2017). *The Bible and Feminism: Remapping the Field*. Oxford: Oxford University Press.
Skon-Jedele, N. (1994). *"Aigyptiaka": A Catalogue of Egyptian and Egyptianizing Objects Excavated from Greek Archaeological Sites, ca. 1100–525 B.C., with Historical Commentary*. Unpublished Ph.D. thesis, University of Pennsylvania.
Snodgrass, A. (1989–1990). "The Economics of Dedication at Greek Sanctuaries." *Scienze dell'Antichità* 3–4: 287–294.
Strøm, I. (1992). "Evidence from the Sanctuaries." In *Greece between East and West, 10th–8th Centuries BC: Papers of the Meeting at the Institute of Fine Arts, New York University, March 15–16th, 1990*, edited by G. Kopcke and I. Tokumaru, 46–60. Mainz am Rhein: Philipp von Zabern.
Sugimoto, D. T. (2014). "The Judaean Pillar Figurines and the 'Queen of Heaven'." In *Transformation of a Goddess: Ishtar-Astarte-Aphrodite*, edited by D. T. Sugimoto, 141–166. Fribourg: Academic Press and Göttingen: Vandenhoeck & Ruprecht.

Tadmor, M. (1982). "Female Cult Figurines in Late Canaan and Early Israel: Archaeological Evidence." In *Studies in the Period of David and Solomon and Other Essays: Papers Read at the International Symposium for Biblical Studies, Tokyo, 5–7 December, 1979*, edited by T. Ishida, 139–173. Winona Lake: Eisenbrauns.

Talalay, L. E. (2005). "The Gendered Sea: Iconography, Gender, and Mediterranean Prehistory." In *The Archaeology of Mediterranean Prehistory*, edited by E. Blake and A. B. Knapp, 130–155. Malden: Blackwell.

Talalay, L. E. and T. Cullen. (2002). "Sexual Ambiguity in Plank Figurines from Bronze Age Cyprus." In *Engendering Aphrodite: Women and Society in Ancient Cyprus*, edited by D. Bolger and N. Servint, 181–195. Boston: ASOR.

Tartaron, T. (2013). *Maritime Networks in the Mycenaean World*. Cambridge: Cambridge University Press.

Tazawa, K. (2014). "Astarte in New Kingdom Egypt: Reconsideration of Her Role and Function." In *Transformation of a Goddess: Ishtar-Astarte-Aphrodite*, edited by D. T. Sugimoto, 103–124. Fribourg: Academic Press and Göttingen: Vandenhoeck & Ruprecht.

Teissier, B. (1984). *Ancient Near Eastern Cylinder Seals from the Marcopoli Collection*. Berkeley, Los Angeles, and London: University of California Press and Beverly Hills: Summa Publications.

Tringham, R. and M. Conkey. (1998). "Rethinking Figurines." In *Ancient Goddesses: The Myths and the Evidence*, edited by L. Goodison and C. Morris, 22–45. Madison: University of Wisconsin Press.

Tsukimoto, A. (2014). "'In the Shadow of Thy Wings': A Review of the Winged Goddess in Ancient Near Eastern Iconography." In *Transformation of a Goddess: Ishtar-Astarte-Aphrodite*, edited by D. T. Sugimoto, 15–32. Fribourg: Academic Press and Göttingen: Vandenhoeck & Ruprecht.

Ulbrich, A. (2008). *Kypris: Heiligtümer und Kulte weiblicher auf Zypern in der kyproarchaischen und kyproklassischen (Königszeit). Alter Orient und Altes Testament 44*. Münster: Ugarit-Verlag.

Ugander, J., L. Backstrom, C. Marlow, and J. Kleinberg. (2012). "Structural Diversity in Social Contagion." *PNAS* 109 (16): 5962–5966.

Valdés, M. (2005). "The Cult of Aglauros (and Aphrodite) in Athens and in Salamis of Cyprus: Reflections on the Origin of the *Genos* of the *Salaminioi*." *Ancient West & East* 4 (1): 57–76.

Vassileva, M. (2014). "Phrygian Bronzes in the Greek World: Globalization through Cult?" In *Melammu: The Ancient World in an Age of Globalization. Max Planck Research Library for the History and Development of Knowledge Proceedings 7*, edited by M. J. Geller, 217–234. Berlin: Edition Open Access.

Villing, A. (2017). "Greece and Egypt: Reconsidering Early Contact and Exchange." In *Regional Stories towards a New Perception of the Early Greek World. Acts of an International Symposium in honour of Professor Jan Bouzek*, edited by A. Mazarakis Ainian, A. Alexandridou, and X. Charalambidou, 563–596. Volos: University of Thessaly Press.

von Graeve, V. (2013). "Das Aphrodite-Heiligtum von Milet und seine Weihgaben." In *Sanktuar und Ritual. Heilige Plätze im archäologischen Befund. Menschen – Kulturen – Traditionen. Studien aus den Forschungsclustern des Deutschen Archäologischen Instituts. Vol. 10*, edited by I. Gerlach and D. Rau, 5–17, Rahden and Westalen: Verlag Marie Leidorf.

Waraksa, E. (2009). *Female Figurines from the Mut Precinct: Context and Ritual Function. Orbis Biblicus et Orientalis 240.* Fribourg: Academic Press and Göttingen: Vandenhoeck & Ruprecht.

Ward, W. A. (1969). "La déesse nourricière d'Ugarit." *Syria* 46 (3–4): 225–239.

Watts, D. J. and S. H. Strogatz. (1998). "Collective Dynamics of 'Small-World' Networks." *Nature* 393: 440–442.

Webb, V. (2016). *Faience Material from the Samian Heraion. Samos XIII.* Wiesbaden: Reichert Verlag.

Weiss, K. (2012). *Ägyptische Tier- und Götterbronzen aus Unterägypten: Untersuchungen zu Typus, Ikonographie und Funktion sowie der Bedeutung innerhalb der Kulturkontakte zu Griechenland.* Wiesbaden: Harrassowitz Verlag.

Wenger, E. (1998). *Communities of Practice: Learning, Meaning, and Identity.* Cambridge: Cambridge University Press.

West, M. L. (1966). *Hesiod, Theogony: Edited with Prolegomena and Commentary.* Oxford: oxford University Press.

West, M. L. (1997). *The East Face of Helicon: West Asiatic Elements in Greek Poetry and Myth.* Oxford: Oxford University Press.

Whitley, J. (2002). "Objects with Attitude: Biographical Facts and Fallacies in the Study of Late Bronze Age and Early Iron Age Warrior Graves." *Cambridge Archaeological Journal* 12 (2): 217–232.

Whitley, J. (2013). "The Cretan Orientalizing. A Comparative Perspective." In *Kreta in der geometrischen und archaischen Zeit. Akten des Internationalen Kolloquiums am Deutschen Archäologischen Institut, Abteilung Athen 27–29 January 2006*, edited by W.-D. Niemeier, O. Pilz, and I. Kaiser, 409–426. München: Hirmer Verlag GmbH.

Wilson, I. (2012). "Judean Pillar Figurines and Ethnic Identity in the Shadow of Assyria." *Journal for the Study of the Old Testament* 36 (3): 259–278.

Wilson-Wright, A. M. (2016). "Athtart: The Transmission and Transformation of a Goddess in the Late Bronze Age." In *Forschungen zum Alten Testament 2. Reihe 90.* Tübingen: Mohr Siebeck.

Winter, U. (1983). *Frau und Göttin: Exegetische und ikonographische Studien zum weiblichen Gottesbild im Alten Israel und in dessen Umwelt. Orbis Biblicus et Orientalis 53.* Göttingen: Vandenhoeck & Ruprecht.

Woodard, R. (2019). "The Disappearance of Telepinu in the Context of Indo-European Myth." In *Hrozný and Hittite: The First Hundred Years. Proceedings of the International Conference Held at Charles University, Prague, 11–14 November 2015*, edited by R. I. Kim, J. Mynářová, and P. Pavúk, 583–602. Leiden: Brill.

Woolley, C. L. (1955). *Alalakh: An Account of the Excavations at Tell Atchana in the Hatay, 1937–1949. Reports of the Research Committee of the Society of Antiquaries of London 18.* London: The Society of Antiquaries.

Wyatt, N. (2002). *Religious Texts from Ugarit*, 2nd Ed. Sheffield: Sheffield Academic Press.

Yener, K. A. (2015). "Material Evidence of Cult and Ritual at Tell Atchana, Ancient Alalakh: Deities of the Transitional Middle-Late Bronze Period." In *From the Treasures of Syria: Essays on Art and Archaeology in Honour of Stefania Mazzoni*, edited by P. Ciafardoni and D. Giannessi, 203–216. Leiden: Netherlands Institute for the Near East.

Ziffer, I. (2010). "Western Asiatic Tree-Goddesses." *Ägypten und Levante* 20: 411–430.

3 Weak and strong ties in the diffusion of coinage during the Archaic period in Greece[1]

John Mooring

The spread of coinage changed Greek society fundamentally. Although this invention was technically not that radical, the implications were far reaching; for instance, increasing labour specialisation as workers did not have to be self-sufficient but were able to pay for goods and services, or the enabling of Athens' direct democracy as citizens were paid for duties such as officeholders and participation in the assembly, council, and courts. In this chapter, the role of networks is explored in the diffusion of this innovation. To analyse the spread of coinage and the significance of networks in this phenomenon, several models are applied. The aim is to build a hypothesis about decisive elements in this development. Subsequently, this hypothesis is tested against the available sources.

Peer Polity Interaction and the diffusion of innovation

In his book *Ancient History from Coins*, Chris Howgego used the Peer Polity Interaction model to think about the spread of coinage, the both unconscious and deliberate competition between Greek *poleis* which can be observed in the widespread adoption of a whole range of phenomena, from military techniques and political structures to particular crafts and tastes.[2] Howgego explained the rapid spread of coinage using the assumption that Greek poleis watched their peers closely and imitated or even surpassed their achievements.

The Peer Polity Interaction model offers useful insights into the swift spread of coinage in the sixth century BCE, in particular the flow of information between equivalent communities. When coining the term Peer Polity Interaction, Colin Renfrew stressed that '[t]he emphasis here is not primarily upon interaction in terms of the exchange of material commodities, but rather in the flow of information of various kinds between the polities'.[3] It is notable that this flow of information, to which the Peer Polity Interaction model was successfully applied for Archaic period developments, took place through small communities. Renfrew mentioned as an example the Greek islands:

> These islands, despite their small size, were self-governing, had a well-defined constitution, were subject to explicit law, issued coinage in

DOI: 10.4324/9780429429217-4

> the case of Naxos and Paros, and were famed for their accomplishments in sculpture. Yet it should be noted that the population of each was generally not more than 5,000 persons [...] The small states in Greece emerged *together*, pulling each other up by the bootstraps, as it were.[4]

Although the Peer Polity Interaction model offers an attractive explanation for the swift diffusion of coinage in the Greek world, it is lacking in detail as far as the dissemination process itself is concerned. For this purpose, the Diffusion of Innovation theory is more appropriate. This theory was developed in communication studies by Everett Rogers. In 1962 he published his seminal book *Diffusion of Innovations* in which he identified and structured the key success factors in the spread and adoption of innovations. He based his work on 405 studies on diffusion, from the controlling by the British Navy of scurvy through citrus juice in the eighteenth century and the introduction of hybrid corn in Iowa in the middle of the twentieth century to the failed diffusion of the Dvorak keyboard.[5] This theory has a sound empirical foundation, which has been updated and extended over the years up to 2003, when the fifth and final edition was published. Rogers' theory has had a tremendous impact on the study of innovations, and some terms Rogers introduced, such as early adopters and laggards, are common expressions in both the academic world and business. Another strong side of this theory is that it is based on research from regions all over the world and throughout time. In this theory, Rogers distinguished four main elements that determine the success or failure of an innovation. Of course, the characteristics of the innovation such as relative advantage, compatibility, and complexity, are key success factors. Other decisive elements in the adoption process are the social system, the innovativeness of the individuals in a certain environment and the communications channels through which an innovation is diffused. I discuss all of these elements in my analysis of the introduction of coinage to the Greek world elsewhere;[6] for this chapter however, I will focus on the communication channels.

Weak ties at work: the dissemination of knowledge

According to the Diffusion of Innovations theory, the decision to adopt an innovation is based on two preceding steps, viz. the dissemination of knowledge about the innovation, and the persuasion of a person to use the innovation.[7] First, knowledge about the innovation must be available. Mark Granovetter's well-known network theory on the strength of weak ties explains how information can rapidly diffuse in a low-density network.[8] Such a low-density network existed in the Mediterranean and Black Sea area in the Archaic period, where people and communities were connected through origin, trade, colonisation, and other ties. The effectiveness of the communication about coinage was boosted as the nodes in the network were people who were alike, or homophilous.[9] That is, they shared the same culture, the same deities and methods of worship, (in the most part) the same language, social status, and usually even the same kind of education. At first sight, it may sound contradictory that these independent states shared the same culture, but the mobility of Greek speaking people in the Archaic period and their colonisation across the Mediterranean encouraged the formation of overarching identities.[10] Local communities were given a veneer of Panhellenism through the presence of travelling individuals such as aristocrats, poets, and experts, for example, urban and territorial planners, lawmakers, architects, sculptors, seers, and other cult experts.[11] A shared Greek culture, even with all kinds of local and regional flavours, had emerged by the sixth century.

Strong ties at work: the process of persuasion

The second step in the diffusion process after the knowledge about coinage had been spread was the process of persuasion. This communication was transmitted through interpersonal contacts or channels. According to Rogers, interpersonal channels are the most effective communication channels in the diffusion process as they involve a face-to-face exchange between two or more individuals. These channels are very effective at overcoming resistance or apathy and are therefore very helpful in the process of persuasion.[12] It is through these trusted channels that potential adopters are persuaded to use the innovation. Diffusion is a social process in which modelling and imitation by potential adopters takes place through contacts of their network. Therefore, social network contacts are vital in the diffusion of an innovation. The role of opinion leaders within the social network is particularly important. In Rogers' words: 'Opinion leaders gain their perceived competency by serving as an avenue for the entrance of new ideas into their system'.[13] Opinion leaders are characterised as more cosmopolite than their followers, having greater social participation and higher socio-economic status.[14] Anne Kandler and Fabio Caccioli developed a simulation framework which explicitly models the adoption decision of individuals with regard to an innovation to establish a causal relationship between individual-level processes and population-level patterns.[15] With this mathematical model, it can be demonstrated that the dynamics of the adoption of an innovation is dependent on the social status of the first adopter; a higher status leads to a quicker adoption.[16] The same model demonstrates that the effect of homophilistic bias is stronger in a small community.[17]

Trust

The setting of small communities with shared norms and values also resulted in and generated something else: increased levels of trust. To explore the role trust played in the spread of coinage, let us look at the creation of trust. A broadly used definition describes trust as 'the psychological state comprising the intention to accept vulnerability based upon positive expectations of the intentions or behaviour of another'.[18] Trust is based upon the expectation that a truster can predict the intentions or behaviour of the trustee. As trust is dependent on the predictability of the behaviour of others, trust will grow in an environment where there is much interaction between the actors. This means that trust increases more quickly in small communities as the number of interactions with the same person is larger and therefore the predictability of this person's behaviour is more substantial. Alongside the number of actors, mutual elements also help to predict future behaviour. When the norms and values of the community members are the same, or homophilous, the predictability of behaviour becomes higher. Recent sociological research confirms that homophily across linguistic, ethnic, and religious characteristics is a critical determinant of trust.[19] From all this it can be deduced that in small Greek communities that were in close contact with other Greek communities with which they shared a common culture, trust levels were comparatively high as the number of interactions between members was fairly high and the behaviour was predictable due to the shared culture. When communities increased in size and number of inhabitants, for example, in the case of classical Athens, abstract systems such as accountability, standardisation of measures and weights, and political governance replaced personal trust as was illustrated by Steven Johnstone.[20]

Coins as fiduciary money

Why was a high level of trust so important in the diffusion of coinage in Greek society? The reason is simple: trust is the only discriminating feature of coinage. Coins can be defined as ***pre-weighted*** pieces of ***(precious) metal*** bearing ***the sign of the issuing authority*** which was produced as a means of payment. Technically, it was easy to produce good-quality silver pieces of the same weight or mark it by means of a stamp. These were not the characteristics that determined the success of coinage. In the sixth century, bullion silver was the most frequently used means of payment in the ancient Near East and Egypt. The most prevalent form was *Hacksilber*, pieces of silver like ingots, rods, and scrap silver.[21] All these pieces with different weights were put together in the scales until the balance levelled out and the right amount was collected to fulfil the amount due.[22] In case of doubt the silver quality could be checked to ensure that the pieces were made of genuine silver and not debased. Coinage was a convenient invention as coins were pre-weighted conforming to the local weight standard and usually the quality of silver was high. This simplified the payment process as the checks on quality and weight were superfluous. However, this was only possible if the users had trust in that coinage. There had to be trust in the issuing government to produce coins from high-quality silver according to a recognisable weight standard, and there had to be trust in the person paying with coins that the coins had not been tampered with. In other words, there had to be trust that the received coins could be used in the future at the same value. From this it can be inferred that coinage was the first fiduciary money.[23]

Before stating my hypothesis, I summarise the building blocks found. In the Diffusion of Innovations theory, two steps in the communication of the innovation are essential. First, the innovation has to be known. Knowledge about coinage was distributed throughout the Archaic Greek world through the weak ties linking the Greeks from all over the Mediterranean and Black Sea area. The second step was taken by the opinion leaders within the Greek communities who persuaded their constituents to adopt coins. This was possible as they possessed strong ties and acted as the trusted experts of the community. Trust was essential in this process as coinage is the first fiduciary money. Elements that contributed to a high level of trust were the shared culture which possibly contained individualistic characteristics and the small scale of Greek communities. The small interacting communities bring us back to the Peer Polity Interaction model.

The role of panhellenic gatherings

In describing the Peer Polity Interaction model, Renfrew proposed that the crucial exchange of information occurred in gatherings of the elites of these polities during panhellenic games and festivals at such centres as Olympia, Isthmia, Nemea, Delphi, and Delos.[24] Starting as local sanctuaries, these places evolved into religious centres, where (elite) representatives from the whole Greek world gathered at specific intervals for celebrations, games and markets. This development first took place in Olympia well before the end of the eighth century and later festivals were modelled on it at Nemea, Delphi, and Isthmia in the first half of the sixth century.[25] The reason to assign this central role in the communication of innovations to the panhellenic sanctuaries was their ability to facilitate the direct communication of hundreds of separate and politically independent polities. Athletes, magistrates, colonists, and traders came to these cult places and created a dense network in which information on all sorts of topics flowed. Kostas Vlassopoulos, in his analysis of the ways various networks functioned beyond the level of the individual polity, considered the Greek world as a so-called world-system, meaning a 'larger than any juridically defined political unit' where he identified three processes at work: the movement of people, the movement of goods and the movement of technology and ideas;[26] it appears that the panhellenic sanctuaries tied all three together. Anthony Snodgrass aptly described Delphi as 'the main central clearing-house for information of a geographical and political kind'.[27] I would like to hypothesise that in a similar vein the idea of coinage spread fast throughout the Greek world as it was passed on during panhellenic festivals. The information the visitors, who belonged to the elites of the *poleis*, brought home from the panhellenic gatherings made a huge impact on their local societies and was quickly adopted by the other inhabitants through the elites' status and the small size of their community.

The question of why coinage was adopted lies outside the scope of this chapter,[28] but some remarks can be made about the necessary context for the successful adoption of this innovation. Coinage spread fast in the second half of the sixth century BCE (see also Figure 3.1). This was a time of growth and change in the Greek world.[29] According to Josiah Ober this epoch was part of the rise and fall of classical Greece, an extended period of efflorescence—'increased economic growth accompanied by a sharp uptick in cultural achievement'.[30] He quantified this through ten proxy indicators of economic growth of which four were related to coins and hoards. The social, economic, and cultural developments were manifold in the late Archaic period, the emerging *poleis* being one of the most distinctive and studied phenomena. Many ancient historians have explained the success of coinage by pointing to certain payments which coins could have facilitated such as the hiring of mercenaries,[31] official payments to be made or received by the government[32] or the purchase of foodstuffs by soldiers.[33] However, Modern Money Theory, a modern version of Chartalism, has made clear that the basis for money lies in the acceptance by governments of 'money-things' in payments to the state.[34] Therefore, the origins of coinage must be sought in the emergence of levies and indirect taxes on import, export, sale, and consumption which took place in the late Archaic period.[35] The minting of coins fitted well in the incremental development of public finance, where, as Henry Kim phrased it, 'coinage should not be seen as the starting point so much as a milestone along a much longer road in the use of money'.[36]

Turning to the sources

As discussed in the first part of this chapter, the contacts, and communications between the elites of several *poleis* sustained by panhellenic gatherings were related to the diffusion of coinage across the Greek world in the Archaic period. But how can this hypothesis be tested? There are publications in which the reasons for minting in a particular (city) state are investigated.[37] However, these reasons cannot be taken as representative for the entire Greek world which consisted of over 1,000 *poleis*. To draw general conclusions for the Archaic period, it is necessary to verify this hypothesis across the complete Greek population. For that, the investigations conducted by the Copenhagen *Polis* Centre directed by Mogens Hansen between the years 1993 to 2003 is invaluable. In the synthesising publication, *An Inventory of Archaic and Classical Poleis* (hereafter '*Inventory*'), a wealth of information about all 1,035 identified *poleis* has been brought together, based on a standardised format with some forty topics. After a regional introduction, all *poleis* are briefly described by a team of archaeologists and historians. The publication is completed by twenty-seven indices that list specific characteristics such as type of constitution, territory size, major divinities, and cults. With the rich data from the *Inventory*, a hypothesis can be tested for the entire Greek world. The only downside of this approach is the omission of non-*polis* information.[38]

The increase in minting poleis

Coinage was invented in Lydia or Ionia in the third quarter of the seventh century BCE. The first coins were made of electrum, an alloy of silver and gold. The provenance of these early electrum coins is difficult to establish as they are anepigraphic and the number of types, differentiated by the images on the coins, is overwhelming. In addition, most of them stem from the market instead of controlled archaeological excavations. Therefore, detailed understanding of the early stage of coinage is still hazy. The situation changed towards the middle of the sixth century when electrum was replaced by silver, and the number of types decreased and stabilised. Minting was taken up in the Greek mainland by cities such as Aegina, Athens, Chalcis, and Corinth, and in the second half of the sixth century coinage spread quickly to all parts of the Greek world. The table of cities minting their own coins by 480 BCE that Robin Osborne assembled is often referred to as an (albeit rather static) illustration of the success of coinage: over 100 mints were listed by Osborne.[39] The *Inventory* has now made it possible to demonstrate the pace of this diffusion instead of only the outcome at the end of the Archaic period. The number of mints increased from 35 in 550 BCE to 111 in 480 BCE. When all *poleis* are taken into account that minted between 550 and 480, so including those that ceased minting in that period, the number increases to 147. This is a spectacular rise of 317% (or even 422% for all *poleis* starting to mint) in 70 years (see Figure 3.1).

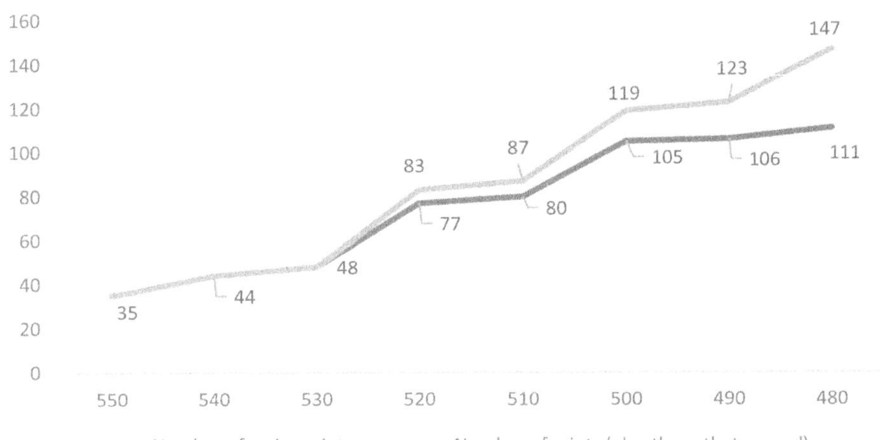

Figure 3.1 Diffusion of coinage in the Archaic period.

The issuing side: elite social networks of theorodokoi and proxenoi

The next step in evidencing the hypothesis is to gain insight in the elite social network relations in Archaic Greece. As a proxy for network contacts, the appointment of *theorodokoi* can be used. Although the appointments of these officials are epigraphically attested only from the fourth century onwards, these institutions were already established in the seventh century.[40] *Theorodokoi* were appointed by a *polis* to host and assist the *theoroi*. The latter were messengers sent out by a state that was about to organise a panhellenic game or festival to announce this event. Even though *theorodokoi* were appointed by the community and, hence, a civic institution, it was a highly regarded position held by the upper-class. The appointment of *theorodokoi* could therefore serve as a materialisation of active participation of an appointing *polis* in the panhellenic activities. The *Inventory* lists the *theorodokoi* of four panhellenic gatherings: the Heraia at Argos, the Asklepiaia at Epidauros, the Pythian Games at Delphi and the Nemean Games. Why the *theorodokoi* of the most important panhellenic gathering, the Olympic Games, were not listed was not explained by Hansen and Nielsen. On the one hand, this is a serious omission, but as the Olympic Games must have been visited by representatives of virtually all Greek communities, perhaps it would not add much to this argument. Even though these sanctuaries can be classified as panhellenic, there seemed to be a certain regional focus: the origins of the *theorodokoi* of the Heraia at Argos were concentrated in Akarnania and Ionia; those from the Asklepiaia at Epidauros from Akarnania, Chalkidike, and Magna Graecia; the ones from the Pythian Games at Delphi from Arkadia and the documented *theorodokoi* of the Nemean Games came from Akarnania, Northern Greece, Asia Minor, and Cyprus.[41]

Another indication of elite social network contacts is found in the *proxenoi*. The earliest epigraphical attestations of *proxenoi* probably date from the beginning of the sixth century BCE, whereas the earliest references in literary sources are found in Pindar.[42] *Proxenos* was the honorary title bestowed upon a citizen of the *polis* who hosted foreign ambassadors and looked after the interests of the other state's citizens. The appointment of a *proxenos*, who mostly came from the upper classes, must have indicated that a *polis* had strong ties with another state. These appointments provided the Greek *poleis*, both large and small, with an enormous network.[43]

To determine the strength of the network contacts in a *polis*, the four types of *theorodokoi* and the *proxenoi* were scored. The *Inventory* makes a distinction between *proxenoi* given and received, meaning the hosting *polis* and the facilitated state, respectively. Therefore, a *polis* could 'earn' six network points, one for each category: for *proxenoi* given, *proxenoi* received and *theorodokoi* at all four places (see, for an example, Figure 3.2). The results tell a clear story: from the total population of 1,035 *poleis* listed in the *Inventory*, only 28% had network contacts through *proxenoi* or *theorodokoi*.[44] This means that on average 28% of the Greek poleis in the Archaic and Classical

Polis	1. Proxenoi received	2. Proxenoi given	3. Theorodokoi at Heraia at Argos	4. Theorodokoi Pythian Games at Delphoi	5. Theorodokoi Asklepiaia at Epidauros	6. Theorodokoi at Nemean Games	Contacts with other poleis (totals)
Akragas	1				1		**2**
Gela	1				1		**2**
Leontinoi					1		**1**
Syrakousai	1				1		**2**
Zankle / Messana					1		**1**
Rhegion					1		**1**
Athenai	1	1	1	1			**4**
Akanthos					1		**1**

Figure 3.2 Example of how network contacts are scored.

periods had network contacts. When looking at the 147 *poleis* which minted in the period from 550 to 480 BCE, 61% had network contacts such as *proxenoi* or *theorodokoi* (compare the left and right hand side of Figure 3.3).[45] In other words, the *poleis* that started minting in the Archaic period were included in elite social networks far more frequently, which strongly suggests that network contacts and minting were interrelated. It also worked the other way around: from the 747 poleis without network relations, only 58 (less than 8%) minted in the Archaic period.

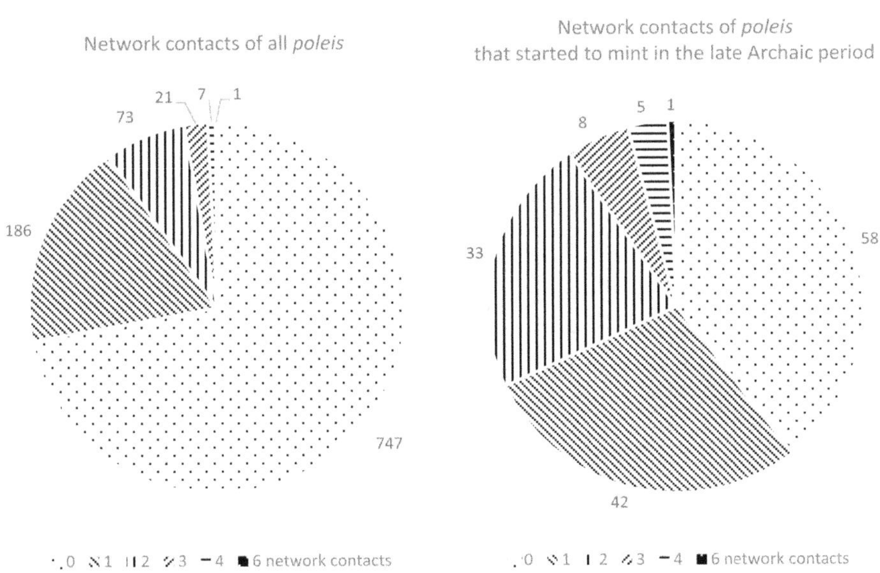

Figure 3.3 Number of network contacts.

The acquiring side: information from Archaic hoards

To further demonstrate the correlation between coinage and elite social network relations, information from coin hoards was analysed alongside the data from the *Inventory*. For that purpose, the data from Coin Hoards (http://coinhoards.org/), which is based on the print publication *Inventory of Greek Coin Hoards* (IGCH), were used.[46] The provenance of the coins found in the 85 hoards from the Archaic period was added to the database.[47] In this way linkages between *poleis* became visible as coins from their mints were brought together in a hoard. For the purposes of this research, the number of coins from a single mint was deemed irrelevant, but instead, the presence or absence of coins from each mint was noted for each hoard. On average a hoard contained coins from 3.6 different mints, 76% of which were from *poleis* that had elite social network contacts. This means that on the user side, deduced from the hoard data, the connection between elite social network contacts and coinage was even stronger than on the production side as demonstrated by the data on *proxenoi* or *theorodokoi*. This is also to be expected as states with extensive elite social network contacts must have witnessed much inbound and outbound travel. It is logical that coins from these *poleis* dispersed through these travels, even though coins were produced for local purposes.[48]

So, both the evidence for *proxenoi* and *theorodokoi* and the occurrences of coins in the Archaic hoards indicate a relationship between elite social network contacts and minting, but is it possible to take this one step further? To determine the correlation between the network contacts of a city and the occurrence of coins from that *polis* in Archaic hoards, I counted the number of hoards in which coins from a specific *polis* were found. Together with the scores on network contacts (see earlier section and Figure 3.2), these numbers are compared; see Figure 3.4 for a sample of the complete list.

Next, the correlation between these data was calculated. There is a significant positive correlation between coinage and elite social network contacts:

Polis	Network contacts	Number of hoards
Akragas	2	1
Gela	2	1
Leontinoi	1	1
Syrakousai	2	1
Zankle / Messana	1	7
Rhegion	1	3
Athenai	4	17
Akanthos	1	7

Figure 3.4 Example of how correlation between network contacts and number of hoards is calculated.

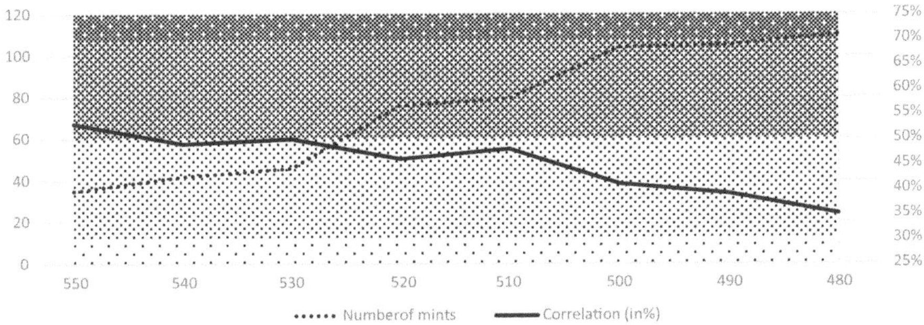

Figure 3.5 Correlation between diffusion of coinage and network contacts.

in 550 BCE there was a moderate positive correlation between the number of elite social network contacts and the number of occurrences in hoards. This correlation decreased over time until there was only a low positive correlation between the two at the end of the Archaic period (see Figure 3.5). This suggests that elite social network contacts were more important at the start of the diffusion of coinage, but less so by the end of this period of time. Maybe this can be explained by the fact that elite social network contacts were important for the diffusion of a relatively unfamiliar and less well-understood invention in 550 BCE, but by the end of the Archaic period, there were so many more mints and users of coinage, that the advantages of coinage were well-known, which made the persuasion to adopt by opinion leaders less necessary.

Two hoards and their network

Whilst the correlation is substantiated by the hoard data, it is thought-provoking trying to get a clearer picture how the elites influenced each other regarding minting. As the type of a coin, that is, its visual appearance is determined by local aspects such as patron deities, regional legends, or puns on the *polis*' name and not by panhellenic aspects; influences of the elites in this area were, as expected, not found.

In addition to its appearance, the weight standard is another important feature of coinage. The weight of the coins, the denominations and how they were divided into fractions are characteristic aspects of a coin. In the Archaic period, some eight different standards can be distinguished. The choice for a certain standard was a deliberate, mostly political decision by the minting authority.[49] When examining the hoard data, it might be possible to distinguish the rationale behind some of these decisions. For example, IGCH 2066, a hoard that contained 1176 coins which were buried in Gela in the second decade of the fifth century BCE, had an interesting composition (see the table in Figure 3.6 and the map in Figure 3.7). The hoard was found in 1956 and was dispersed, but the details of 850 pieces were documented.

Number in Hansen and Nielsen (2004)	Mint	Region	Weight standard	Theorodokoi at Heraia at Argos	Theorodokoi Asklepiaia at Epidauros	IGCH 2065	IGCH 2066
9	Akragas	Sikelia	Attic		x	0	798
17	Gela	Sikelia	Attic		x	0	505
33	Leontinoi	Sikelia	Attic (?)		x	0	1
47	Syrakousai	Sikelia	Attic		x	0	51
51	Zankle / Messana	Sikelia	Euboian and Attic		x	9	5
68	Rhegion	Italia and Kampania	Euboian		x	3	2
361	Athenai	Attika	Attic	x		20	353
559	Akanthos	Chalkidike	Attic		x	4	6

Figure 3.6 Network contacts and the distribution of coins in IGCH 2065 and 2066.

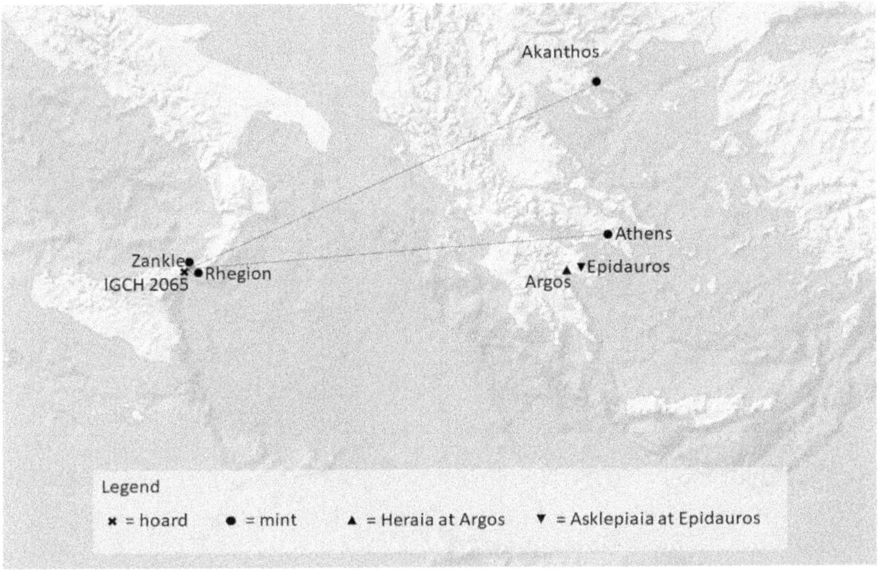

Figure 3.7 Coin hoard IGCH 2066 and the provenance of the coins.

The hoard contained coins from eight different mints: Gela, Akragas, Leontinoi, Syrakousai, Zankle, Rhegion, Akanthos, and Athens. Except for Athens, all these *poleis* had appointed *theorodokoi* for the Asklepiaia at Epidauros. Another hoard from Sicily, IGCH 2065, found in Messina (ancient Zankle/Messana) was much smaller, but again contained coins from Zankle (Messana), Rhegion, Akanthos, and Athens (see the table in Figure 3.6 and the map in Figure 3.8).

Except for the coinage from Rhegion, all coins were on the Attic standard. This is remarkable because the Attic standard became dominant in the fifth century BCE, after Athens exploited its leading role in the Persian Wars, well after the burial date of IGCH 2065 and 2066. Although Athens was involved in the foundation of colonies in Northern Greece and Italy, there is no evidence that strong ties between Athens and Gela, Akragas, Leontinoi, Syrakousai, Zankle, or Akanthos existed. The fact that Gela, Akragas, Leontinoi, Syrakousai, Zankle, and Akanthos minted on the standard of Athens is therefore remarkable and suggests some otherwise invisible connection.

Possibly, the connections between these cities were the people working at the mint. Travelling artisans are well-known from the Archaic period,[50] and Ionian die cutters were presumably active in the West.[51] It is therefore not an unfounded suggestion that the people working at the mint in Athens may have been involved somehow in the minting at Gela, Akragas, Leontinoi, Syrakousai, Zankle, or Akanthos and were accountable for the introduction of the Attic standard there. However, other explanations have

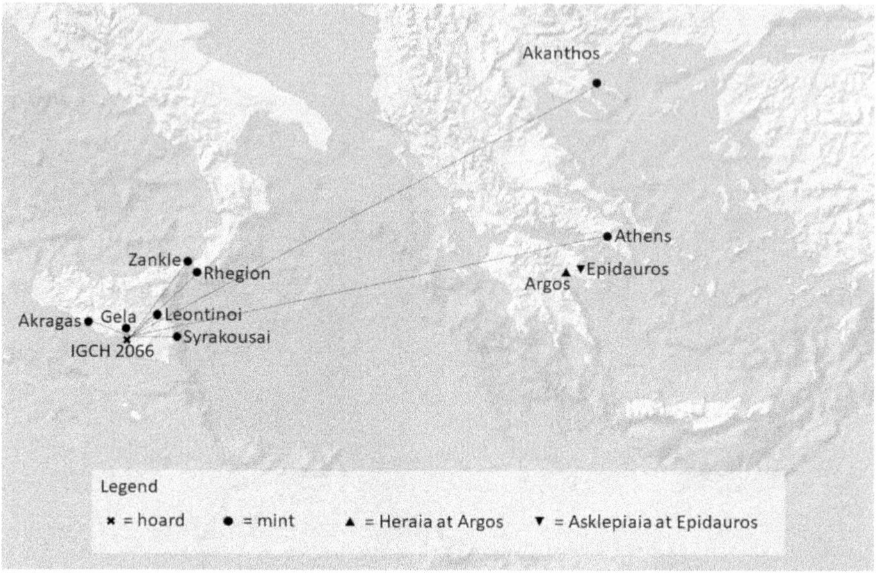

Figure 3.8 Coins hoard IGCH 2065 and the provenance of the coins.

been suggested; for example, that some mints readily accepted orders from neighbouring rulers, or acted as collective mints within a network of cities.[52] Either way, in the process of minting, imitation and cooperation between *poleis* must have taken place regularly. From there it is only a small step to encourage peer cities' enthusiasm regarding the usefulness and value of minting coinage, and to offer help with initiating this process.

Conclusion

Returning to the main theme of this book, I have argued that the diffusion of coinage was stimulated by the elite social network contacts between *poleis* sustained by panhellenic gatherings. These contacts were alternating weak and strong ties; after all, 'weak ties provide the bridges over which innovations cross the boundaries of social groups; the decision making, however, is influenced mainly by the strong-ties network in each group'.[53] During the panhellenic gatherings the participants formed a temporary community with a strong tie network structure, and when they returned to their home communities, they acted like random weak ties: bringing new knowledge about coinage into their home environment. Because the local elites were important and respected members of the community that we can classify as opinion leaders, after knowledge of coinage was spread, it was the strong level of trust or strong ties between the elite opinion leaders bringing and advocating the innovation, and the individuals in the home community accepting it, that was responsible for the adoption of coinage. Members of local elites, visiting panhellenic gatherings and returning to their homes, were both the horizontal and the vertical axes of transmission: they were more exposed to all forms of external communication and thus were more cosmopolite—gathering new information across the peer-to-peer horizontal axis; and because they had a higher social status and therefore were able to act as a force of change in their home communities—that is, they were able to pass new information and encourage new behaviour across the vertical axis of transmission, means that the local elites can be clearly identified as the engine of the innovation that caused coinage to spread so widely and so profoundly across the Archaic Greek world.

Notes

1. I wish to warmly thank Dr Anna Collar, Prof Jan Paul Crielaard, and an anonymous reader from Routledge for their invaluable suggestions and comments. Of course, I remain completely responsible for any remaining mistakes.
2. Howgego (1995) 16.
3. Renfrew (1986) 8.
4. Renfrew (1986) 11. On the size of the communities, see also Snodgrass (2006) 245.
5. In the last version published in 2003, Roger estimated the number of diffusion publications to be more than 5,200.
6. This will be in my dissertation at the Vrije Universiteit Amsterdam in which I use the Diffusion of Innovation theory to analyse the introduction of coinage in the Greek world. Special attention will be given to the role of trust and the differences between the Near East, Egypt and Greece; Mooring, *Graeca fides* (forthcoming).
7. Rogers spends a whole chapter on what he calls the Innovation-Decision Process. However, it is good to be careful of presupposing rational behaviour in all steps of this case-study.
8. Granovetter (1983).
9. Rogers (1995) 286.
10. Malkin (2011) 209.
11. Crielaard (2017) 392 and Malkin (2011) 222; see also the comment on Malkin by Antonaccio (2013) 248–249 and note 9 in Daniels, this volume.
12. Rogers (1995) 194.
13. Rogers (1995) 293–294.
14. Respectively generalisations 8–9, 8–11 and 8–12. See Rogers (1995) 294.
15. Kandler and Caccioli (2016) 176.
16. Kandler and Caccioli (2016) 185.
17. Kandler and Caccioli (2016) 193.
18. Lyon, Möllering and Saunders (2015) 3.
19. Gleave, Robbins and Kolko (2010) 210.
20. Johnstone (2011).
21. For some overview of the used money items in the Near East, see Powell (1978), Snell (1995) and Appendix 2 Pre-Greek Coinage in Schaps (2004). A recent discussion on *Hacksilber* can be found in Heymans (2018). See for the bigger picture Rahmstorf (2016).
22. For more technical details, see Hafford (2005).
23. Seaford is one of the very few scholars on ancient coinage, who drew the same conclusion; Seaford (2004) *passim*.

24. Renfrew (1986) 16. On the role of Iron Age sanctuaries as 'wide bridges' across which information and innovation could pass, see also Daniels, this volume.
25. Osborne (1996) 243–244.
26. Vlassopoulos (2007) 17. Note that Vlassopoulos is rather vague which networks he is referring to: 'What makes Croton, Aegina, Athens, Samos, and Persia all components of the same world-system is not their direct contacts; it is rather the networks that already existed in the sixth century [...] that link these places together into a 'system'; *ibidem*.
27. Snodgrass (2006) 247.
28. See also note 6.
29. E.g., Ian Morris tried to quantify this through archaeological data; Morris (1998) 35.
30. Ober (2015) 2. Note that Ober designated with the term 'classical Greece' not the (art) historian period from 480 to 330 BCE, but the rougher 'layman use' of the term for the period ranging from 1,000 to 300 BCE; Ober (2015) XVIII.
31. Cook (1958) 261.
32. Kraay (1964) 89.
33. Schaps (2014) 35.
34. Wray (2014) 28.
35. Van Wees (2013) 23.
36. Kim (2001) 9.
37. To give a few examples which are neither representative nor exhausting: Abdera in Kagan (2006), Aegina in Figueira (1981) and Kroll (2011), Athens in Davis (2011) and Kroll (2011), Cyrene in Van Alfen (2016), Delphi in Picard (2005), Eretria in Van Alfen (2015), Gela in Holle (1978), Ichnai in Wartenberg (2015), Paros in Sheedy (2000), Siris and Pyxoes in Horsnæs (2011), multiple Boiotian *poleis* in Mackil and Van Alfen (2006) and satrapal coinages in Bodzek (2014).
38. This information can be found in Mackil (2013) which among others covers the cooperation of several entirely independent communities such as the Boiotian federation in minting coinage. It goes without saying that the communication on coinage between these communities was not limited to panhellenic gatherings.
39. Osborne (1996) 253–255.
40. Hansen and Nielsen (2004) 103 and Malkin (2011) 20.
41. It is intriguing as to why Akarnania was so well represented at all four sanctuaries.
42. Herman roughly sketches the 'thousands of inscriptions on stone or lead [...] covering a temporal span from the seventh century B.C. to the second century A.D.'; Herman (1987) 130. Hansen and Nielsen cite two inscriptions from the late seventh or early sixth century BCE; Hansen and Nielsen (2004) 98. Perlman mentions inscriptions from the late sixth century BCE in which foreign athletes at Epidauros are mentioned; Perlman (2000) 59.
43. Hansen and Nielsen (2004) 100.
44. See left pie chart in Figure 3.3, where all the parts with 1 or more network contacts are added:

    186 + 73 + 21 + 7 + 1 = 288 *poleis* = 288 / 1,035 = 28%.
45. See right pie chart in Figure 3.3, where all the parts with 1 or more network contacts are added:

    42 + 33 + 8 + 5 + 1 = 89 *poleis* = 89 / 147 = 61%.
46. The original *Inventory of Greek Coin Hoards* (IGCH), edited by Margaret Thompson of the American Numismatic Society, Otto Mørkholm of the Danish cabinet in Copenhagen and Coin Kraay of the Heberden Coin Room in the Ashmolean Museum, Oxford, was published in 1973 by the American

Numismatic Society for the International Numismatic Commission. The work contains inventory listings of 2387 hoards, covering the whole of the ancient Greek numismatic world. Unfortunately, it was not possible to incorporate material from *Coin Hoards* (10 volumes), published by the Royal Numismatic Society, and more recently by both the Royal Numismatic Society and the American Numismatic Society, as these publications are not available in the Netherlands.

47. From nine of these hoards, the provenance of the coins was uncertain, and four contained only non-Greek coins.
48. Kraay (1964) 84–85.
49. Psoma (2016) 107 and Van Alfen (2018) 347. See, for some examples, Van Alfen (2015) and Van Alfen (2016).
50. See note 10.
51. Van Alfen (2018) 346.
52. Fischer-Bossert (2018) 62. Peter van Alfen even attributes similarities between Eretrian and Athenian coinage to rivalry and competition; Van Alfen (2015) 274.
53. Granovetter (1983) 219.

Bibliography

Antonaccio, C.M. (2013), 'Networking the Middle Ground? The Greek Diaspora, Tenth to Fifth Century BC' in *Archaeological Review from Cambridge,* 28.1: 237–251.

Bodzek, J. (2014), 'Remarks on the origins of Achaemenid Satrapic Coinages' in *Folia numismatica: supplementum ad Acta Musei Moraviae* 28.1: 3–10.

Cook, R.M. (1958), 'Speculations on the Origins of Coinage' in *Historia* 7: 257–262.

Crielaard, J.P. (2017), 'Skilled in the Muses' lovely gifts: lyric poetry and the rise of the community in the seventh-century Aegean', in X. Charalambidou and C. Morgan (eds.), *Interpreting the seventh century BC. Tradition and innovation* (Archaeopress Archaeology, Oxford): 382–392.

Davis, G. (2011), *Law, money, and the transformation of Athens in the sixth century B.C.E.* (unpublished thesis for the degree of Doctor of Philosophy, Macquarie University).

Figueira, T.J. (1981), *Aegina. Society and politics* (The Ayer Company, Salem).

Fischer-Bossert, W. (2018), 'Die-sharing and the transfer of dies: evidence for orders and shipments of large sums of money in the ancient Greek world (5th to 3rd centuries BC)' in B. Woytek (ed.), *Infrastructure and distribution in ancient economies. Proceedings of a conference held at the Austrian Academy of Sciences, 28–31 October 2014* (Austrian Academy of Sciences, Vienna): 59–66.

Gleave, E., B. Robbins, and B. Kolko. (2010), 'Trust in Uzbekistan' in *International Political Science Review* 33.2: 209–229.

Graeca fides (forthcoming). *PhD thesis, Vrije Universiteit Amsterdam.*

Granovetter, M. (1983), 'The Strength of Weak Ties: A Network Theory Revisited' in *Sociological Theory* 1: 201–233.

Hafford, W.B. (2005), 'Hanging in the Balance: Precision Weighing in Antiquity' in *Expedition* 47.2: 35–37.

Hansen, M.H. and T.H. Nielsen. (eds.) (2004), *An inventory of archaic and classical poleis. An investigation conducted by the Copenhagen Polis Centre for the Danish National Research Foundation* (Oxford University Press, Oxford).

Herman, G. (1987), *Ritualised friendship and the Greek city* (Cambridge University Press, Cambridge).

Heymans, E.D. (2018), *Argonauts of the eastern Mediterranean: The early history of money in the eastern Mediterranean Iron Age (Unpublished PhD* thesis, Tel Aviv University).

Holle, B.F. (1978), *Historical considerations on the origins and the spread of Greek coinage in the archaic age* (unpublished PhD thesis, University of Michigan).

Horsnæs, H.W. (2011), 'Coinages of indigenous communities in Archaic Southern Italy – the mint as a means of promoting identity?' in M. Gleba and H.W. Horsnæs (eds.), *Communicating identity in Italic Iron Age communities* (Oxbow, Oxford): 197–209.

Howgego, C.J. (1995), *Ancient history from coins* (Routledge, London/New York).

Irwin, K. and N. Berigan. (2013), 'Trust, Culture, and Cooperation: A Social Dilemma Analysis of Pro-Environmental Behaviors' in *The Sociological Quarterly* 54: 424–449.

Johnstone, S. (2011), *A history of trust in ancient Greece* (University of Chicago Press, Chicago/London).

Kagan, J.H. (2006) 'Small change and the beginning of coinage at Abdera' in P.G. van Alfen (ed.), *Agoranomia. Studies in money and exchange presented to John H. Kroll* (The American Numismatic Society, New York): 49–60.

Kandler, A. and F. Caccioli. (2016), 'Networks, homophily, and the spread of innovations' in T. Brughmans and A. Collar and F. Coward (eds.), *The Connected Past. Challenges to Network Studies in Archaeology and History* (Oxford University Press, Oxford): 175–197.

Kim, H.S. (2001), 'Archaic coinage as evidence for the use of money' in A. Meadows and K. Shipton (eds.), *Money and its uses in the ancient Greek world* (Oxford University Press, Oxford): 7–21.

Kraay, C.M. (1964), 'Hoards, Small Change and the Origin of Coinage' in *Journal of Hellenic Studies* 84: 76–91.

Kroll, J.H. (2011), 'Minting for export: Athens, Aegina, and others' in T. Faucher and M.-C. Marcellesi and O. Picard (eds.), *Nomisma. La circulation monétaire dans le monde grec antique. Acte du colloque international. Athènes, 14-17 avril 2010 = Bulletin de Correspondance Hellénique Supplément 53* (École française d'Athènes, Athens): 27–38.

Lyon, F., G. Möllering, and M.N.K Saunders. (2015), 'Introduction. Researching trust: the ongoing challenge of matching objectives and methods' in F. Lyon and G. Möllering and M. N. K. Saunders (eds.), *Handbook of Research Methods on Trust* (Edward Elgar Publishing, Cheltenham): 1–22.

Mackil, E. (2013), *Creating a common polity. Religion, economy, and politics in the making of the Greek Koinon* (University of California Press, Berkeley/Los Angeles).

Mackil, E. and P.G. Van Alfen. (2006), 'Cooperative coinage' in P.G. Van Alfen (ed.), *Agoranomia. Studies in money and exchange presented to John H. Kroll* (The American Numismatic Society, New York): 201–246.

Malkin I. (2011), *A small Greek world: networks in the ancient Mediterranean* (Oxford University Press, Oxford).

Morris, I. (1998), 'Archaeology and archaic Greek history' in N. Fisher and H. van Wees (eds.), *Archaic Greece: new approaches and new evidence* (Duckworth, London): 1–91.

Ober, J. (2015), *The rise and fall of classical Greece* (Princeton University Press, Princeton).

Osborne, R. (1996), *Greece in the making, 1200–479 BC* (Routledge, London/New York).

Perlman, P. (2000), *City and sanctuary in ancient Greece. The Theorodokia in the Peloponnese = Hypomnemata. Untersuchungen zur Antike and zu ihrem Nachleben* 121 (Vandenhoek and Ruprecht, Göttingen).
Picard, O. (2005), 'Les XPHMATA d'Apollon et les débuts de la monnaie à Delphes' in *Topoi: Orient-Occident* 12.13: 55–68.
Powell, M.A. (1978), 'A contribution to the history of money in Mesopotamia prior to the invention of coinage' in B. Hruška and G. Komoróczy (eds.) *Festschrift Lubor Matouš* 2 (Eötvös Loránd University of Budapest, Budapest): 211–243.
Psoma, S.E. (2016), 'Choosing and changing monetary standards in the Greek world during the Archaic and the Classical periods' in E.M. Harris and D.M. Lewis and M. Woolmer (eds.), *The ancient Greek economy: markets, households and city-states* (Cambridge University Press, Cambridge/New York): 90–115.
Rahmstorf, L. (2016), 'From 'value ascription' to coinage: a sketch of monetary developments in Western Eurasia from the Stone to the Iron Age' in C. Haselgrove and S. Krmnicek (eds.), *The archaeology of money. Proceedings of the workshop 'Archaeology of Money', University of Tübingen, October 2013* (Leicester Archaeology Monographs, Leicester): 19–42.
Renfrew, C. (1986), 'Introduction: Peer Polity Interaction and socio-political change' in C. Renfrew and J.F. Cherry (eds.), *Peer Polity Interaction and socio-political change* (Cambridge University Press, Cambridge): 1–18.
Rogers, E.M. (1995), *Diffusion of innovations* (fourth edition) (The Free Press, New York).
Schaps, D.M. (2004), *The invention of coinage and the monetization of ancient Greece* (The University of Michigan Press, Ann Arbor).
Schaps, D.M. (2014), 'War and peace, imitation and innovation, backwardness and development: the beginnings of coinage in ancient Greece and Lydia' in P. Bernholz and R. Vaubel (eds.), *Explaining monetary and financial innovation. A historical analysis = financial and monetary policy studies* 39 (Springer, Cham): 31–51.
Seaford, R. (2004), *Money and the early Greek mind. Homer, philosophy, tragedy* (Cambridge University Press, Cambridge).
Sheedy, K. (2000), 'Archaic Parian coinage and Parian marble' in D. Schilardi and D. Katsanopoulou (eds.), *Paria lithos* (Paroikia, Paros): 117–120.
Snell, D.C. (1995), 'Methods of exchange and coinage in the ancient Western Asia' in J.M. Sasson (ed.), *Civilizations of the Ancient Near East* III (Simon & Schuster Macmillan, New York): 1487–1497.
Snodgrass, A.M. (1986), 'Interaction by design: the Greek city state' in C. Renfrew and J.F. Cherry (eds.), *Peer polity interaction and socio-political change* (Cambridge University Press, Cambridge): 47–58. Reprinted in A.M. Snodgrass, (2006), *Archaeology and the emergence of Greece* (Cornell University Press, Ithaca): 234–257.
Van Alfen, P.G. (2015), 'The Chalkid(ik)ian beginnings of Euboian coinage' in U. Wartenberg and M. Amandry (eds.), ΚΑΙΡΟΣ. *Contributions to Numismatics in honor of Basil Demetriadi* (The American Numismatic Society, New York): 255–283.
Van Alfen, P. (2016), 'The beginnings of coinage at Cyrene: weight standards, trade, and politics' in M. Asolati (ed.), *Le monete di Cirene e della Cirenaica nel Mediterraneo. Problemi e prospettive Atti del V Congresso Internazionale di Numismatica e di Storia Monetaria Padova, 17-19 marzo 2016 = Numismatica Patavina* 13 (Esedra editrice, Padova): 15–32.

Van Alfen, P.G. (2018), 'The limits of cooperation: prolegomena to a die study of the archaic silver coinages of Phokaia' in C. Ünal and A. Ersoy, C. Gürbıyık and B. Kasalı (eds.), *Port cities of the Aegean world. Coins, seals and weights* (Manisa Celal Bayar Üniversitesi Yayınları, Manisa): 342–352.

Van Wees, H. (2013), *Ships and silver, taxes and tribute. A fiscal history of Archaic Athens* (I.B. Tauris, London/New York).

Vlassopoulos, K. (2007), 'Beyond and Below the *Polis*: Networks, Associations, and the Writing of Greek History' in *Mediterranean Historical Review* 22.1: 11–22.

Wartenberg, U. (2015), 'Thraco-Macedonian bullion coinage in the fifth century B.C.: the case of Ichnai' in U. Wartenberg and M. Amandry (eds.), *ΚΑΙΡΟΣ. Contributions to Numismatics in honor of Basil Demetriadi* (The American Numismatic Society, New York): 347–364.

Wray, L.R. (2014), From the State Theory of Money to Modern Money Theory: An Alternative to Economic Orthodoxy. Levy Economics Institute Working Paper 792. Retrieved on March 10, 2021 from https://papers.ssrn.com/sol3/papers.cfm?abstract_id=2407711

4 Strong ties and deep habits

The Samothracian diaspora in network perspective

Sandra Blakely and Joanna Mundy[1]

Spreading Secrets: networks, mysteries, and the diffusion of ideas

How can secret practices cast light on strong ties? Strong ties are characterised by deep affinity, such as that shared among family members or colleagues in daily interaction with each other, with common cultural background, identities, and shared interest. Trust, cooperation, commitment, and emotional attachment are notably greater for strong ties than for weak ones. Strong ties translate into increased innovation and information transfer, as individuals are more likely to accept knowledge from people they trust and who, because of shared cultural context, are able to anticipate the fit of the information in the recipient's system.[2] As the recipient uses the new knowledge in her or his domain, innovation emerges, precisely because of the distinctions which remain between the domains of the source and the recipient, distinctions which lie beyond the boundaries of their shared experience.

If the social outcome of strong ties is the diffusion of knowledge, practices sealed by secrecy would logically seem to promise little for their exploration. The mystery religions of the Greek and Roman Mediterranean, however, offer paradoxically apt case studies. The mysteries were defined by the silence required of initiates regarding procedures and propositional content, so that information flow is by definition forbidden. What did flow, and flow broadly, were patterns of affiliation among Greeks and Romans scattered in city states around the Mediterranean's shores. Those patterns were explained at the mythic level as the effect of travelling charismatic founders. A social network analysis offers explanation at the historical level, building on the evidence available for the initiates themselves, their travels, and local adaptations of the rites as they put their mysteric affiliations to social use. For ancient studies, the model of strong ties, foregrounding the role of the affective, the symbolic, and shared constituencies of cult, city, and professional peers, casts light on the complex of dynamics through which the ancient mysteries spread, and offers a framework for integrating data from far beyond the cult site with the 'black box' of the initiation itself. The

DOI: 10.4324/9780429429217-5

mysteries, in turn, offer a contribution to social network analyses in ancient contexts, distinguished from the analysis of living cultures by problematically fragmentary and contestable datasets.

This chapter takes up the mystery cult of the Great Gods of Samothrace as a case study of the use of strong ties to investigate the spread of a cult sealed by secrecy. It begins with the typology of the mysteries in the ancient world, positions Samothrace among them, and introduces the epigraphic record relevant to the rites. These inscriptions record a range of categories through which individuals were connected to the rites, their cities, and other initiates. Each of these categories is considered both in terms of the typology of the ancient mysteries and the phenomenon of strong ties, including juridical weight, perceptual intensity, kinship, and the generative force of secrecy as a cultural practice. Analysis of the Samothracian network reveals several nodes of high betweenness, at Rhodes, Rome, and Thera. Comparison of the Rhodian dynamics with evidence from Delos, Mylasa, and cities of the western Black Sea reveals exemplars of innovation which build on strong ties, and help explain the spread and resilience of the rites. The outcomes confirm the expectation from contemporary studies of the usefulness of strong ties in tracing innovation and adaptation. They also recommend a two-part response to the challenges of social network analysis in ancient contexts, pairing a micro-scale, granular attention to the historical context of individual pieces of evidence with the macro-scale concern for typologies which were as emic to the cultures under investigation as the technologies of analysis are foreign.

Defining the mysteries

Mystery religions have long resided in the interstices between *polis* religion, voluntary cults, magic liminality, and charismatic antecedents to Christianity.[3] Their terminology comes from the words used by Greek and Romans themselves; their categorical elusiveness arises in part from the scholarly trends of the nineteenth and twentieth centuries, drawn to an apologetic comparison with Christianity, as well as from the sheer number of phenomena categorised as 'mysteries'. These include the rites attributed to Orpheus, a mythic musician and founder of rituals: 'Orphic' texts, engraved on gold and bone and small enough to be held in the hand, suggest a ritual focus on the afterlife.[4] That focus is shared with the great mystery cult of Attica, celebrated at Eleusis, where thousands of people at a time gathered into a massive hall to witness 'things shown, things said, and things performed'.[5] And mysteries focused on Dionysos, the god of wine and social inversion, seem from the literary evidence to have led followers to the boundaries of the acceptable, while Roman silverware and wall paintings reflect its usefulness as a marker of elite status.[6]

Among the mysteries were dozens of local cults which remained specific to their communities, their priesthoods provided by prominent local families. Those which rose to panhellenic prominence foreground various models of diffusion, in both ancient sources and modern scholarship. Charismatic reformers loom large, such as Dionysos himself in Euripides' *Bacchae*, an analogy that carried into the spread of the early Christian church. Pausanias describes the missionary figure Methapos, who came from Athens to Boiotia to found the mysteries of the Kabeiroi at Thebes: he was a hierophant, literally 'one who makes manifest the sacred', a musical priest, and an expert in the mysteries, establishing rites for Hermes, Demeter, and Kore as well as Kabeiroi (Pausanias 4.1.7). And no fewer than three missionary founders were linked to the mysteries of Andania.[7] Caucon was the first, known variously as a son of Poseidon, or Arcas, or Lykaon, all mythic figures (Pausanias 4.1.5).[8] He appeared in a dream to the generals of the Argives and the Thebans on the eve of the battle of Leuctra, promising victory and giving enigmatic instructions for the discovery of a buried bronze urn containing a tin scroll inscribed with the mysteries (Pausanias 4.26).[9] The second founder was Lycus, the son of Pandion, a primordial king of Athens; their names, 'Wolf' and 'All gods', condense cultural notions of bravery, untamed power, and a pantheon of indeterminable boundaries. The third was Mnasistratus, a wealthy Messenian in the early imperial period, recognised in inscriptions that commemorate his re-establishment of the mysteries.[10]

Divine authority, liturgical texts, metaphoric force, and priestly charisma cluster around these tales of the spread and renewal of mystery cults.[11] Explanations focused on followers rather than founders have emphasised personal psychology and flashes of transcendence, while apologetic models which saw the rites as antecedents for Christianity have emphasised universal longings for salvation in a deracinated Hellenistic world.[12] Other models emphasise the solidarity and identity afforded the military men who celebrated Mithras, a god whose origins were imaged as Persian, but whose rites were rooted in the Roman world.[13] Analogous arguments are made for the sailors and merchants who brought Isis from Egypt and the eastern shores of the Mediterranean to Delos and on to Rome.[14] Scholarship long categorised the cults of Mithras and Isis as 'oriental', a title which threatens to obscure the fact that neither god was celebrated with mysteries in their regions of origin: an act of cultural borrowing and implicit mobility underwrites both phenomena as experienced in the West.

The Samothracian rites are among those mysteries whose initiates came from across the Mediterranean world, far beyond the Thracian coast and the mountainous island that gave them their name. They informed literary productions from the fifth century BCE through the end of antiquity, and inspired local celebrations in Asia Minor, the Black Sea, Egypt, and Rome.[15] Those who craved the rites' promises needed to travel to the island only once, where initiation and *epopteia*, the 'viewing' which was the highest level of instruction, could be experienced on a single day. The motivation for initiation seems in some cases to be nearly happenstance, a stopover on an already determined trip, such as Ovid *en route* to the Black Sea, the mythic Argonauts on their way to Colchis, or the Roman administrators initiated along with their households.[16] Against this we may balance the miles travelled to reach the sanctuary, the inherent risks of sea travel, and the status accorded to mystery initiation.[17] The reach of the rites and their reputation directs our attention to the range of data that suggest how Samothrace became part of the cognitive and experiential landscape of the shared Greek and Roman Mediterranean. A generous portion of those data bring the ancient individuals, rather than the mythic founders, into view, as they integrated their Samothracian experience into their social webworks. Decrees recording *theoria*, an ambassadorial status to the island's festivals, and *proxenia*, a commemoration of service granted to the citizens of one city by a citizen of another, index its integration with civic institutions, and the forging of relationships among Hellenistic *poleis*. Altars and shrines to the Samothracian gods shaped the landscapes of religion far from the island; processions and sacrifices were public performances of Samothracian connections; associations forged cohesive sub-*polis* communities. These monuments, performances, and institutions did not demand travel to the island; that the idea of the rites themselves travelled, however, is clear. Samothrace was second only to Eleusis in its reputation among the ancient mysteries, but the reasons for its expansiveness remain obscure. Eleusis' reputation may be linked to Athenian power and ambition, and analogies may be found in the patronage of the Macedonians and Ptolemies on Samothrace. But the Greeks who colonised Samothrace never achieved the international prominence of Athens in historic, political, or economic endeavours, and Macedonian and Ptolemaic enthusiasms signal more than explain the value accorded the cult. The social and historical dynamics behind the success of the rites—materialised in architectural dedications from competing rulers, an abundant epigraphic record, and a long literary presence—remain as mysterious as the ritual experience itself.

A Samothracian network?

Network analysis is a promising approach for the available data and historical questions regarding Samothrace, as well as current conversations about the role of strong ties and complex contagion in the spread of ideas. Its pursuit also highlights the caveats attendant on statistical analyses in antiquity.[18] Samothrace offers a relatively robust epigraphic record. Of over 900 individuals, 545 may be connected to one of 109 sites away from the island.[19] Graham has characterised epigraphy as 'low hanging fruit' for ancient social network analysis: it combines the geospatial specificity of archaeological materials with the personal voices of the ancient world, functioning within cultural institutions which shaped human interactions towards the fulfilment of specific needs.[20] Among these are economic goals, personal relationships, inter-*polis* collaboration, care for the dead, and the traffic in praise. The creation of an inscription was itself a social practice reflecting the needs for memory, identity, and social mobility. Epigraphic networks have informed analyses of religious innovation in the Roman world, the evolution of political power in the Classic Maya, and ancient trade and economic patterns.[21] The extensive map that emerges from Samothracian inscriptions reflects a multiplicity of possible motivations (Figure 4.1). Both literary and archaeological evidence confirm the prestige of successful pilgrimage to the island, enrolling one's self alongside legendary and royal initiates including Pythagoras, Agamemnon, and Odysseus along with the Argonauts, Ptolemies, and Thracian kings.[22] Equally well known is the promise of the rites to keep its initiates safe in sea travel, materialised in monumental and epigraphic form, and appearing in literary evidence as an ecological fantasy of nature sympathetic to the faithful initiate. The *pompilos* fish would guide their vessels, violent winds would cease, cooperative breezes would rise, and the Dioskouroi, as shining celestial lights, would dispel the darkest storm.[23] The pragmatic civic counterpart of these myths may be found in networks of collaboration, communication, and mutual non-aggression among those who affiliated themselves with the island.[24] De Ligt and Gabrielsen propose that festivals and fairs were frameworks for interactions that fostered interpersonal and interpolis trust, and thus enabled the long-distance trade that tied together the scattered centres of the ancient Mediterranean.[25] This renders the geospatial spread of Samothracian affiliates a measure of the potential for those promises to be realised, bridging mythic images and

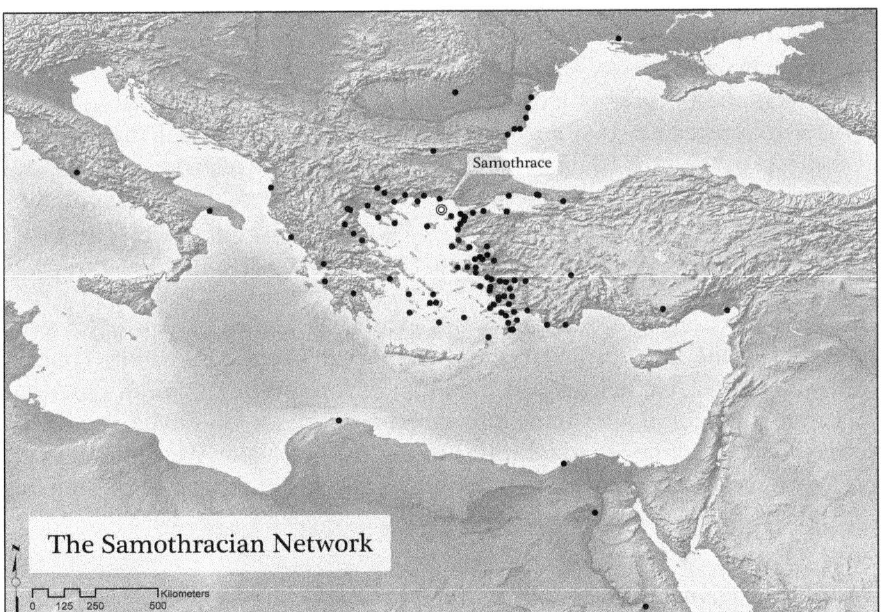

Figure 4.1 Sites attested in Samothracian inscriptions.

Source: Created by Michael C. Page.

ritual promises with social interactions that enabled success. This model of the Samothracian network resonates with Fafchamp's solidarity networks in pre-industrial societies, in which symbolic claims play a significant role in defining and ensuring social arrangements, and punishment strategies help assure compliance.[26] The symbolic claims of Samothrace added the gravitas of divine promises to the long-established cultural habitus of *proxenia* and *theoria*, the self-advertisement of the epigraphic habit, and the specifically Greek practices of mystery cults, in which the demands of secrecy index the mastery of cultural norms. The punishment would come from the gods themselves, who surrendered to the sea those who profaned the rites.[27]

All of these factors, rooted in the shared cultural norms of the rites' participants, recommend an analysis oriented to strong ties; the alignment, in turn, between strong ties and cultural innovation encourages a focus on betweenness centrality. Network analyses are capable of measuring multiple forms of centrality: degree, indicating the number of connections; closeness, the proximity of one entity to another; and betweenness, the proportion of shortest paths that cross a node. Of these three, betweenness centrality has the most critical role in information flow and innovation.[28] Nodes of high betweenness are literally 'in the thick of things'; they become either bottlenecks or bridges for the flow of information, able to withhold or distort data in the process of transmission. The distortions to which they give rise are modifications that fit the information to their own best interests. Betweenness centrality is thus a predictor of innovation. In business contexts, it often correlates with the number of patents secured, and financial performance; in research collaboration networks, betweenness centrality has greater impact on preferential attachment than degree or closeness centrality.[29] Nodes of high betweenness in the Samothracian network may index both a significant role in the diffusion of the rites beyond the island, and loci where we may investigate most clearly the uses to which the rites were put. These nodes reflect models of complex contagion, in which the diffusion of ideas is enabled by alignment with existing cultural traditions. Their reception, rather than disrupting the new environment, energises the social forces of memory, reciprocity, and trust that characterise strong ties.[30] Three nodes exhibiting higher than usual degree centrality in the Samothracian graph—Rhodes, Thera, and Rome—offer a case study of the challenges and potential of interpreting a graph based on epigraphic data, and an avenue into the micro-, meso-, and macro-scales of connectivity which local adaptations to the rites enabled. We begin with an exploration of the epigraphic record on Samothrace itself, and the strong ties engendered in the interactions they record.

Abundance, lacunae, and deep ties: Samothracian inscriptions and experience

The data for this analysis come from 236 inscriptions, 171 of which come from the island itself: they are dated largely to the second century BCE through second century CE, often through letter form. They represent less than half of the lifetime of the rites, which were celebrated for a thousand years, from the sixth century BCE into the fourth century. They also begin some 150–200 years after the first signs of interest in the rites from beyond the island, materialised in epigraphic evidence from Thracian Seuthopolis and the Macedonian monumentalisation of the site itself.[31] Those from the island record initiation, *epopteia*, and grants of *theoria* and *proxenia*, providing a snapshot of groups assembled from different and very distant locations. Inscriptions off the island thank the gods for salvation at sea, or record local priesthoods, temples, and organisations which recognised the Samothracian gods. These two types of inscription let us make a provisional gauge of the weight accorded to the Samothracian rites in different centres around the Mediterranean, and confirm face-to-face encounters on the island for representatives from those sites.[32] As historical sources, the inscriptions are as sparse in detail as they are numerous by the standards of an ancient site.[33] Very few offer iconography, and none any narrative; with a single exception, no divine names are confirmed, as the gods of the rites are referenced only euphemistically as *theoi megaloi*.[34] Individual names generally occur only once, and with rare exception are not known from elsewhere in the historical record. They do not, therefore, provide the kind of information about person to person connection which is the heart of sociological network analysis. City names, however, do. Individual cities appear with varying frequency: 2% appear on more than ten inscriptions, 55% on only one document. Each city may have from one to forty individuals associated with it, names associated with a range of civic and ritual roles. The number of cities with many representatives is very small: only three sites have thirty to forty individuals, while over 50% have two, one, or no names connected to them. Given the ability to measure the comparative intensity of engagement with the network as a whole, we have elected to have cities take the place of actors in the Samothracian network, with individuals acting as proxies for their home towns. This approach has the unfortunate impact of disqualifying nearly half of the names preserved in inscriptions relevant to Samothrace; only 545 of the 950 preserved names along with their home cities. A simple numerical analysis of the data, moreover, would seem to recommend a focus on weak ties, the network emphasis which, as Centola and Macy have noted, offers little measure of the social dynamics of information diffusion and behavioural adoption.[35]

The potential, however, of these materials for network analysis, and particularly for a focus on strong ties, is significant. Laconic as the inscriptions are, they are relational: they record interactions among individuals under the aegis of the gods. These are the interactions which, as Sindbæk noted, must be synthesised for archaeological materials that have no textual component.[36] The relationships they record are largely of three types: initiations, the visits of *theoroi*, and grants of *proxenia*. The strength of these relationships may be gauged by their contractual nature, their resonance with civic forms, the requirements of secrecy, the intensity of the initiatory experience, and the social stakes of honour, profit, and competition.

Initiation as a strong tie: moral contracts, ritual space, civic consequences, and affective intensity

Mystes, in their sole visit to the island, entered into a contractual relationship with the god: the reward for their discretion regarding the rites would be safety in travel at sea.[37] There are moral dimensions to the contract, which required both a ritually achieved purity and a commitment to future behaviour. Purity at Samothrace was accomplished through confession of sins, consistent with practices familiar in many other mysteries as well as non-mysteric sites.[38] Eleusinian initiates bathed in the sea *en route* to the sanctuary;[39] Mithraic initiates were purified with honey poured on their hands as well as their tongues, a catharsis of both acts and speech.[40] Orphic purification was a matter of social performance in day-to-day life, including the wearing of white clothing, vegetarianism, and avoidance of moments of birth and death.[41] The act of confession at Samothrace fulfils this requirement through an interaction between two individuals, the candidate and a priest whose title was 'Koies'.[42] Lehmann suggests that the exchange took place at the point of entrance into the Hieron, a building likely used for *epopteia*, the highest degree of initiation.[43] Here the priest would stand on one raised stone, the initiate on the other, separated from each other by a flaming torch, inserted in a third stone intended for that purpose. The position of interrogator and confessor echoes that used in civic contexts in courts of law: at the Athenian Areopagus, those seeking magistracies or serving as witnesses would stand on one stone, while being questioned by a magistrate standing on the other.[44] The dramatic visual impact of two torch-lit faces, conversing across a burning flame on a dark Aegean night, thus replayed the legal procedures traditional in Greek city-states. The semantic importance of lit torches for the island's experience is underscored by the depiction of flaming torches before doorways on four initiate lists from the second to the first centuries BCE; the images are rare exceptions to the aniconism which characterises most Samothracian inscriptions.[45] Fragments from within the Hieron confirm the architectural realisation of this image in the form of torches which functioned as frames for the curtain which screened off access to the *sanctum sanctorum*, the point of highest revelation. The initiate would have seen these shortly after the conclusion of the interrogation that granted him or her entrance to the building.[46] The embodied, interactive moment of the initiate's confession was thus intensified in monumental form at the point of transition into the culminating moment of the rites, a visual image capable of functioning as metonymic for initiation itself; at the same time, it inscribed the island's ties into the vocabulary of civic procedure.

The confession rendered the life history of the initiate immediately relevant to their bond with the island's gods. As a recitation of errors acknowledged, it foregrounded the candidate's own imperfection; realised through person-to-person interaction, it emphasised asymmetrical relationships and relational hierarchies, first between initiate and priest, second in the bond between men and gods.[47] That imbalance of power intensified the contract of the rites, as it made the divine capacity for punishment especially threatening. That the gods disciplined impiety was a familiar theme in the Greek world: that violators of the mysteries were among their targets was equally well known.[48] Miltiades was seized by panic when he violated the Parian sanctuary of Demeter Thesmophoros: he leaped from the wall, twisted his leg, and eventually died of the infection that set into the wound.[49] Battos, founder of Libyan Cyrene, was emasculated by the women who tended the rites of Demeter Thesmophoros.[50] Those who offended the Samothracian gods could expect destruction at sea, given the prominence of this promise in the rites, and the divine propensity for sinking the ships of those who dishonoured them.[51]

Civic authorities translated that divine wrath into legal proceedings against those who profaned the mysteries. Athens preserves the most compelling evidence for these, a reflection of the centrality of Eleusis to Athenian identity and prestige.[52] It was the magistrates and citizens of Athens, not the priests of Eleusis, who secured punishment of the guilty parties.[53] Slabs of stone and bronze were set up around the Eleusinion and on the Acropolis at Athens, inscribed with the names of those who had violated the rites; penalties for the guilty included fees, confiscation and sale of property, exile, and execution.[54] Such offences fell into the legal category of *asebeia*, impiety, applicable to more ritual practices than the mysteries alone.[55] The prominence of some so charged underscores the political force of rites:[56] among the charged were Aeschylus, the dramatist; Andokides, a speech writer included in the Alexandrian Canon of logographers; philosopher Diagoras of Melos; and Alcibiades, a notorious politician and military advisor of the aristocratic Alkmaeonid clan.[57]

Initiation was experienced but once in a lifetime: the strength of its social ties could not emerge from frequent or regular interactions. It was characterised, however, by an emotional and affective intensity that would contribute to the sense of community.[58] Aristotle claimed that the rites were a matter of experience rather than propositional content, in which it was not necessary to learn, but to suffer: οὐ μαθεῖν τί δεῖν, ἀλλὰ παθεῖν.[59] Plato describes a 'kinship of souls and bodies' συγγένεια ψυχῶν καὶ σωμάτων produced by participation at Eleusis.[60] A range of mechanisms for this intensity has been identified within the general typology of mystery cults, including preparatory stages, mind-altering substances, and mythic violence.[61] Preparations included fasting, rhythmic noises, exhaustion, pain, and sensory deprivation.[62] These would enable the confusion appropriate for the claim of Sopatros Rhetor that he came out of the telesterion 'feeling a stranger to himself',[63] and for Plato's suggestion that initiates were no longer aware of their own identities.[64] Similarly dislocating is the role proposed for mind-altering substances. Psychotropic substances have long been claimed and debated for Eleusis, and ancient poets and Christian fathers both mocked and feared Dionysos' wine as an instrument of intimacy with gods as well as fellow celebrants.[65] Authors from the fifth century BCE through the second century CE confirm a blend of terror, amazement, and blissful happiness as part of the rites. Plato claimed the new initiate experienced trembling and pain, joy, and happy apparitions;[66] Aelius Aristides describes Eleusis as the most frightening and resplendent of all that is divine for men, a blend of hope and terror.[67]

The Samothracian rites have particular evidence for two techniques of mysteric emotionality: shining lights at the culminating moment of the rites, and a complex engagement with music and dance.[68] Within the context of mysteries more broadly, such lights assume multiple functions. They may mark the emotional transition from fear into joy; they offer a transcendent identity for the initiates themselves, when they are conceptualised as dancing choruses of stars.[69] The lights invoke a range of mythological narratives, including the birth of Ploutos, 'conspicuous star, true light for a man'[70] or Dionysos Iakhos, the 'light bearing star of the night time mystic rituals',[71] and 'chorus leader of the stars, master of night-time voices'.[72] Bright lights could also replay the lightning flash that birthed Dionysos, the fertilising thunderstorms appropriate for his control over fertility, or the death of those struck by lightning. Such death brought benefits, in the form of purification from primordial crimes and the transformation of the victim into an object of ritual celebration.[73] Analogously broad semantics are associated with the lights of Samothrace. A second–first-century BCE epitaph for Isidorus of Athens boasts that its owner was an initiate at both Eleusis and Samothrace, and that on the island he encountered the 'doubly sacred light of Kabiros'.[74] The Kabeiroi are often equated to Dioskouroi, the divine twins whom Varro and Nigidius Figulus assigned a leading role in the rites.[75] Nigidius claimed that their translation into stars was the topic of the mysteries themselves, and emphasises their role saving those in peril on the sea.[76] The lights on Samothrace thus condensed hopes of conquering death with the promise of salvation at sea which was specific to the island's rites and framed, in monumental form, the highest moment of revelation. The form of those lights invested the embodied experience of initiates with mythic significance, as they carried in their hands the torches metonymous with the salvific light of the ritual. This would intensify their experiential connection to the light in a way similar to the star choruses and torch tossing of Eleusis. Those initiates familiar with more than one mystery, as Isidorus was, would recognise in those flames the deeply rooted cultural phenomenon of which they became a part.

The dances that were part of the mysteries dissolved the boundaries of the self and reinforced the boundaries of the group at social, divine, and cosmic interfaces. Lucian wrote that he knew of no mysteries that did not include dance, and that those who divulged the mysteries were said to 'dance them out'.[77] The experience is social: the mania of the Bacchantes and followers of Cybele are group experiences, analogous to the star choruses that articulate cosmic order through human movement.[78] The ability of dance to convey mythic tradition is a notable advantage in initiatory contexts. The girls who 'danced the bear' at Brauron engaged in a kinetic realisation of the aetiological myth of the rites, assuming animal identities as part of their rite of passage.[79] Dionysos' maenads danced at the edge of sanity, civilised life, and their own identities, perceiving the world through divinely altered eyes; analogous madness beset the dancers for Kotyto and for Cybele.[80] The *daimones* who clamour about the shores of Samothrace are mythic enactments of strong ties, both in the form of their dance and their association with the civic priesthoods of Asia Minor. Their dances are those in which the movement of bodies alters the perception of the mind, yielding a group experience of temporarily shared identity. Strabo compares it with the frenzy of Dionysos, the war dances that re-enact the birth of Zeus, and the uproar of Cybele's rites. The dancers have multiple names; all are corporate groups, appropriate models for the socio-genetic force of dance.[81] Those called Korybantes were said to have given their name to both the island and its town, and are part of the divine genealogy of the place.[82] Diodorus Siculus identifies Korybas as the child of Cybele and a Samothracian prince, who took the Samothracian rites to Asia Minor.[83] Dionysius Scytobrachion connects Korybantes to cultural flow in the other direction, claiming that they were the sons of the mother of the gods whose mysteries were celebrated on the mainland. Their mother settled them on the island, along with other colonists, where the identity of their father was the secret communicated in the mysteries.[84] If Scytobrachion's Korybantes were imagined, as Diodorus', to descend from the Samothracian royal house, their role in the rites would confirm the prestige of most elite stratum of the island's inhabitants. A bond of this type between Korybantic performers and social privilege is consistent with the historical and epigraphic evidence for civic priesthoods in Asia Minor. Here individuals of the best families purchased and executed the rites for the mystic mother of the mountain, imagined as Cybele, Artemis, Hekate, or Dindemene, and were conceptualised as Kouretes, Korybantes, or Bacchants.[85] The island's dancers thus tie the rites into a broadly spread and profoundly flexible cultural type, one with strong roots in the region from which Samothrace's initiates, *theoroi* and *proxenoi* most emanated.

From the mystic to the civic: strong ties and recurring festivals

The Asia Minor priesthoods exemplify the permeability of the boundary between the mystic and the civic: names associated with personal madness in myth, and with ritual performance in the mysteries, translate into civically ordered festival performances.[86] They also, in contrast to the singularity of initiations, reflect the repeated and regular practices which are recognised generators of strong ties of trust. The festivals of Samothrace share in this quality. The ties generated in initiation relied on the intensity of experience, the intersection with law, and the engagement with the cultural symbols, narratives and practices, including the commitment to secrecy, which were shared by Greeks from the Black Sea to Athens. All of these counter the challenge of creating strong ties through an experience accessed only once in a lifetime. Samothracian festivals, in contrast, were events whose function, in establishing Samothrace's position among the city states of the Greek world, was best answered by their recurrence. The festivals brought together representatives of cities from around the Greek world to the island. These pilgrims were designated *theoroi*, that is, 'watchers', a reflection of their responsibility to observe the festival. *Theoroi* represented not themselves, but their entire city-state. The activities of the festival were non-initiatory, public events, typically including a sacrifice, marching in procession, offering a gift, and viewing an event, likely the Dionysia, a celebration of Dionysos, which would have been held in the sanctuary's theatre. An appointment as *theoros* was a civic honour, one acquired by the well-connected and influential of their cities.[87] The Samothracian festivals would have presented these individuals with opportunities to create and strengthen their own personal networks beyond their city's boundary.[88]

The names and ethnics of Samothrace's *theoroi* are preserved in inscriptions which record representatives of up to seventeen different cities at a given event.[89] Several family groups may be identified among the Samothracian *theoroi*, suggesting that the epigraphic snapshots we obtain of a single festival may reflect longer lasting patterns of connectivity.[90] *Proxenia* grants were often made to those who came and performed the duties of the *theoroi*, as recorded on thirteen block grants on Samothrace, dated to the second to first centuries BCE, and one monument base.[91] *Proxenia* was the city's formal recognition and praise of the good intentions, eagerness, and personal resources the recipient had already manifested. In return he received the right to sail into and out of the harbour without paying fees, *asylia* (freedom from seizure) in times of war and peace, *ateleia* (exemption from taxes and fees), access to the *boule*, front seats in the theatre, and being first among the *hiera* (sacred rites).[92] Several Samothracian grants specify that *proxenia* was given to the recipient's descendants, a transmission within family lines that would further strengthen these ties, grounding them in the local kinship lines of Samothrace's scattered affiliates.[93] *Theoroi* also became initiates, combining the occasion of their civic duty with the personal benefits of the rites.[94] The number of *theoroi*-initiates foregrounds the permeability of personal, mystic experience with the fulfilment of a civic role; so too does the depiction of young women performing for the festival on the Hall of Choral Dancers, located inside the area of the sanctuary itself.[95] It would be furthered by the shared experience of dramatic performances of Samothracian myth in both the festivals and the mysteries. The epigraphic habit, as material realisations of civic and individual pride, emphasises the status of these appointments, and the benefits of advertising one's own access to sealed rites. They reflect a Greek analogue for sociological analyses of the practice of secrecy, which emphasise its function as a performance of status, built around the question of knowledge: one must simultaneously announce that one possesses knowledge, but refrain from revealing the knowledge itself. This demands acts of discretion which engage the symbolic structures and performances unique to every society: the modalities of secrecy are a uniquely strong index of specific cultural identities.[96]

Initiation, *theoria,* and *proxenia*, taken together, reflect strong ties of several types: contractual force, affective intensity, family lines, and the cultural *habitus* of secrecy. The bridges between civic and ritual are many, manifested in law and civic priesthoods as well as appointment as *theoroi* and *proxenoi.* Each of these ties, moreover, mediates an individual with a supra-human entity, the gods themselves or the states the humans represent. To the degree that the festivals helped build the social infrastructures of maritime success, all three also underwrote the promises central to Samothrace's fame. The cities who engaged with these ties did so, however, with varying degrees of intensity, and in response to the articulation of strong ties within their own communities. Network graphs help identify the most promising places to investigate these ties, in the form of loci of high betweenness. These are sites which, according to the theory and practice of network analysis, promise indices of high adaptation, innovation, and therefore the strongest role in the diffusion of Samothracian affiliation.

Investigating emergence: three degrees of betweenness

The betweenness centrality graph of our Samothracian data shows one node, Rhodes, notably stronger than the rest; it is followed at a second level by Ephesos, Thera, and Rome, with Chios and Kyzikos at a third level of strength (Figure 4.2).[97] The Degree modularity graph offers a visualisation of two factors—the number of names identified per site, and the sub-neighbourhoods of nodes which interact with each other most frequently (Figure 4.3). The graph for Samothrace suggests strong patterns of connectivity among several top performers. Rhodes and Ephesos appear in the same neighbourhood, Chios, Kyzikos, and Halikarnassos another; Rome and Thera are outliers, a visual indication that the cultural ramifications of Samothrace were less relevant in connecting those sites to each other than they were in connecting them to the broader range of participants.

The high betweenness of Rome and Thera arises from very different dynamics than those of Ephesos, and points towards the role of weak rather than strong ties in their visits to the island. Rome and Thera are positioned at the very edges of the graph, and gain their comparatively high betweenness by providing the sole bridge between the outlying clusters of four cities each and the rest of the cities. The four to which Thera gives access are attested in a single inscription, all of which but Alexandria are otherwise unattested, and thus suggests the potential for weak ties to increase the betweenness of a node.[98] Rome, analogously, provides the connector for four cities which appear only once in Samothrace's record. The three inscriptions that record their names reflect the 'boatload' phenomenon: initiation at Samothrace

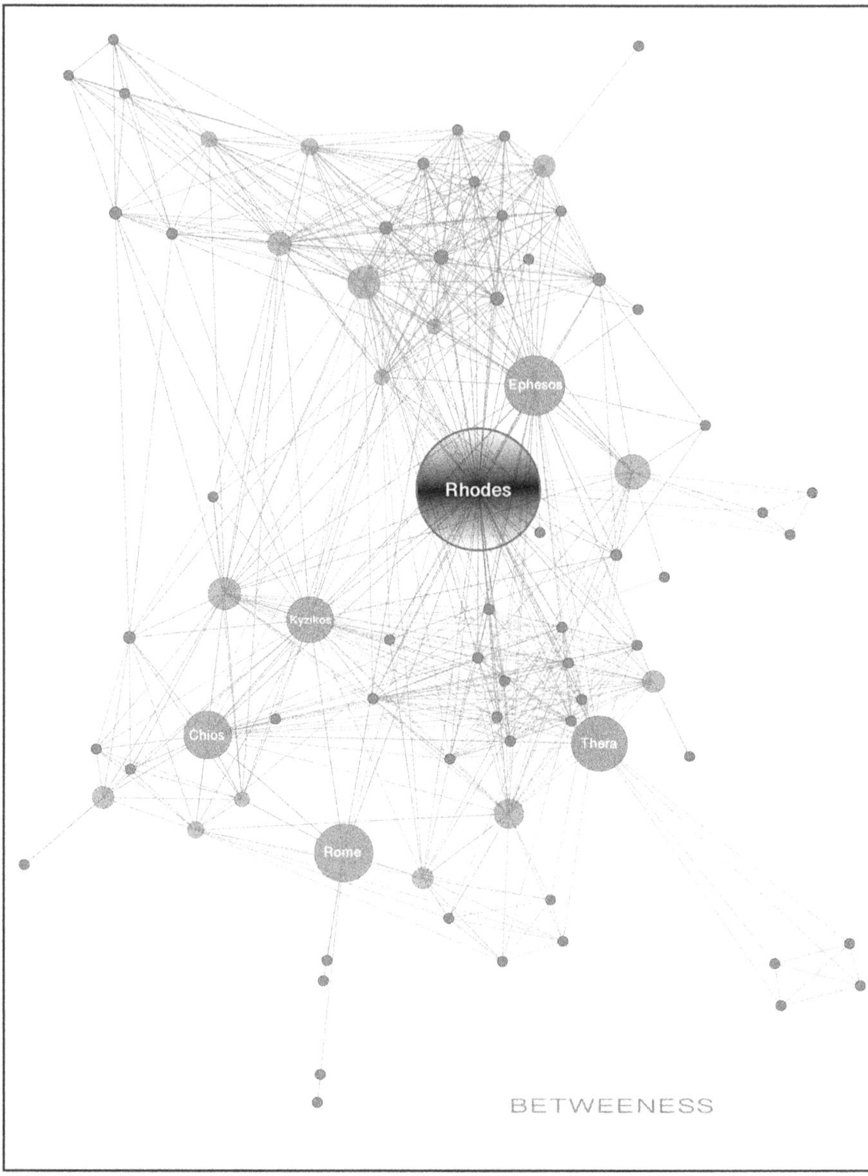

Figure 4.2 Betweenness centrality graph; Samothracian network.

Source: Created by Michael C. Page.

from at least the third century BCE was available on demand, when a ship came into port.[99] These inscriptions come from the crew of a war galley, a boat carrying a proconsul's household, and the stopover on the island of a collection of eminent Romans, as they returned from a pirate-fighting expedition. Knidos appears in a single inscription, on which is recorded the

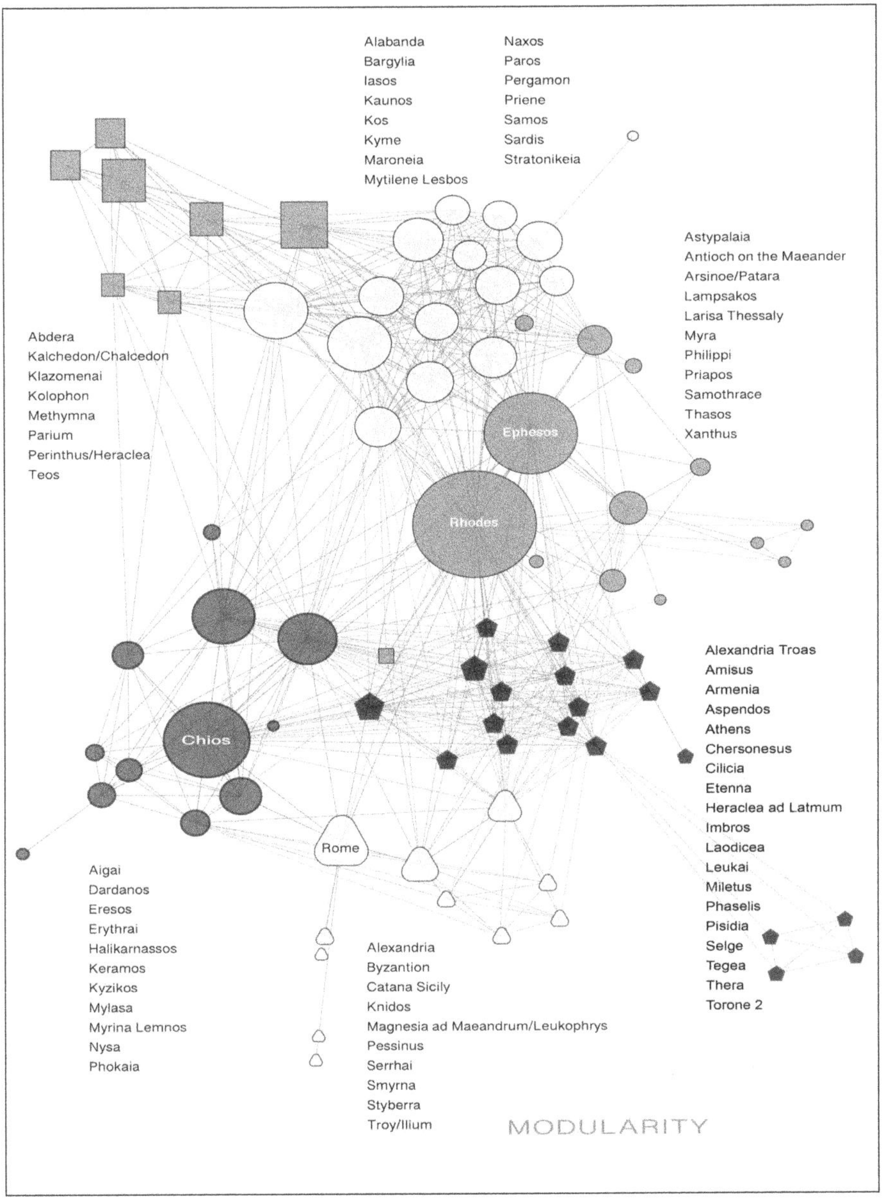

Figure 4.3 Degree modularity graph; Samothracian network.

Source: Created by Michael C. Page.

crew of a war galley who became *mystes* and *epoptes* along with their trireme commander (*trierarch*), vice captain (*epiplous*), admiral of the fleet (*nauarch*), and naval officer (*grammateus*); they are described as συμπλεύσαντες, those who sail together. Romans were initiated alongside them, and commemorated their achievement in Latin.[100] An inscription from May–June CE 186 offers the sole record of initiates from Pessinus and Smyrna, alongside Romans, in what seems to be the initiation of P. Iulius Geminius Marcianus, the proconsul of Asia, along with his slaves and household.[101] And the only Sicilian initiate on Samothrace is preserved on a stele from September 4, 100 BCE, recording the initiation of prominent Romans on their return from the campaign of M. Antonius against Cilician pirates: included in the list are Lucius Tullius, the uncle of Cicero, M. Fanius, praetor of 80 BCE and Plebian aedile.[102] Romans were a familiar presence on the island as initiates, *epoptes* and patrons; they were also exceptionally inventive, sowing Samothrace into Roman founding myths, topography, and institutions.[103] Their Samothracian network, however, is created without any of the civic or private institutions through which the Greeks enabled its spread. Many of its initiates share their experience only with other Romans, so that their time in the emotional and aesthetic intensity of the rites did not immediately connect them with the broader Aegean world.

Ephesos, in contrast, appears on seven inscriptions recording festival attendance, with up to seventeen different cities present at a single time, as well as one inscription of initiation: in these festivals, several connections emerge repeatedly.[104] Ephesian representatives shared festivals with their counterparts from Alabanda and from Rhodes four times; with individuals from Kyme, Lampsakos, and Kyzikos twice.[105] Given that *proxenia* and *theoria* could run in family lines, handed down from father to son, these repeated meetings suggest the potential for strong ties based on the interests of the leading families of the city through multiple generations. The decree culture of the Hellenistic world turned these patterns within kin groups into matters of public display, materialising the position of the Samothracian rites in the Hellenistic marketplace of honours.

Deep ties and local habits: priests and associations at Rhodes and beyond

Rhodes repeats this engagement with Samothracian festivals; its exceptional betweenness centrality arises as well, however, from the range of local institutions and habits through which it engaged with the island's gods. Rhodian representatives appear alongside the *theoroi* from sixteen other cities on a *proxenia* grant dated to 150 BCE, and with their counterparts from Ephesos, Kyzikos, Alabanda, Lampsakos, and possibly Athens on a *theoros* list of the second-first century BCE.[106] A Rhodian residing in Chalcedon represented that city as a *theoros* in a third-century BCE inscription, along with representatives from eight other cities.[107] Rhodians also shared their initiations with individuals from around the Mediterranean, from Antioch to Priapos and Rome.[108] These combine to make Rhodes one of the highest scoring sites in terms of names alone, with some fifty individuals claiming Rhodes as their site of origin. The inscriptions also reflect how often Rhodians' time on the island coincided with other cities with relatively high levels of engagement, notably Kyzikos and Ephesos.[109] Materials from Rhodes itself, moreover, provide significant evidence for the inclusion of Samothracian gods in celebrations at home. Priests are recorded on three separate inscriptions; on Karpathos, closely tied into the Rhodian orbit, a single third-century BCE inscription records thirty-nine individuals who held the priesthood in succession, suggesting a perennial, well-established institution.[110] The Rhodian inscriptions list the Samothracian gods next to Dioskouroi and Korybantes, suggesting an association on Rhodes with other gods linked to the mysteries. The Karpathian inscription may have come from the temple of Poseidon Porthmios, a context that suggests a focus on Samothrace's maritime promises.[111]

None of these detail the priests' responsibilities; the records of Samothracian priesthoods from nine other sites suggest possibilities, which index the adaptation and innovation that characterise nodes of high betweenness. These innovations integrate the Samothracian gods into a range of deeply rooted civic and economic patterns, including calendars, topography, performances, and economic concerns.[112] In the West Black Sea cities of Tomis, Istros, and Dionysopolis, the cities themselves distribute the Samothracian priesthoods. A second-century BCE inscription from Istros records the city's grant of the Samothracian priesthood to Dionysios, son of Strouthion, in recognition of his interest-free, year-long loan to the city, which enabled its purchase of grain.[113] The priesthood would belong to his family in perpetuity, δεδόσθαι δὲ αὐτῶι καὶ ἐκγόνοις ἀεὶ; his honour would be inscribed on the base of his statue, and on a stone to be placed near the altar or temple of the Samothracian gods.[114] Both family and civic organisations are again engaged in Tomis, where an inscription details the regulations of an association for Samothracian initiates.[115] Priesthoods may be purchased; they are held for life, though heritability is not confirmed. Those who purchase them are responsible for providing, on the seventh of the month of Apatouria (Pyanepsion, at the end of October), materials for the *mystes* to perform their celebrations and a procession as well as libations. The *mystes* also grant wreaths. This links the Samothracian gods to the city's calendar, topography, and engagement with the Hellenistic market of honours. Dionysopolis offers similar dynamics: here again, the city is in charge of the Samothracian priesthood, which it grants to Akornion, the ambassador of King Burebista who conquered the north and west Pontic cities in the late 50s and early 40s BCE.[116] Akornion's ambassadorial clout translated into the ritual sphere, with four priesthoods recorded on a single inscription: among these, the Samothracian seems the most recently acquired. It would last for his lifetime, and required him to make sacrifices and pay for the processions of *mystes* and citizens. Here, the role of the city in granting the priesthood, and the celebrations which juxtaposed citizens and initiates, manifest the care to integrate the city and the mysteries. Associations of *mysteis* reflect the translation of a singular experience into a habit carried out within the contexts of the city, and put on display through periodic processions that highlight without dissolving the boundary between those who were initiated and those who were not.

These priests were active in engaging Samothrace with their communities: Dionysios and Akornion suggest a pattern of high political standing among those who eager to accept this charge. Inscriptions in Asia Minor and on Delos offer more articulate case studies for how Samothracian priesthoods and votives mediated international connectivity through alignment of strong local traditions with emerging political needs. In the sanctuary of the local moon god in Sinuri near Mylasa in Caria, the annual wreathing of the priest of the Samothracian gods was used to date two inscriptions regarding the purchase and leasing of land by the sanctuary.[117] This practice, whereby temples would purchase private lands and then lease them back to their original owners, was widely practised in Mylasa, and bears analogy to the priesthood sales as well as to Trajanic alimentary schemes that provided the sanctuary with a steady revenue stream.[118] The use of the Samothracian priesthood to provide a date constitutes a significant semantic link, tying the idea of the gods into the practice that itself wove the sanctuary into the solvency of the entire community. The temple at which these were found is situated on a small promontory overlooking the fertile valley which was both the economic base of the community and its major route towards Mylasa.[119] Mylasa is one of the sites whose engagements with Samothrace materialise in inscriptions both on and off the island: it appears with Kyzikos on a *theoros/proxenos* inscription of the second century BCE, though no record of initiates from the city is preserved.[120] Reasons for a local commitment to Samothrace may lie in the complicated Hellenistic fortunes of the city, which was controlled by Ptolemies in the third century, then Seleucids, until it was liberated by Philip V of Macedon. Romans appear as 'common benefactors' in Caria as early as 188 BCE.[121] The Macedonians, Seleucids, Ptolemies, and Romans who in turn controlled Caria all had a notable presence on and enthusiasm for Samothrace. An integration of the gods into civic calendars, landscapes, and economies provided a point of contact legible by each of the powers who crossed through the revolving doors of the regions' political power over three turbulent centuries.

Clearer evidence comes from Delos, where an enterprising Athenian, Helianax, cites his status as a priest of Poseidon Aisios and of Theon Megalon Samothraikon Dioskouron Kabeirion in his dedication of a temple to Mithradates Eupator Dionysos, Mithradates VI of Pontos, on behalf of the people of Athens and Rome, in 102/101 BCE. The temple was built in the temenos of the Samothracian gods on Mount Kynthos, a location which afforded an exceptional view of the harbour and the sea.[122] His priesthoods reflect, on the one hand, a concern for maritime safety: Poseidon Aisios, an 'auspicious Poseidon', suggests the maritime success that made both Delos and Rhodes prosper.[123] The inclusion of Dioskouroi, whose powers to save at sea were attributed to Samothrace by the Hellenistic period, echoes the juxtaposition of Dioskouroi and Samothracian gods on two Rhodian priests' inscriptions.[124] That maritime focus materialises, in the Samothrakeion, in votives of naval and maritime implements, including a trident and wooden and metal anchors.[125] A Rhodian origin has been suggested for Helianax. He may descend from the same family who made a dedication at Delos ca. 169 BCE to the sister of the Pontic king Pharnakes and his successor Mithradates IV of Pontus, and the elements of his name suggest but do not confirm Rhodian ties.[126] His concern for Rome as well as Athens is consistent with the status Mithradates VI claimed until 89 BCE, 'Friend and ally of the Roman People'. The Romans on Delos would have been the merchants and *negotiatores* who linked Rome to the commercial centres of Asia. Why Helianax chose the Samothracian sanctuary for his dedication remains unclear. His status as priest offers the simplest logic; Sanders and Catling conclude that no convincing argument, however, has been found for interest on Mithradates' part in Samothrace, an indifference which sets him apart from other Hellenistic kings. It is easy, however, to imagine either an Athenian or Rhodian family who prospered through trade with the Pontic regions, and as a result chose to honour its king on Delos.[127] The inclusion of Kabeiroi with the Samothracian gods suggests resonance with Athenian perspectives: the name is otherwise absent from epigraphic evidence for the cult, with the single exception of another Athenian voice, the mime Isidoros Nikostratou, whose epitaph records his Samothracian encounter with the doubly sacred light of the Kabeiroi.[128] Roman interests are well reflected in the rites as well. The abundance of Roman initiates includes many who, like the initiate and proconsul P. Iulius Geminius Marcianus, were magistrates stationed in Asia. By the 90s BCE, they in many ways dominated the entire island of Rhodes, its political as well as its commercial institutions.[129] Helianax's choice of location suits more than his own convenience: the Samothracian gods provided a framework that mediated Roman, Athenian, and possibly Rhodian interests with the object of his dedication, precisely because of Samothrace's status as a token of the shared interests and cultural symbols which characterise strong ties. Those Samothracian connections combined to let him honour the Pontic king by positioning him in the interstices of the contemporary trading powers on which Delian fortunes relied.

An analogously detailed insight into the network dynamics engaging the Samothracian gods emerges at Rhodes in the evidence for *koina*.[130] Rhodes was the centre of gravity for these associations, with two hundred known from the third century BCE to the second century CE: Samothracian gods appear in ten of these.[131] *Koina* were private groups, whose inscriptions record sacrifices, feasts, mutual assistance, the lending of money, the distribution of honours, and provision for burial of its members. Each group had a distinctive proper name: these are notoriously agglutinative, such as the *Soteriastai Asklapiastai Poseidaniastai Herakleiastai Athanaistai Aphrodisiastai Hermaistai Matron Theron koinon*.[132] The names are human as well as divine. Those human names, turned into adjectives, allowed the founders themselves to be onomastically celebrated alongside the cluster of divinities who signalled the orientation of the group members. The groups also had internal organisation, and a clearly defined membership, which included foreigners, Rhodians, and even women. Early scholars emphasised the cultic and professional roles of these groups: they have seemed faithful imitations of the *polis*, which made accessible to outsiders the privileges which Rhodian citizens had from the *polis* itself. The *koina* were generators of strong ties: from their shared activities emerged commensality, camaraderie, and trust,[133] and Aristotle declared that they were formed for sacrifice and social intercourse.[134] They were also, moreover, deeply tied to the honour-generating machinery which was essential to the functioning of the Hellenistic world of cities, commerce and political exchange. The tombstone of Hagetor, a *hegemon* and *strategos* (leader and general) of the late third or early second century BCE, listed no fewer than twenty-seven associations—a number far in excess of the norm, but indicative of the responsiveness of associations to the need for honour and belonging.[135] An inscribed altar base from the eastern necropolis of Rhodes records the membership of the entombed, whose name is not preserved, in five different *koina* in as many different cities. These enabled him to create for himself a personal network that crossed the island, from the southernmost to the northernmost end.[136] The associations are at one level mesoscopic, functioning as aggregation spaces between the level of the family and the *polis*.[137] They are also capable of connecting individuals to functional organisations across the boundaries of the *polis*, or indeed an island or geospatial region. They were profoundly responsive to the needs of their social environments, models of the complex contagion which contribute to node strength.

The centrality of the sea in Rhodes' economic and political success made maritime gods natural inclusions in the *koina*, and has provided the dominant explanation for the Samothracian gods among them; Poseidon appears in five associations.[138] Rhodians needed to maintain their dominance as the gateway for eastern Mediterranean trade, and the merchants of protection against seaborne predation.[139] That dominance was built on Rhodes' fleet of superior warships, a chain of offshore naval bases, and alliances with Rome and the Hellenistic rulers of the east. A constant shortage, however, was a sufficient number of skilled mariners to man the fleet. *Koina* provided a mechanism for the naval aristocracy to draw such individuals to Rhodes and keep them there.[140] Many *koina* were founded by men of prominent families, with impressive records of maritime service; the *koina* created common ground between an immigrant, eminently mobile workforce, and local, elite descent group that sought their services. The sacrifices, feasts, and concern for securing burial grounds built community and trust, deepened by the tie to favouring divinities, and so created initiative for such workers to make Rhodes their permanent home.

A preserved subscription list of one Samothracian affiliated *koinon*—the *Samothraikiastan Aristobouliastan Hermaistan ton sun Ktesiphonti*—offers vivid confirmation of the multi-ethnic, international dynamics of a Rhodian association.[141] The founder is a foreigner, Ktesiphon from the Chersonesos, and of thirty-two members listed, twenty-four originate outside Rhodes, from locations including Ladokeia, Alexandria, Ephesos, Kyzikos, Chios, Cilicia, and Armenia. Their names suggest interaction with a community as broad spread as that assembled in the Samothracian festivals, but who interacted with greater frequency. The association itself may reflect individuals of significantly different levels of wealth, as members contributed different amounts towards the funds for the purchase of land for their private cemetery. Among the place names are several—Ephesos, Kyzikos, and Chios—which emerge with particular betweenness in the Samothracian graph. Of the three, Chios' betweenness is the most heavily shaped by this stone itself. It appears otherwise only on one *theoros/proxenos* grant, which it shares with Kyzikos and other less-connected cities, and its *mystes* shared their experience with the well-connected Romans.[142] The inscription is at once a caveat on how much a single document may impact network analysis, and an intriguing index of the gravitational force of Samothrace for group formation in sites away from the island. The associations of *mystes* in Tomis, Dionysopolis, and Kallatis acted analogously, bringing together those who shared the dual boundaries of initiation and citizenship. Their analogue in Rhodes may or may not have been initiates, but shared with their Black Sea counterparts the use of Samothrace as a marker of identity.

The Samothracian gods appear in four of Rhodes' associations comprised of shipmates and fellow soldiers. Rostovtzeff, Laundy, and Berthold saw these organisations as manifestations of the desire to perpetuate the bravery, fellowship, and patriotism that emerged from shared time on campaigns.[143] Gabrielsen notes their responsiveness in addition to the perennial need for skilled seamen in the Rhodian economy. Samothrace appears in the title for two groups of *mesoneoi*, rowers, and two of *synstrateusamenoi*, those who fought together.[144] These are the *Samothraikiastai kai Lemniastai synstrateusamenoi*, and *Samothraikiastai Nikostrateioi synmystai synstrateusamenoi*.[145] The Samothracian element would further their effectiveness as advertisement of the group's skills. The image of Samothrace would resonate, at the simplest level, with the local prevalence of maritime gods; Samothrace's promise of maritime safety, additionally, would offer a divine counterpart to Rhodes' dominance in the business of protection against maritime predators. And, the Samothracian gods would provide in Rhodes, as they did at Mylasa and Delos, gods who were among the favourites of their important international trading partners, Ptolemies as well as Romans. The *koina* that declared themselves to have shared initiation as well as labour at sea recall the phenomenon of boatloads of initiates on Samothrace. Particularly dramatic among these is a first century BCE inscription, dated to the years in which Romans and Greeks joined in efforts to resist piracy. A group of Chians arrived in their *lestophylakika ploia*, anti-pirate ships; they record, as well as their own initiation, the names of the commander (*archon*) and junior officer (*dekatarchos*) under whom they sailed.[146] *Koina* which declared themselves fellow rowers and fighters sent a powerful message about their cohesion and camaraderie, a quality eminently marketable in Rhodes or any port. Samothracian initiation would underscore those ties with the invocation of the particular bond that arose, as Plato, Aristotle, and others noted, from the intensity and solemnity of the mysteries, as well as the common baseline of worthiness and commitment to future behaviour which those rites entailed.[147]

Conclusions: caveats, chronology, and complex contagion

The mysteries of Samothrace were at their core relational—the forging of a contract between men and gods—and the contagion of the rites could be fundamental to the realisation of its promises. This casts down the gauntlet to approach Samothrace from a macroscopic perspective, and seek to model the dynamics of relationships that were shaped by a shared familiarity with the mysteries. Network analysis, focused on the graphing of human interactions, offers a methodology for this task. The outcomes suggest historical patterns for the spread of the cult rooted in complex contagion, adaptations at the local level to multiple factors including economic and political need, deep-rooted local institutions, and the various Hellenistic modalities for generating honour, memory, and visibility. Three patterns have emerged repeatedly in these case studies, based in chronology, the quality of our data, and the relationship between macro and microscopic analyses.

The chronological frame of our data, the third to first centuries BCE, has foregrounded the multiple political realities to which Samothracian rites could respond, as regional control moved from one Hellenistic power to another and finally to Rome. Roman reception of the rites has characteristically been approached as relatively *sui generis*, and unrelated to its spread into the Eastern Mediterranean. The Roman and Greek mythopoeic impulses for Samothrace are indeed wildly distinct from each other. The cases of Mylasa, Delos, and Rhodes, however, suggest that precisely the capacity for the same cult to be read as needed by different audiences could shape its capacity to prove useful in those locations, and itself become part of local habitus.

The lacunous nature of our record raises the caveats familiar in digital and statistical approaches to the ancient world. Our graphs cannot be considered valuable for their historical accuracy: they are not Rankean maps of 'how it actually was' *wie es eigentlich gewesen ist*. They are uniquely useful, however, for the macroscopic perspective they afford on the data we have to work with. As a synopsis of the attested patterns of interaction, the graphs help us select case studies out of all the locations that entered into a relationship with the island, and through it with the broader Mediterranean world. Those that come to the fore are those which have the most helpful data for exploring those ties. That data reflect local epigraphic habits and the accidents of preservation as well as the adaptation of the cult. The betweenness graph foregrounds not raw numbers of individuals and monuments, but an index of how actively those sites seem to have enabled the flow of information and ideas. The institutions that cluster on Rhodes—priesthoods, temples, and organisations—represent the kinds of cultural traditions that may enable complex contagion, connecting local calendars, topographies, performances, and patronage to the island's reputation. Comparison with the Black Sea cities highlights the extent to which there is not a deterministic relationship between local organisations and network ties: the Samothracian rites could generate different responses in different places, even when functioning in analogous institutions. The Black Sea cities, for all their priesthoods, temples, and organisations of *mystes*, are far behind Rhodes in their connectivity. At Rhodes, Delos, and Mylasa, however, the epigraphic records offer insights into some of the stories behind these patterns. These take microhistoric form, in the ambitious priests eager to gather crowns from their *koina*, mesoscopic, in the local institutions that enabled connectivity, and macroscopic, when a Delian priest deploys his Samothracian access to tie Athens, Rome, Rhodes, and Mithradates into a coherent credit to himself.

The permeability between these different strata of action is a repeated pattern in these case studies: the line between individual initiation and civic representative was often crossed, on the island as well as in the poleis. This suggests that the emotion, memory, ambition, and ritual intensity that the rites could engage would flow naturally into the spread of the cult as traced through civic institutions and private associations. Our quantitative analyses of epigraphic data, in the end, thus return us to the individual motivations—whether salvation, solidarity, or enthusiastic conviction—that inform the long investigation of the initiatory cults. What the graphs offer, however, is a more nuanced sense of the great variation of social, political, and economic contexts in which these motivations functioned, and how they connected Greeks and Romans to each other.

Notes

1. The authors thank Emory's Center for Digital Scholarship for their support of this project: special thanks are due to Sara Palmer, Digital Text Specialist, and Michael Page, Geographer. https://scholarblogs.emory.edu/samothraciannetworks/.
2. Tortoriello, Reagans and McEvily 2012; Leyden, Link and Siegel 2013; Friedkin 1982.
3. Bremmer 2010; Eidinow 2011; Edmonds 2013: 71–95; Kindt 2009; Naiden 2013; Taylor and Vlassopoulos 2015.
4. Edmonds 2013: 71–94; Smith 1990.
5. Cosmopoulos 2015: 22–24.
6. Gruen 1990: 34–78; Holzhausen 2008; Hutchinson 1991; Hearnshaw 1999.
7. Robertson 1988: 254
8. Poseidon, Aelian *De Natura Animalium* 1.24; Arcas, scholia to Homer *Odyssey* 3.366; Lykaon, Apollodoros *Library* 3.97

9. Ribichini 2016
10. Pirenne-Delforge 2010.
11. Walters 1893; Farnell 1907: 208; Bremmer 2014: 87; Bowden 2007.
12. Fox 2005, McLoughlin 2005, Merkur 2005, Gifford 2005; Margry 2004; Stark and Bainbridge 1980; Smith 1990.
13. Stoba (this volume); Price 2012; Gordon 2007.
14. Glomb et al. 2018; cf. Tilly 2010; Wortham 2006; Bremmer 2014: 110–114;
15. Lewis 1958; Cole 1984.
16. Cole 1989: 1580–1587.
17. Maillot 2012; Bremmer 2014.
18. Davies 2015; Brughmans 2010, 2013; Collar 2007, 2013; Graham 2014; Knappett 2011; Taylor and Vlassopoulos 2015.
19. Robert 1963: 51; Dimitrova 2008: 1–6; Mack 2015: 330–332.
20. Graham 2014.
21. Collar 2013; Scholnick, Munson and Macri 2013; Munson and Macri 2009; Graham 2014; Remesal et al. 2014.
22. Pythagoras' initiation appears in third/fourth centuries CE accounts, Porphyry *Vita Pythagorae* 17 and Iamblichus, *de Vita Pythagorica* 146; Thracian royalty, Dimitrova 2008:115–119; for the presence of Ptolemaic dedications on Samothrace, Cole 1984: 22; Burkert 1993: 185; Agamemnon and Odysseus, *Scholia Laurentiana to Apollonius Rhodius Argonautica* 1.917–18; Argonauts, Apollonius of Rhodes *Argonautica* 1.918; Diodorus Siculus 5.48.4–50.1; Skytobrachion *BNJ* 32 F 14; Diodorus 4.43 and 4.48.5.
23. Aristophanes *Pax* 276–86; scholia to Aristophanes *Pax* 277–278; Theophratus *Characteres* 25.2; Callimachus Epigram 47 (*Anthologia Palatina* 6.301); Automedon Epigram (*Anthologia Palatina* 11.346); Apollonius of Rhodes *Argonautica* 1.915–21; Valerius Flaccus *Argonautica* 2.431–442; Orphic *Argonautica* 467–72, ed. Dottin; Scholia Laurentiana to Apollonius of Rhodes *Argonautica* 1.917–918; Athenakon *BNJ* 546 F 1b; Scholia Parisina to Apollonius Rhodius *Argonautica* 1.918; Cicero *de Natura Deorum* 3.37.89; Diogenes Laertius 6.2.59; Ovid *Tristia* 1.10.45–50; Aelian *de Natura animalium* F 90, Suidas s.v. ἀκμῇ καλεῖ; Aelian *de Natura animalium* 15.23; Athenaeus *Deipnosophistae* 7.283a; 10.421d; Lucian Epigram 15 (*Anthologia Palatina*. 6.164).
24. Blakely 2016.
25. De Ligt and de Neeve 1988; Gabrielsen 2007.
26. Fafchamps 1992.
27. Horace shudders to set foot on a vessel carrying a man who revealed the rites, *Odes* 3.20: cf. Ovid, *Ars Amatoria* 2.601–604.
28. Bavelas 1948; Freeman 1978; Latora and Marchiori 2007.
29. Caldarelli and Catanzaro 2012: 87; Abbasi, Hossain and Leydesfdorff 2012; Brandes 2001.
30. Centola and Macy 2007; Centola 2018: 7–8, 21–29, 90–91, 137–138.
31. Cole 1984: 17–21; Rabadjiev 2017.
32. Blakely 2016.
33. Dimitrova 2008: 79; 330–333.
34. For epigraphic confirmation of the Kabeiroi among the Samothracians, Dimitrova 2008: 90.
35. Centola and Macy 2007; Centola 2018: 7–8, 21–29, 90–91 137–138.
36. Sindbæk 2013.
37. Safety at sea as the Samothracian benefit, Lewis 1958: 102–111; Blakely 2018b, 2019.
38. Seaford 1994: 275 and n.3; Hemberg 1950: 108–110; Lehmann 1969a: 19; Parker 1983: 281–307; Lupu 2009: 14–21; Chaniotis 2009.
39. Bremmer 2014: 5; Ephoros *BNJ* 70 F 80, Polyaenus 3.11.2; Hesychius α 2727; *IG* I^3 84.35–36; *IG* II^2 847.20; scholia to Aeschines 3.130.

40. Bremmer 2014: 136; Porphyry *De antro* 15.
41. Edmonds 2013: 77–82, 195–247; Bremmer 2014: 66–67; white clothing, Euripides *Cretans* F 472.9–19, Herodotus 2.81.2, Aelian *Varia Historia* 12.32; vegetarianism, Euripides *Hippolytus* 952–54.
42. Hesychius s.v. κοίης.
43. Lehmann 1969b: 54 and n. 8.
44. Lehmann 1969a: 18–20; 1969b: 54–56.
45. *IG* XII.8.190–192; Dimitrova 2008: 140–144, no. 58; *IG* XII 8.190, Dimitrova 2008: 119, no. 47; *IG* XII 8. 189, Dimitrova 2008 137–139, no. 57; *IG* XII 8.188, Dimitrova 2008: 135–137, no. 56.
46. Lehmann 1969a: 31–32; Lehmann 1969b: 136; Blakely 2018a.
47. Chaniotis 2006, 2011, 2012.
48. Parker 1983: 179; Mylonas 1961: 224–226; see Livy 31.14.6–8, Aelian F 46 D–K, Pausanias 9.25.8–10.
49. Herodotus 6.134.2.
50. Aelian F 44 Hercher; see also the account of Erysichthon and Lykourgos, Furley 2015.
51. For Samothracian secrecy, ancient sources collected in Lewis 1958: 100–102; for Samothracian promises of protection in seafaring, Lewis 1958: 102–111; for divine vengeance executed on sea voyages, Aeschylus *Seven Against Thebes* 602; Aeschylus *Agamemnon* 338–342; Aeschylus *Persians* 797–782; Euripides *Electra* 1347–56; Xenophon *Cyropaedia* 8.1.25; Lysias *Against Andokides* 6.19; Andokides *On the Mysteries* 137–39; Horace *Odes* 3.2.25–29.
52. Gagné 2009: 218–219; Furley 1996: 34–3.
53. Gagné 2009: 224–227.
54. Furley 1996: 31, 45–46 Gagné 2009: 221–224; Pritchett 1953, 1956.
55. O'Sullivan 1997; Rubel 2000: 49–119; Meijer 1981; Filonik 2013.
56. Gomme, Andrewes and Dover 1970: 284–5; Furley 1996.
57. Aeschylus: Aristotle, *Nicomachean Ethics* 3.1.17 (1111a8–10); Gagné 2009: 52 n. 220. Andokides: ps–Lysias *Against Andokides* (Oration 6), Isokrates *De bigis* (*Oration* 16), Diodorus Siculus 13.2.3–5; Plutarch, *Life of Andokides* (834B–835B), Bremmer 1995: 74–76; Gagné 2009: 222 and n. 63; Rubel 2000: 157–177; Diagoras of Melos, Krateros *BNJ* 342 F 16; Melanthios *BNJ* 326 F 3b; Aristotle *Nicomachean Ethics* 3.1.17 (1111a); Heraclides Ponticus F 170 (Wehrli); Clement of Alexandria *Stromata* 2.14.60.3; scholia on Aristophanes *Birds*1073; Livy 31.14; Alcibiades, Plutarch *Life of Alcibiades* 19–22; Andokides *On the Mysteries* 1.11–13; Xenophon *Hellenica* 1.4.14f; Thucydides 6.27–29. On three comedic writers accused of sacrilege, see Furley 1996: 131–146, Bremmer 1995: 77; for blasphemers who were themselves mocked in comedies, and cursed by the citizens, see Gagné 2009: 229 and n. 101–102.
58. *Epopteia*, a second level of initiation, was available at Eleusis and at Samothrace; at Samothrace this could be experienced on the same visit as initiation itself: Cole 1984: 27 and n. 216. Bremmer 2012: 387–388; Chaniotis 2006, 2011; Riedweg 1987: 1–69; Ustinova 2012: 121–123.
59. fr. 15 Rose.
60. *Epistulae* 7.334b7.
61. Ustinova 2013:115–118; Burkert 1987: 77; cf. Seaford 1981: 253–254.
62. Ustinova 2013: 113 and n. 64; Burkert 1987: 102 notes that Hesychius defines flagellation as purification, καθαρθῆναι μαστιγωθῆναι. One day of the Eleusinian initiation was devoted to fasting; similar deprivations are attested for the cult of Isis in Apuleius *Metamorphoses* 11.23.
63. *Rhetores Graeci* 8:114–115.
64. *Phaedrus* 250A.

65. Rinella 2000: 70–78; Bizzoto 2018; Bennett and Bentley 1999; Seaford 2006: 29–33.
66. *Phaedrus* 249–251.
67. *Orationes* 22.2, 24.28; Plutarch compares Eleusinian initiation to what the soul feels at the point of death, characterised by nervous movement, terror, amazement and physical trembling, described as φρίσσειν and φρίκη (F 178), initiates experience joy, confusion, hope and πτόησις, an 'anxious excitement' (*Moralia* 943c; Gagné 2009: 219–220; Lada-Richards 1999: 90–98; Bremmer 2014: 13 and n. 92; Seaford 1994: 284; Riedweg 1987: 64–67; Lada-Richards 1999: 90–92. The hand-held inscriptions of Orphic initiates confirm contemplation of the bliss to come, a hope shared by Eleusinian initiates as well; see Seaford 1994: 288 and n. 79; cf. Seaford 1981: 257
68. Bremmer 2014: 14 and n. 48, 105 and n. 111, and 121; Richardson 1974: 26; Seaford 1981: 254, 258; Seaford 1994: 276 and n. 10–11; Blakely 2019; Plutarch F 178, *Moralia* 81e; Dio Chrysostom 12.33; Hippolytus *Refutatio omnium haeresium* 5.8; Mesomedes Hymn 5 (Bremmer 2014: 12); Firmicus Maternus *De errore profanarum religionum* 22; Posidippus 43; Maximus of Tyre *Dissertationes* 39.3, Himerius *Orationes* 67.9; Apuleius *Metamorphosis* 11.23.7.
69. Seaford 1994: 276 and n 10–11. For dancing choruses, Csapo 2008. A night-long search for Kore at Eleusis concluded with the tossing of torches; an analogous search for a missing girl, Harmonia, has been posited as part of the Samothracian rites: Richardson 1974: 24–26; Mylonas 1961: 261–272; Marconi 2010: 125–128. Seaford 2005 offers analogy between the light shining in darkness and the personages of light seen in near death experiences across a range of cultural contexts.
70. ἀστὴρ ἀρίζηλος, ἐτυμώτατον ἀνδρὶ φέγγος (Pindar *Olympian* 2.53), see Seaford 1981: 256 and n. 46.
71. Aristophanes *Frogs* 342–3.
72. Sophokles *Antigone* 1146–52; see scholia ad loc, which alludes to the role of the god among the stars as part of a tale told in the rites: cf. Euripides *Ion* 1074ff, Seaford 2005: 604 and n 15–16.
73. Mendelsohn 1992: 110 and n. 20; 117 and n. 47. Fragments of Aeschylus' *Bassarae* suggest that mystic light could invoke Apollo Helios as a subterranean sun: Seaford 2005.
74. Dimitrova 2008: No. 29, p, 83–90; Karadima and Dimitrova 2003.
75. Blakely 2018a.
76. Nigidius F 91 Swoboda; Arnobius *Adversus nationes* 3.42; Varro *De Lingua Latina* 5.10.5–58;
77. *De saltatione* 15; Lada-Richards 2007: 27, 178 and notes 37–39; Hardie 2004.
78. Edmonds 2006; Henrichs 1978; Bremmer 1984; Bremmer 2014: 100–103; Ustinova 2013: 114–115.
79. Kahil 1977; Ekroth 2003; Nielsen 2009.
80. Bremmer 1984: 276–284; Furley 2015; Bundrick 2005: 47–48.
81. Strabo 10.3.7; 10.3.19–21.
82. Diodorus Siculus 5.48.4–50.1; Priscian 544–56, *GGM* II. 195); Dionysius Periegesis 524, *GGM* II 135; Scholia to Dionysius Periegesis 524, *GGM* II 450; Anon. Paraphrasis Dionysii Periegeis 524, *GGM* II 416; Eustathius, *Commentarii in Dionysii 'Periegesin'* 524, *GGM* II 317; scholia to Aristophanes *Pax* 277–78, Suidas s.v. ζηρυνθία, Nonnos *Dionysiaca* 13.393–407.
83. Cf. Pherekydes *BNJ* 3 F 48.
84. *BNJ* 32 F 7.
85. Bremmer 1984: 273, 284; Holzhausen 2008: 55; Henrichs 1978; Dignas 2002: Rogers 2013. There is a significant disjuncture between the literary and material evidence for Samothracian dance. Lehmann proposed a setting for the ecstatic telestic dances known to Plato in the building called the Anaktoron, where a

round stage seemed a likely setting for the 'thronosis' ceremony, in which frenzied dancers induced madness in an initiate. That round stage proved, on further analysis, to be a lime kiln: Nock 1941: 577–578; Wescoat 2017: 61 n. 66. The Hall of Choral Dancers offers an iconographic confirmation of a role for dance on the site: a frieze of up to 920 dancers circles the top of the building: Marconi 2010. While they carry the instruments—kithara, auloi, and tympanon—appropriate for ecstatic dance, their movements are orderly and restrained, more appropriate for the choral performances of summer festivals: Marconi 2010: 121: cf. Statius *Achilleid* n1.830–832 on stately Samothracian dances.

86. Dignas 2002.
87. Rutherford 2013: 4–9, 11, 51–70; Dimitrova 2008: 72–74; Cole 1984: 48–56; Cf. Kerenyi, who defined *theoria* as the human vision of the god and the god of the worshippers: 1942:100–121. For the honours received on their return, Androtion of Athens, *BNJ* 324 F 36. For members of a religious gene among the theoric delegations, Rutherford 1998; for their inviolability en route, Rutherford 1995.
88. For the role of *theoria* in strengthening panhellenic identity among the participating city states, Ker 2000; Rutherford 2007.
89. 17 cities, IG XII.8.170, Dimitrova 2008 no. 5, 28–31.
90. Mack 2015: 330; Dimitrova 2008: 72–74
91. Mack 2015: 330; Dimitrova 2008: 72–74, 253, and nos. 1, 2, and likely 3, 5–13i; p. 18, 72. The inscriptions on blocks which from a building are dated to the second and first centuries BCE; for the addition of *proxenia* to *theoria*, Rutherford 2013: 23–27; Mack 2015: 330.
92. Third and second century BCE inscriptions show grants of *proxenia* in response to friendship (IG XII.8.151, 157, 159), provision for the city's needs from their own resources (151, 152, 157), and goodwill (151); this exemplifies Mack's *proxenos* paradigm (2015: 22–89); an analogous eagerness underlies those who seek initiation in IG XII.8.156, see Dimitrova 2008: 170; Mikalson 2016: 294–295. For *eisploun*, *ekploun*, *asylia*, *ateleia*, 152; 155; 156; access to the *boule*, 152; front seats, 152; first among the *hiera*, 158.
93. Mack 2015: 330 and n. 55; Dimitrova 2008: no. 1, lines 9–10 and no. 8, lines 7–8; cf. *IG* XII.8.152, 153, 155.
94. Dimitrova 2008: 74 and inscriptions nos. 14, 15, 16, 17, 19; Rutherford 2013: 283.
95. Marconi 2010; Dimitrova 2008: 72–74; Rutherford 2013: 282.
96. Beidelman 2008; Blakely 2008.
97. For creation of the database, Blakely 2016.
98. Dimitrova 2008: no. 31, p. 92–93; see also SEG 39.737 and *IG* XII(3) Suppl. No. 1337, discussed in Cole 1984: 61–64, 155, no. 32.
99. Lehmann 1969a: 10–11; Cole 1984: 39; Lewis 1958: 104.
100. Dimitrova 2008: no. 61, pp. 145–146; for the ritual role accorded to crews who sailed with theoric delegations, Jordan 1975: 169.
101. Dimitrova 2008: no. 63, pp. 148–150.
102. Dimitrova 2008: no. 66, pp 152–53.
103. Blakely 2018a; Popkin 2015; Wescoat 2013; Cole 1989.
104. Dimitrova 2008: nos. 1, 2, 4, 5, 23, 24, 50; Matsas and Dimitrova 2006: no. 3, p 129–130; *SEG* 39.737; 23. A marble stele at Ephesos, dated to the mid first century CE, records the use of a Samothrakeion as the collection place for building taxes from fishermen near the harbour, likely within or near the Fishery Customs house: Cole 1984: 80 and inscription no. 49, p. 164–165.
105. Alabanda: Dimitrova 2008: nos. 4, 3, 5, Matsas and Dimitrova 2006: no. 3, p 129–130. Rhodes: Dimitrova 2008: nos. 5, 23, 50, and, with caveat, Matsas and Dimitrova 2006: no. 3, p 129–130. Kyme, Dimitrova 2008: nos. 5 and 1; Lampsakos, Dimitrova 2008: nos. 3 and 24; Kyzikos, Dimitrova 2008: no. 23 and Matsas and Dimitrova 2006: no. 3.

106. Dimitrova 2008: no. 23.
107. Dimitrova 2008: no. 22, *theoroi* from Teos, Methymna, Samos, Kolophon, Klazomenai, Kaunos, Abdera, Parios and Kalchedon; line 70, Diogenes Menyllou Rhodios, Διογένης Μενύλλου Ῥόδιος.
108. Dimitrova 2008: 50, Xanthos and Ephesos; Dimitrova 2008: 51, Antioch, Priapos, Thasos and Arsinoe; Dimitrova 2008: 57, Maroneia, Kyzikos and Rome. Dimitrova 2008: 5, *proxenia/theoria*; Dimitrova 2008: 23, *theoria*.
109. With Ephesos, Dimitrova 2008: 5, 23, 50, Matsas and Dimitrova 2006: no. 3; with Kyzikos, Dimitrova 2008: 23, 57, Matsas and Dimitrova 2006: no. 3.
110. Rhodes, see Cole 1984: no. 33, 2nd or 1st c BCE; no. 34, undated; no. 41 from Lindos, ca. 215 BCE; Morelli 1959: 153 and Cole 1984: 74.
111. Cole 1984: p. 69 and no. 42, *IG* XII(1) no. 1034, Hiller von Gaertringen 1893: 391 no. 6. This was recovered from Tristomo on Karpathos, in the ruins of a church at the east end of the bay, near the honorary decree for Nikagoras, which came from the sanctuary of Ποτειδὰν Πόρθμιος, a Poseidon concerned with maritime transit. The irregularity of the writing suggests that names were added as new priests assumed their position.
112. Cole 1984: 69–75, on priests from Olbia, Tomis, Dionysopolis, Istros, Kalchedon, Delos, Rhodes, Lindos on Rhodes, Karpathos, and Mylasa.
113. Cole 1984: 140–141, no. 4.
114. The existence of that temple is confirmed in two additional inscriptions, Cole 1984: 72, 139–140, nos. 2 (*IOlb* no. 67) and 3 (*SEG* 16 (1959) no. 430. Cole argues that this seems to be a significant moment in the expansion of the cult.
115. Sokolowski 1969: 87. Priesthoods of the Samothracian gods are also attested at Olbia and Kalchedon: for Olbia, Cole 1984: 139 no. 1, 2nd c BCE; for Kalchedon, Cole 1984: 161, no. 44.
116. Cole 1984: 71–73; Suetonius *Caesar* 44.6; Dana 2006; Davis and Paunov 2012; Lica 2000: 73–75, 82–83.
117. Cole 1984: 69–70, 166, no. 50 and 51.
118. Dignas 2000; Williamson 2013.
119. Williamson 2013.
120. Dimitrova 2008: 41, no. 12; Hemberg 1950: 221 proposes that there was also a temple to the Samothracian gods in Mylasa, though Cole notes that none have been found.
121. Cole 1984: 69–70 notes that the month in which the crowning occurred bears a Macedonian name; Dignas 2000: 118 and n. 6 notes Reger's suggestion that these inscriptions, usually dated to the second half of the second or start of the first century, may be as early as the end of the 3rd century BCE, making them contemporary with the earliest evidence for the Samothracian gods on Rhodes.
122. Constantakopoulou 2017: 74–75; Bruneau 1970: 379–99.
123. The cult title appears first here on Delos after Athenians take control in 166, and it was a life priesthood: ID 1562, 1581, 1582. Bruneau 1970: 379–99.
124. Blakely 2018a.
125. Constantakopoulou 2017: 75; Bruneau 1970: 396.
126. IDelos 1555, cited in Sanders and Catling 1990: 331, nn. 17–19.
127. Sanders and Catling 1990; cf. Ballesteros-Pastor 2006.
128. Cole 1984: 25; Sanders and Catling 1990 n. 16.
129. Verdejo Manchado and Antela-Bernárdez 2015; Dimitrova 2008: 148–150, no. 63.
130. *Samothraikiastai mesoneioi*, Cole no. 35, line 10, IG XII.1.43; *Samothraikiastai kai Lemniastai symstrateusamenoi*, Cole no. 35, line 14, IG XII.1.43, Rubensohn 1892: 234–235, Gabrielsen 1997: 124; Smith 1972; *Samothraikiastai mesoneioi Damasileioi*, Pugliese Carratelli 1939–1940: p. 185, from Hiller *Rhodos* 834; *Samothraikiastai [Niko?s]trateioi synmystai synstrateusamenoi*, Cole no. 38a,

Pugliese Carratelli 1939–1940 no. 13; *Samothraikiastai Soteriastai Aristobouliastai Apolloniastai*, Cole no. 36, von Gaertringen 1893: 389–90, no. 3; *Samothraikiastan Aphrodisiastan koinon*, Cole no. 37; *koinon Samothraikiastai* Cole no. 38; Mauri 1925 no. 43; *koinon Samothraikiastai Aphrodisiastai Borboritai*, Cole no. 43; *Heroistai Samothraikiastai*, Cole no. 53, Fraser and Bean 1954: 62–63; *Samothraikiastai Aristobouliastai Hermaistai Panathenaistai hoi syn Ktesiphonti*, SEG 39–373.

131. Maillot 2015; Gabrielsen 2001.
132. Gabrielsen and Thomsen 2015; Arnaoutoglou, personal communication, cited in Liddell and Low 2015: 272 n. 23; Gabrielsen 2017.
133. For an overview of the history of the study of *koina*, Taylor and Vlassopoulos 2015; Gabrielsen and Thomsen 2015.
134. *EthNic* I 1160a20–22.
135. Gabrielsen 2001: 227 and n. 56; Pugliese Carratelli 1939–1940: 165, no. 19.
136. Gabrielsen 2017: 9.
137. Ismard 2015.
138. Gabrielsen 2001: 219–224; 2007: 194 and n. 60; 2017: 17, 37. A maritime function may be suggested for Aphrodite who appears alongside Hermaistai, Haliadai, and Haliastai, Pugliese Carratelli 1939–1940: p. 151 no. 6 lines 16 and 20; cf. Aphrodisiastai Soteriastai, no. 7 line 6. Morelli 1959: 126–134, 153 suggests that associations organised under the aegis of the Dioskouroi and of Hermes likely were connected with mariners, and may themselves represent Samothracian initiates; particularly likely for the latter is the koinon of Soteriastai Asklapiastai Poseidaniastai Herakeliastai Athanaistai Aphrodisiasti Hermaistai Matron Theon (IG 162), or the Hermaistai Thersmophoriastai (IG 157).
139. Gabrielsen 2013.
140. Gabrielsen 1997: 126–127; 2001: 227; 2017: 35–36.
141. Maillot 2015: 142; Gabrielsen 2001: 229; Kontorini 1989: 73, no. 10; cf. Gabrielsen 1992; Harland 2015: 26.
142. *Theoros/proxenos* grant, Dimitrova 2008: 35–37 no. 9; initiations, Dimitrova 2008: 103–104, no. 38, and 174–176, no. 89.
143. Gabrielsen 2001: 223, 1997: 125.
144. as Meseoneoi, IG XII.1.43, Cole 35 line 10 and 14; *Samothraikiastai mesoneoi Demasileioi*, Pugliese Carratelli 185; *synstrateusamenoi*, *Samothraikiastai kai Lemniastai synstrateusamenoi*, Cole 35, IG XII.1.43, line 15, 19–21; *Samothraikiastai Nikostrateioi synmystai synstrateusamenoi*, Cole 38a, PC 153 no. 13. Cole 1984: 85 for the debate over the meaning of '*mesoneoi*', whether this was a reference to a proper name or a group of rowers; Morelli 1959: 153–154 and n. 6 cites Carratelli's thesis that they were the sailors of the crews of the ships that carried Rhodian mystes to Samothrace.
145. *Lemniastai* seems simply an ethnic, though the Rhodians knew of the mysteries on Lemnos: see Rubensohn 1892: 234–235, Smith 1972.
146. The inscription is dated to 80–60 BCE, the years in which Greek cities joined Rome in antipirate patrols: Dimitrova 2008: 122–125; Skarlatidou 1993.
147. See note 31.

Bibliography

Abbasi, A., L. Hossain, and L. Leydesfdorff (2012) 'Betweenness Centrality as a Driver of Preferential Attachment in the Evolution of Research Collaboration Networks', *Journal of Informetrics* 6: 403–412.

Ballesteros-Pastor, L. (2006) 'Los cultos de Mitrídates Eupátor en Delos: Una propuesta de interpretacíon', *Habis* 37: 209–216.

Bavelas, A. (1948) 'A Mathematical Model for Group Structure', *Applied Anthropology* 7: 16–30.

Beidelman, T.O. (2008) 'African and Classical Secrecy and Disclosure: The Kaguru of East Africa and the Ancient Greeks', *Electronic Antiquity* 12.1: 129–160.

Bennett, J.W. and R. Bentley (1999) 'Pride and Prejudice: The Story of Ergot', *Perspectives in Biology and Medicine* 40: 333–355.

Bizzoto, J. (2018) 'The Hypothesis on the Presence of Entheogens in the Eleusinian Mysteries', *Medicina Historica* 2(2): 85–93.

Blakely, S. (2008) 'Introduction: Ancient Mysteries, Modern Secrets', *Electronic Antiquity* 12(1): 1–22.

Blakely, S. (2016) 'Beyond Braudel: Network Models and a Samothracian Seascape', in L. Mazurek and C. Concannon (eds.), *Across the Corrupting Sea: Post-Braudelian Approaches to the Ancient Eastern Mediterranean*, London: Routledge, 17–38.

Blakely, S. (2018a) 'Starry Twins and Mystery Rites: From Samothrace to Mithras', *Acta Antiqua Academiae Scientarium Hungaricae* 58: 427–463.

Blakely, S. (2018b) 'Sailing with the Gods: Serious Games in an Ancient Sea', *Thersites* 7: 108–146.

Blakely, S. (2019) 'Hermes, Kyllene, Samothrace and the Sea', in J. Miller and J. Strauss-Clay (eds.), *Tracking Hermes, Tracing Mercury*, Oxford: Oxford University Press, 271–292.

Bowden, H. (2007) 'Cults of Demeter Eleusinia and the Transmission of Religious Ideas', *Mediterranean Historical Review* 22(1): 71–83.

Brandes, U. (2001) 'A Faster Algorithm for Betweenness Centrality', *The Journal of Mathematical Sociology* 25(2): 163–77.

Bremmer, J. (1984) 'Greek Maenadism Reconsidered', *Zeitschrift für Papyrologie und Epigraphik* 55: 267–286.

Bremmer, J. (1995) 'Religious Secrets and Secrecy in Classical Greece', in H.F. Kippenberg and E.T. Lawson (eds.), *Secrecy and Concealment: Studies in the History of Mediterranean and Near Eastern Religions*, Leiden: Brill, 61–78.

Bremmer, J. (2010) 'Manteis, Magic, Mysteries and Mythography: Messy Margins of *Polis* Religion?', *Kernos* 23: 13–35.

Bremmer, J. (2012) 'Initiation into the Eleusinian Mysteries: A "Thin" Description', in J. Bremmer (ed.), *Mystery and Secrecy in the Nag Hammadi Collection and Other Ancient Literature: Ideas and Practices*, Leiden: Brill, 375–397.

Bremmer, J. (2014) *Initiation into the Mysteries of the Ancient World.* Berlin: de Gruyter.

Brughmans, T. (2010) 'Connecting the Dots: Towards Archaeological Network Analysis', *Oxford Journal of Archaeology* 29(3): 277–303.

Brughmans, T. (2013) 'Thinking through Networks: A Review of Formal Network Methods in Archaeology', *Journal of Archaeological Method and Theory* 20(4): 623–662.

Bruneau, P. (1970) *Recherches sur les cultes de Delos a l'epoque hellenistique et a l'epoque imperial*, Paris: De Boccard, 465–504.

Bundrick, S. (2005) *Music and Image in Classical Athens.* Cambridge: Cambridge University Press.

Burkert, W. (1987) *Ancient Mystery Cults.* Cambridge: Harvard University Press.

Burkert, W. (1993) 'Concordia Discors: The Literary and the Archaeological Evidence on the Sanctuary of Samothrace', in N. Marinatos and R. Hägg (eds.), *Greek Sanctuaries: New Approaches*, London: Routledge, 178–191.

Caldarelli, G. and M. Catanzaro (2012) *Networks: A Very Short Introduction*, Oxford: Oxford University Press.
Centola, D. and M. Macy (2007) 'Complex Contagions and the Weakness of Long Ties', *American Journal of Sociology* 113(3): 702–734.
Centola, D. (2018) *How Behavior Spreads: The Science of Complex Contagions*, Princeton: Princeton University Press.
Chaniotis, A. (2006) 'Rituals between Norms and Emotions: Rituals as Shared Experience and Memory', *Kernos Supplement* 16: 211–238.
Chaniotis, A. (2009) 'Ritual Performances of Divine Justice: The Epigraphy of Confession, Atonement, and Exaltation in Roman Asia Minor', in H.M. Cotton et al. (eds.), *From Hellenism to Islam: Cultural and Linguistic Change in the Roman Near East*, Cambridge: Cambridge University Press, 115–153.
Chaniotis, A. (2011) 'Emotional Community through Ritual: Initiates, Citizens, and Pilgrims as Emotional Communities in the Greek World', in A. Chaniotis (ed.), *Ritual Dynamics in the Ancient Mediterranean: Agency, Emotion, Gender, Representation*, Stuttgart: Steiner Verlag, 264–290.
Chaniotis, A. (2012) 'Constructing the Fear of Gods: Epigraphic Evidence from Sanctuaries of Greece and Asia Minor', in A. Chaniotis (ed.), *Unveiling Emotions: Sources and Methods for the Study of Emotions in the Greek World*, Stuttgart: Franz Steiner Verlag, 205–234.
Cole, S.G. (1984) *Theoi Megaloi: The Cult of the Great Gods at Samothrace*. Leiden: Brill.
Cole, S.G. (1989) 'The Mysteries of Samothrace During the Roman Period', *Aufstieg under Niedergang der römischen Welt II* 18(2): 1564–1598.
Collar, A. (2007) 'Network Theory and Religious Innovation', *Mediterranean Historical Review* 22(1): 149–162.
Collar, A. (2013) *Religious Networks in the Roman Empire*. Cambridge: Cambridge University Press.
Constantakopoulou, C. (2017) *Aegean Interactions: Delos and its Networks in the Third Century*. Oxford: Oxford University Press.
Cosmopoulos, M.B. (2015). *Bronze Age Eleusis and the Origins of the Eleusinian Mysteries*. Cambridge: Cambridge University Press.
Csapo, E. (2008) 'Star Choruses: Eleusis, Orphism, and New Musical Imagery and Dance', in M. Revermann and P. Wilson (eds.), *Performance, Iconography, Reception: Studies in Honour of Oliver Taplin*, Oxford: Oxford University Press, 262–290.
Dana, D. (2006) 'The Historical Names of the Dacians and Their Memory: New Documents and a Preliminary Outlook', *Studia Universitatis Babes Bolyai* 51: 99–127.
Davies, J. (2015) 'Retrospect and Prospect', in C. Taylor and K. Vlassopoulos (eds.), *Communities and Networks in the Ancient Greek World*, Oxford: Oxford University Press, 239–256.
Davis, P. Evgeni Paunov (2012) 'Imitations of Republican denarii from Moesia and Thrace', in E. Paunov and S. Filipova (eds.), *ΗΡΑΚΛΕΟΥΣ ΣΩΤΗΡΟΣ ΘΑΣΙΩΝ. Studia in honorem Iliae Prokopov sexagenario ab amicis et discipulis dedicate*, Veliko Tarnovo: Faber, 389–413.
De Ligt, L. and P.W. de Neeve (1988) 'Ancient Periodic Market: Festivals and Fairs', *Athenaeum* 66: 391–416.
Dignas, B. (2000) 'The Leases of Sacred Property at Mylasa: An Alimentary Scheme for the Gods', *Kernos* 13: 117–126.

Dignas, B. (2002) 'Priestly Authority in the Cult of the Corybantes at Erythrae', *Epigraphica Anatolica* 34: 29–40.

Dimitrova, N. (2008) *Theoroi and Initiates in Samothrace: The Epigraphical Evidence*. Princeton: American School of Classical Studies at Athens.

Edmonds, R. (2006) 'To Sit in Solemn Silence? "Thronosis" in Ritual, Myth and Iconography', *The American Journal of Philology* 127(3): 347–366.

Edmonds, R. (2013) *Redefining Ancient Orphism: A Study in Greek Religion*. Cambridge: Cambridge University Press.

Eidinow, E. (2011) 'Networks and Narratives: A Model for Ancient Greek Religion', *Kernos* 24: 9–38.

Ekroth, G. (2003) 'Inventing Iphigeneia? On Euripides and the Cultic Construction of Brauron', *Kernos* 16: 59–118.

Fabiano, D. (2010) '"Ho fuggito il male, ho trovato il meglio": le punizioni dei non iniziati nell'aldilà greco', *Archiv für Religionsgeschichte* 12: 149–165.

Fafchamps, M. (1992) 'Solidarity Networks in Preindustrial Societies: Rational Peasants with a Moral Economy', *Economic Development and Cultural Change* 41(1): 147–174.

Farnell, L.R. (1907) *The Cults of the Greek States*, Oxford: The Clarendon Press.

Filonik, J. (2013) 'Athenian Impiety Trials: A Reappraisal', *Dike* 16: 11–96.

Fox, J. (2005) 'New Religious Movements', in J.R. Hinnells (ed.), *The Routledge Companion to the Study of Religion*, London & New York: Routledge, 323–337.

Fraser, P.M. and G.E. Bean (1954) *The Rhodian Peraia and Other Islands*. Oxford: Oxford University Press.

Freeman, L.C. (1978) 'Centrality in Social Networks: Conceptual Clarification', *Social Networks* 1(3): 215–39.

Friedkin, N.E. (1982) 'Information Flow through Strong and Weak Ties in Intraorganizational Social Networks', *Social Networks* 3: 273–285.

Furley, W.D. (1996) *Andocides and the Herms: A Study of Crisis in Fifth-Century Athenian Religion*. London: Institute of Classical Studies.

Furley, W.D. (2015) 'Narrative in a Late Hymn to Dionysos (P.Ross. Georg.I.11)', in A. Faulkner and O. Hodkinson (eds.), *Hymnic Narrative and the Narratology of Greek Hymns*, Leiden: Brill, 121–138.

Gabrielsen, V. (1992) 'The Status of Rhodioi in Hellenistic Rhodes', *Classica et Mediaevalia* 43: 43–69.

Gabrielsen, V. (1997) *The Naval Aristocracy of Hellenistic Rhodes*. Aarhus: Aarhus University Press.

Gabrielsen, V. (2001) 'The Rhodian Associations and Economic Activity', in Z. Archibald (ed.), *Hellenistic Economies*, London: Routledge, 215–244.

Gabrielsen, V. (2007) 'Brotherhoods of Faith and Provident Planning: The Non-public Associations of the Greek World', *Mediterranean Historical Review* 22(2): 183–210.

Gabrielsen, V. (2013) 'Rhodes and the Ptolemaic Kingdom: The Commercial Infrastructure', in K. Buraselis, M. Stefanou, and D.J. Thompson (eds.), *The Ptolemies, the Sea and the Nile: Studies in Waterborne Power*, Cambridge: Cambridge University Press, 66–81.

Gabrielsen, V. (2017) 'A New Inscription Attesting to Associations from the Necropolis of Rhodes, with an Appendix by Nicos Christodoulides', *Beiträge zur alten Geschichte, Papyrologie und Epigraphik* 32: 15–40.

Gabrielsen, V. and C. Thomsen (2015) 'Introduction: Private Groups, Public Functions?', in V. Gabrielsen and C.A. Thomsen (eds.), *Private Associations and the Public Sphere*, Denmark: Det Kongelige Danske Videnskabernes Selskab, 7–25.

Gagné, R. (2009) 'Mystery Inquisitors: Performance, Authority, and Sacrilege at Eleusis', *Classical Antiquity* 28(2): 211–247.

Gifford, P. (2005) 'Religious Authority: Scripture, Tradition, Charisma', in J.R. Hinnells (ed.), *The Routledge Companion to the Study of Religion*. Routledge: London & New York, 379–392.

Glomb, T., A. Mertel, Z., Pospilsil, Z. Schon, and A. Chalupa (2018) 'Ptolemaic Military Operations Were a Dominant Factor in the Spread of Egyptian Cults across the Early Hellenistic Aegean Sea', *Plos One* https://doi.org/10.1371/journal.pone.0193786

Gomme, A.W., A. Andrewes, and K.J. Dover (1970) *A Historical Commentary on Thucydides, Vol. IV*. Oxford: Clarendon Press.

Gordon, R. (2007) 'Institutionalized Religious Options: Mithraism', in J. Rüpke (ed.), *A Companion to Roman Religion*, Malden: Blackwell, 392–405.

Graham, S. (2014) 'On Connecting Stamps – Network Analysis and Epigraphy', *Les nouvelles de l'archéologie* 135: 39–44.

Gruen, E.S. (1990) *Studies in Greek Culture and Roman policy*. Brill: Leiden.

Hardie, A. (2004) 'Muses and Mysteries', in P. Murray and P.W. Wilson (eds.), *Music and the Muses: The Culture of Mousike in the Classical Athenian City*, Oxford: Oxford University Press, 11–37.

Harland, P. (2015) 'Associations and the Economics of Group Life: A Preliminary Case Study of Asia Minor and the Aegean Islands', *Svensk Exegetisk Arsbok* 80: 1–37.

Hearnshaw, V. (1999) 'The Dionysiac Cycle in the Villa of the Mysteries: A Re-reading', *Mediterranean Archaeology* 12: 43–50.

Hemberg, B. (1950) *Die Kabiren*. Uppsala: Almqvist & Wiksells Boktryckeri.

Henrichs, A. (1978) 'Greek Maenadism from Olympias to Messalina', *Harvard Studies in Classical Philology* 82: 121–160.

Hiller von Gaertringen, F. 1893. 'Die Samothrakischen Götter in Rhodos und Karpathos', *Mitteilungen des Deutschen Archäologischen Instituts, Athenische Abteilung* 18: 385–394.

Holzhausen, J. (2008) 'Poetry and Mysteries: Euripides' Bacchae and the Dionysiac Rites', *Electronic Antiquity* 12(1): 41–72.

Hutchinson, V.J. (1991) 'The Cult of Dionysos/Bacchus in the Graeco-Roman World: New Light from Archaeological Studies', *Journal of Roman Archaeology* 4: 222–230.

Ismard, P. (2015) "Playing with Scales' in the Classical City: The Case of the Marathonian Tetrapolis', in C. Taylor and K. Vlassopoulos (eds.), *Communities and Networks in the Ancient Greek World*, Oxford: Oxford University Press, 80–97.

Jordan, B. (1975) *The Athenian Navy in the Classical Period*. Berkeley: University of California Press.

Karadima-Matsa, C. and N. Dimitrova (2003) 'An Epitaph for an Initiate at Samothrace and Eleusis', *Chiron* 33: 335–345.

Kahil, L. (1977) 'L'Artémis de Brauron: Rites et Mystère', *Antike Kunst* 20(2): 86–98.

Ker, J. (2000) 'Solon's "Theoria" and the End of the City', *Classical Antiquity* 19(2): 304–329.

Kerényi, K. (1942) *Die antike Religion: Eine Grundlegung*. Amsterdam: Akademische Verlagsanstalt Pantheon.

Kindt, J. (2009) '*Polis* Religion – A Critical Appreciation', *Kernos* 22: 9–34.

Knappett, C. (2011) *An Archaeology of Interaction: Network Perspectives on Material Culture and Society*. Oxford: Oxford University Press.

Kneppe, A. and D. Metzler (eds.) (2003) *Die emotionale Dimension antiker Religiosität*. Münster: Ugarit-Verlag.

Kontorini, V. (1989) *Ανεκδοτες Επιγραφες Ροδου II*. Athens: Kardamitsa.

Kreuz, P.-A. (2009) 'Monuments for the King: Royal Presence in the Late Hellenistic World of Mithridates VI', in J.M. Højte (ed.), *Mithridates VI and the Pontic Kingdom*, Aarhus: Aarhus University Press, 131–144.

Lada-Richards, I. (1999) *Initiating Dionysos: Ritual and Theatre in Aristophanes' Frogs*. Oxford: Clarendon Press.

Lada-Richards, I. (2007) *Silence Eloquence: Lucian and Pantomime Dancing*. London: Duckworth.

Latora, V. and M. Marchiori (2007) 'A Measure of Centrality Based on Network Efficiency', *New Journal of Physics* 9(2007): 188.

Lehmann, K. (1969a) 'The Epopteia and the Function of the Hieron', in P.W. Lehmann (ed.), *Samothrace: The Hieron, Text II*, Princeton: Princeton University Press, 3–50.

Lehmann, P.W. (1969b) *Samothrace: The Hieron. Text II*. Princeton: Princeton University Press.

Lewis, N. (1958) *Samothrace Vol. 1: The Ancient Literary Sources*. New York: Pantheon Books.

Leyden, D.P., A.N. Link, and D.S. Siegel (2013) 'A Theoretical Analysis of the Role of Social Networks in Entrepreneurship', *Research Policy* 43: 1157–1163.

Lica, V. (2000) *The Coming of Rome in the Dacian World*, Konstanz: Universitätsverlag.

Liddell, P. and P. Low (2015) 'An Honorific "Gnome of the Koinon" of the Phrikyladai. A New Inscription from Liverpool', *Annual of the British School at Athens* 110: 263–284.

Lupu, E. (2009) *Greek Sacred Law: A Collection of New Documents*. Leiden: Brill.

Mack, W. (2015) *Proxeny and Polis: Institutional Networks in the Ancient Greek World*. Oxford: Oxford University Press.

Maillot, S. (2012) 'Le formalization des réseaux de mobilité méditerranéens. Remarques sur les associations à l'epoque hellénistique', in L. Capdetrey and J. Zurbach (eds.), *Mobilités grecques. Mouvements, réseaux, contacts en Méditerranée, de l'Époque archaïque à l'Époque hellénistique*, Paris: De Boccard, 235–260.

Maillot, S. (2015) 'Foreigners' Associations and the Rhodian State', in V. Gabrielsen and C.A. Thomsen (eds.), *Private Associations and the Public Sphere: Proceedings of a Symposium held at the Royal Danish Academy of Sciences and Letters, 9–11 September 2010*. Denmark: Det Kongelige Danske Videnskabernes Selskab, 136–182.

Marconi, C. (2010) '*Choroi*, *Theoriai* and International Ambitions: The Hall of Choral Dancers and its Frieze', in O. Palagia and B.D. Wescoat (eds.), *Samothracian Connections: Essays in Honor of James R. McCredie*, Oxford: Oxbow Books, 106–135.

Margry P.J. (2004) 'Global Network of Divergent Marian Devotion', in C. Partridge (ed.), *Encyclopedia of New Religions: New Religious Movements, Sects and Alternative Spiritualities*, Oxford: Lion Publishing, 98–102.

Matsas, D. and N. Dimitrova (2006). 'New Samothracian Inscriptions Found Outside the Sanctuary of the Great Gods', *Zeitschrift für Papyrologie und Epigraphik* 155: 127–136.

Mauri, A. (1925) *Nuova silloge epigrafica di Rodi e Cos.* Firenze: F. Le Monnier.

McLoughlin, S. (2005) 'Migration, Diaspora and Transnationalism: Transformations of Religion and Culture in a Globalizing Age', in J.R. Hinnells (ed.), *The Routledge Companion to the Study of Religion*, London & New York: Routledge, 526–549.

Meijer, P.A. (1981) 'Philosophers, Intellectuals and Religion in Hellas', in H.S. Versnel (ed.), *Faith, Hope and Worship: Aspects of Religious Mentality in the Ancient World*, Leiden: Brill, 216–262.

Mendelsohn, D. (1992). '*Synkeraunoo*: Dithyrambic Language and Dionysiac Cult', *Classical Journal* 87(1): 118–122.

Merkur, D. (2005) 'Psychology of Religion', in J.R. Hinnells (ed.), *The Routledge Companion to the Study of Religion*, London & New York: Routledge, 164–181.

Mikalson, J. (2016) *New Aspects of Religion in Ancient Athens: Honors, Authorities, Esthetics and Society.* Leiden: Brill.

Morelli, D. (1959) *I Culti in Rodi.* Pisa: Goliardica.

Munson, J.L., and M.J. Macri (2009) 'Sociopolitical Network Interactions: A Case Study of the Classic Maya', *Journal of Anthropological Archaeology* 28: 424–438.

Mylonas, G.E. (1961). *Eleusis and the Eleusinian Mysteries.* Princeton: Princeton University Press.

Naiden, F.S. (2013) 'Recent Study of Greek Religion in the Archaic through Hellenistic Periods', *Currents in Biblical Research* 11(3): 388–427.

Nielsen, I. (2009) 'The Sanctuary of Artemis Brauronia. Can Architecture and Iconography Help to Locate the Settings of the Rituals?' in T. Fischer-Hansen and B. Poulsen (eds.), *From Artemis to Diana: The Goddess of Man and Beast*, Copenhagen: Museum Tusculanum Press, 83–116.

Nock, A.D. (1941) 'A Cabiric Rite', *American Journal of Archaeology* 45(4): 577–581.

O'Sullivan, L.L. (1997) 'Athenian Impiety Trials in the Late Fourth Century B.C.', *Classical Quarterly* 47: 136–152.

Parker, R. (1983) *Miasma.* Oxford: Clarendon Press.

Pirenne-Delforge, V. (2010) 'Mnasistratos, The 'Hierophant' at Andania (*IG* 5.1.1390 and *SYLL.*[3] 735)', in J. Dijkstra, J. Kroesen, and Y. Kuiper (eds.), *Myths, Martyrs, and Modernity: Studies in the History of Religions in Honour of Jan N. Bremmer.* Leiden: Brill, 219–235.

Popkin, M.L. (2015) 'Samothracian Influences at Rome: Cultic and Architectural Exchange in the Second Century B.C.E.', *American Journal of Archaeology* 119(3): 343–373.

Price, S. (2012) 'Religious Mobility in the Roman Empire', *The Journal of Roman Studies* 102: 1–19.

Pritchett, W.K. (1953) 'The Attic Stelai, Part I', *Hesperia* 22(4): 225–299.

Pritchett, W.K. (1956) 'The Attic Stelai, Part II', *Hesperia* 25(3): 178–328.

Pugliese Carratelli, G. (1939–1940) 'Per la storia delle Associazioni in Rodi Antica', *Annuario della Scuola archeologica italiana di Atene* 1(2): 147–200.

Rabadjiev, K. (2017) 'Theoi Samothrakes at Pontos Euxeinos', *Archaeologia Bulgarica* 21(2): 11–25.

Rauh, N. (1993) *The Sacred Bonds of Commerce: Religion, Economy and Trade Society at Hellenistic Roman Delos, 166–87 B.C.* Amsterdam: J.C. Gieben.

Remesal, J., A. Díaz-Guilera, B. Rondelli, X. Rubio-Campillo, A. Aguilera, D. Martín-Arroyo, A. Mosca, and G. Rull (2014) 'The EPNet Project: Production and Distribution of Food during the Roman Empire: Economics and Political Dynamics', in *Information Technologies for Epigraphy and Cultural Heritage: Proceedings of the First EAGLE International Conference*, Rome: Sapienza Università Editrice, 455–464.

Ribichini, S. (2016) 'Covered by Silence: Hidden Texts and Secret Rites in the Ancient Mystery Cults', in G. Colesanti and L. Lulli (eds.), *Submerged Literature in Ancient Greek Culture 2: Case Studies*. Berlin: De Gruyter, 161–176.

Richardson, N.J. (1974) *The Homeric Hymn to Demeter*. Oxford: Clarendon Press.

Riedweg, C. (1987) *Mysterienterminologie bei Platon, Philon und Klemens von Alexandrien*. Berlin: de Gruyter.

Rinella, M.A. (2000) *Pharmakon. Plato, Drug Culture and Identity in Ancient Athens*. Lanham: Lexington Books.

Robert, L. (1963) 'Review of Karl Lehmann', in P.M. Fraser (ed.), *Samothrace. Excavations Conducted by the Institute of Fine Arts, New York, Vol. 2 Part 1: The Inscriptions on Stone*. New York: Pantheon Books 1960, *Gnomon* 35(1): 50–79.

Robertson, N. (1988) 'Melanthus, Codrus, Neleus, Caucon: Ritual Myth as Athenian History', *Greek Roman and Byzantine Studies* 29(3): 201–261.

Rogers, G.M. (2013) *The Mysteries of Artemis of Ephesos: Cult, Polis, and Change in the Graeco Roman World. Synkrisis*. New Haven: Yale University Press.

Rubel, A. (2000) *Stadt in Angst: Religion und Politik in Athen während des Peloponnesischen Krieges*. Darmstadt: Wissenschaftliche Buchgesellschaft.

Rubensohn, O. (1892) *Die Mysterienheiligtümer in Eleusis und Samothrake*. Berlin: R. Gaertner's Verlagsbuchhandlung.

Rutherford, I. (1995) 'Theoric Crisis', *Studi e materiali di storia delle religioni* 61: 276–292.

Rutherford, I. (1998) 'The Emphikleidai of Sicilian Naxos', *Zeitschrift für Papyrologie und Epigraphik* 122: 81–89.

Rutherford, I.C. and F. Franco. 'Theoria', in H. Cancik, H. Schneider and C.F. Salazar (eds.), *Brill's New Pauly*. Consulted online on 25 May 2021 http://dx.doi.org.proxy.library.emory.edu/10.1163/1574-9347_bnp_e1209940

Rutherford, I. (2007) 'Theoria and Theatre at Samothrace. The Dardanos of Dymas', in P. Wilson (ed.), *The Greek Theatre and Festivals: Documentary Studies*. Oxford: Oxford University Press, 279–293.

Rutherford, I. (2013) *State Pilgrims and Sacred Observers in Ancient Greece: A Study of Theoria and Theoroi*. Cambridge: Cambridge University Press.

Sanders, G.D.R. and R.W.V. Catling (1990) 'From Delos to Melos: A New Fragment of I Delos 1562', *Annual of the British School at Athens* 85: 327–332.

Seaford, R.A.S. (1981). 'Dionysiac Drama and the Dionysiac Mysteries', *Classical Quarterly* 31(2): 252–275.

Seaford, R.A.S. (1994) 'Sophokles and the Mysteries', *Hermes* 122(3): 275–288.

Seaford, R.A.S. (2005) 'Mystic Light in Aeschylus' *Bassarai*', *Classical Quarterly* 55(2): 602–606.

Seaford, R.A.S. (2006) *Dionysos*. London: Routledge.

Sindbæk, S. (2013) 'Broken Links and Black Boxes: Material Affiliations and Contextual Network Synthesis in the Viking World', in C. Knappett (ed.), *Network Analysis in Archaeology: New Approaches to Regional Interaction*. Oxford: Oxford University Press, 71–94.

Scholnick, J. B., J.L. Munson, and M.J. Macri (2013) 'Positioning Power in a Multi-relational Framework: A Social Network Analysis of Classic Maya Political Rhetoric', in C. Knappett (ed.), *Network Analysis in Archaeology: New Approaches to Regional Interaction*, Oxford: Oxford University Press, 95–124.

Skarlatidou, E.K. (1993) 'Κατάλογος μυστών και Εποπτών από τη Σαμοθάκη', *Horos* 8(9): 153–172.

Smith, D.R. (1972) 'Hieropoioi and Hierothytai on Rhodes', *L'antiquité Classique* 41(2): 532–539.

Smith, J.Z. (1990) *Drudgery Divine: On the Comparison of Early Christianities and the Religions of Late Antiquity*, Chicago: University of Chicago Press.

Sokolowski, F. (1969) *Lois sacrées des cités grecques*. Paris: De Boccard.

Stark, R. and W.S. Bainbridge (1980) 'Networks of Faith: Interpersonal Bonds and Recruitment to Cults and Sects', *American Journal of Sociology* 85(6): 1376–1395.

Taylor, C. and K. Vlassopoulos (2015) 'Introduction: An Agenda for the Study of Greek History', in C. Taylor and K. Vlassopoulos (eds.), *Communities and Networks in the Ancient Greek World*, Oxford: Oxford University Press, 1–31.

Tilly, C. (2010) 'Cities, States and Trust Networks', *Theory and Society* 39: 265–280.

Tortoriello, M., R. Reagans, and B. McEvily (2012) 'Bridging the Knowledge Gap: The Influence of Strong Ties, Network Cohesion, and Network Range on the Transfer of Knowledge between Organizational Units', *Organization Science* 23(4): 1024–1039.

Ustinova, Y. (2012) 'Madness into Memory: Mania and Mneme in Greek Culture', *Scripta Classica Israelica* 31: 109–132.

Ustinova, Y. (2013) 'To Live in Joy and Die with Hope: Experiential Aspects of Ancient Greek Mystery Rites', *Bulletin of the Institute of Classical Studies* 56(2): 105–123.

Verdejo Manchado, J. and B. Antela-Bernárdez (2015) 'Pro-Mithridatic and Pro-Roman Tendencies in Delos in the Early First Century BC: The Case of Dikaios of Ionidai (*ID* 2039 and 2040)', *Dialogues d'Histoire Ancienne* 41(1): 117–126.

von Gaertingen, F.H. (1893) 'Die Samothrakischen Götter in Rhodos und Karpathos', *Mitteilungen des Deutschen Archäologischen Instituts (Athen. Abt.)* 18: 385–394.

Walters, H.B. (1893) 'Odysseus and Kirke on a Boeotian Vase', *Journal of Hellenic Studies* 13: 77–87.

Wescoat, B.D. (2013) 'Insula Sacra: Samothrace between Troy and Rome', in M. Galli (ed.), *Roman Power and Greek Sanctuaries: Forms of Interaction and Communication*, Athens: Scuola Archeologica Italiana di Atene, 45–81.

Wescoat, B.D. (2017) *Samothrace: The Monuments of the Eastern Hill, Vol. 9: Text*. Princeton: American School of Classical Studies at Athens.

Williamson, C. (2013) 'Public Space beyond the City: The Sanctuaries of Labraunda and Sinura in the Chora of Mylasa', in C.P. Dickenson and O.M. van Nijf (eds.), *Public Space in the Post-Classical Greek City*, Leuven: Peeters, 1–36.

Wortham, R.A. (2006) 'Urban Networks, Deregulated Religious Markets, Cultural Continuity and the Diffusion of the Isis Cult', *Method & Theory in the Study of Religion* 18(2): 103–123.

Part II

Storytelling: the narratives that bind

5 Ritual ties, 'portable communities', and the transmission of common knowledge through festival networks in the Hellenistic world

Christina G. Williamson[1]

Festivals are central to the formation of communities. Some of the earliest evidence that we have for collective human action implicates recurring ritual gatherings. The prehistoric ceremonial site Göbekli Tepe predates the known proto-urban agglomerations by several millennia and serves to remind us that festivals, in whatever form they take, fulfil an innate drive among humans to celebrate together, and to cooperate in ways that transcend mutual exploitation or meeting basic physical needs (e.g. hunting, gathering, trade, and market). Before there were cities, there were festivals.[2] Festivals also shaped emerging cities, certainly in the Greek world. It has been argued that panhellenic games, Olympia in particular, were part of what held this world together during the wave of colonisation in the Archaic period.[3] Festivals attracted interaction at various scales, even on the tiny island of Ikaros (Failaka Island, Kuwait—Figures 5.1 and 5.2), where a third century BCE Seleukid inscription calls for new athletic and music contests in support of a pre-urban population far removed from the centre of the Greek world.[4]

According to Duffy and Waitt, 'festivals help sustain narratives of belonging through bringing people together to share participating in various activities, but are also an exercise in remembering the past'.[6] This is especially the case with festivals in the Hellenistic world, which certainly engaged people from far and wide, and in which the past played a significant role, whether historically authentic or fabricated to suit the times. Figure 5.2 shows over 160 new or newly reorganised mondial festivals starting in the Hellenistic era and continuing in the early imperial period, and this is by no means complete amidst ongoing discoveries and re-assessments of inscriptions, coinage, and other evidence of festival cultures.

These festivals tended to model themselves on the great panhellenic festivals, often with 'crowned' (stephanitic) games declared equal in status to their mainland counterparts, starting with the 'isolympic' Ptolemaia at Alexandria in the third century BCE. Scholars have long noted a surge in such festivals in the Greek world after the conquests of Alexander the Great.[7] Until recently, however, this was largely attributed to a presumed decline of the *polis*, as cities were subsumed into empires, with increased personal

DOI: 10.4324/9780429429217-7

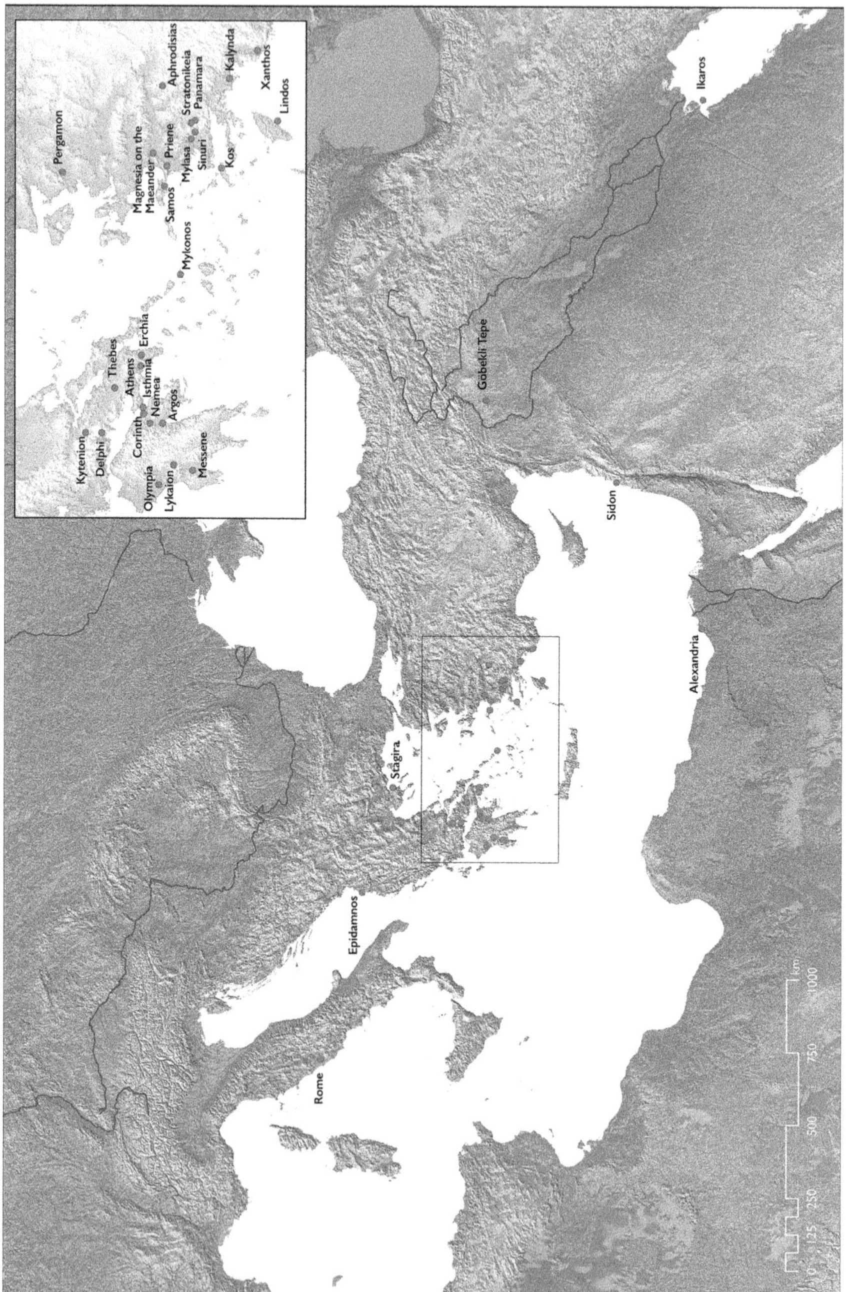

Figure 5.1 Places discussed in the text.

Source: Author.

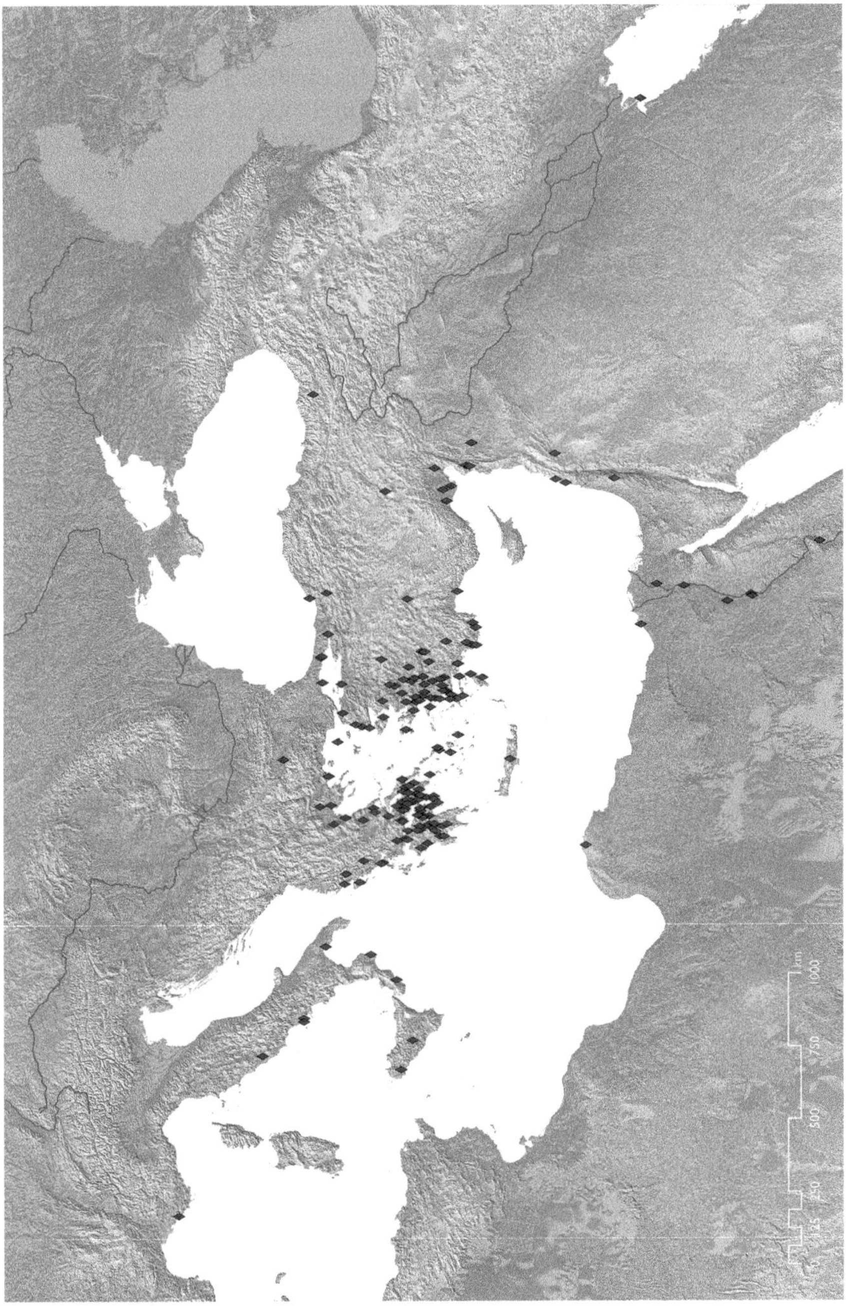

Figure 5.2 Inter-urban festivals in the Hellenistic and early imperial world, based on Chaniotis 1995, and the Groningen and Mannheim databases.[5]

Source: Author.

mobility; at the same time, mounting anxieties in turbulent times presumably fostered an introspective turn to the past and a revival of civic festivals as a result.[8] While this latter may be observed, recent studies demonstrating the vibrancy of cities and their inter-connectivity in this era lead us to consider alternative interpretations of this phenomenon. In our approach, we view festivals as a fundamental channel through which the Greek world became increasingly linked.[9] One of our larger aims is to trace the development of inter-urban festivals in order to fully interpret this phenomenon and how it came about. Meanwhile, the present chapter is a probe into the kinds of connections that were being established, the preconditions for such connectivity to take place, and especially why festivals were so effective at reproducing common knowledge.

Network theory provides a promising approach as it examines the dynamics of relationship structures by modelling people (or places or things) as nodes and their links as ties.[10] Granovetter's leading study posits that innovation spreads through networks via 'weak ties', or sporadic contacts, rather than through 'strong-tie' networks, where everyone knows everyone else, and where it is presumed that knowledge is common, that is, reciprocal and hence static.[11] Weak ties, on the other hand, ideally possess different ('new') knowledge and therefore have a strong brokering potential, filling the disconnect, or 'structural hole', among networks. Ideas common to (strong tie) Network A may thus be shared via (weak tie) Broker B to (strong tie) Network C, where they then rapidly spread due to the dense nature of strong-tie connectivity, eventually becoming common there as well. This moment is designated as 'phase transition' and is key to 'information cascades', where information rapidly flows across multiple networks simultaneously as individual members themselves further become brokers within their own strong-tie networks.[12] Watts and Strogatz, moreover, used statistical analysis to demonstrate the importance of random links, that is, weak ties, in providing shortcuts between nodes across an extended network, thereby creating a 'Small World' in which everyone is only a few handshakes away.[13] Important parameters in their model are the parity of nodes, the stasis of the strong-tie network, the external location of innovation, and especially the randomness of weak-tie contacts. This model has been applied in numerous situations, from analyses of transport systems to the spread of infectious diseases. Problematic, however, is the notion of parity and random contacts, since human relationships are rarely entirely equal or arbitrary.[14] Phase transition is much less likely to occur across asymmetrical network relations, as divisions of race, class, and particularly diverging belief systems can be real social barriers.[15] The success of knowledge transfer thus relies on the level of trust between parties, as well as what is at stake, that is, the importance of information being shared.

Concerning festival networks, the weak-tie model is attractive as trans-regional brokers may be identified through the many delegations (theoroi), athletes, and musicians.[16] We envisage models in which festivals are hubs that connect nodes, either of entire cities, or travelling individuals who represent their home town abroad, with emerging clusters and temporal waves of inter-urban interaction as a result. The problem arises when we force these individual actors into the weak-tie model, while the evidence informs us that they were not perceived as filling any kind of structural hole, but were part of a carefully planned mission, reinforcing a network that was believed to be in place already, one principally framed through terms of religion. The significance of a festival was often measured by the antiquity of cult, but also through festival time, the unique capacity of festivals to fluidly and instantaneously connect the present to a deep and heroic past, crossing meanwhile tangible and intangible boundaries.[17] Communities increasingly believed themselves to be connected to each other through perceived kinships rooted in mythological ancestry. Such ties were often intentionally construed by magistrates, and festivals were the conduits through which these networks were shaped, through the agents of travelling delegations, but also athletes and musical performers. Finally, the efficacy of collective ritual in producing and perpetuating common knowledge provided a coordinating mechanism that helped ensure the tightness and continuity of communal identity, so clearly evident at festivals through common feasting, sacrificing, but also the shared values of honour and public awards. The frequency and expanding geographic scope of festivals across the Graeco-Roman world created a plethora of 'portable communities' that helped spread this common knowledge and should be considered through a globalising lens.

Festivals thus present a different kind of network connectivity than that which current network theory was designed to address, oriented as it is to modern, urban, and largely anonymous societies; some of its basic tenets therefore need to be reassessed. This chapter cannot hope to cover all of the areas, but in what follows, festival connectivity will be discussed in relation to the element of festival time, the formulation of ritual ties, and the production of common knowledge. First, however, a brief excursion into festivals in the modern world for the sake of comparison will highlight important issues in addressing these 'portable communities'.

'Portable communities'—a comparative study

Gardner coined the term 'portable community' to depict bluegrass festival culture in the United States, in which individuals in 'community-starved' environments travel to festivals to find and create community with other like-minded individuals.[18] These festivals generally last some days and take place in small towns in the mountains, far from urban sprawl; Gardner's case study was in the Rocky Mountains, but the movement began in the Appalachians in the 1950s and 1960s. Participants usually pitch a tent and wander freely in and out of each other's campsites, sharing food and engaging in spontaneous music sessions, parallel to the main venue. Past and present are mixed as traditions are blended with a sense of nostalgia and inclusive belonging.

These festivals are open to anyone who cares to join, thus creating a 'community of memory' felt by most to be missing in quotidian life; bluegrass music is as Gardner states 'a romantic reaction against modernity'.[19]

Yet despite the seemingly informal nature, these are highly regulated events that follow a logic of their own. A distinctive, albeit invisible, hierarchy among participants is measured by knowledge and experience: full inclusiveness is first enjoyed by the bluegrass performers themselves, who either possess or are in the process of acquiring the expertise of producing music in the 'right way'; then come the regulars, the insiders who frequent the festivals and have a deep knowledge and appreciation for the music and musicians, even though they might not play themselves; following this are the incidental visitors or tourists, who possess little knowledge of either the music or the non-verbal rules of engagement, but who are coaxed into the logic of the festival. Gardner attributes the wide attraction of these festivals to the lack of social interaction, bonding, and resonance in the everyday mainstream lives of the participants.[20] Besides the remote far-away locations at which these festivals take place, their strong normative character, familiar pattern of organisation, and frequency together especially make these festivals, wherever they are held, such effective repeaters—each one ideally serving to reify and reproduce this portable community.

Like the bluegrass festivals, major festivals in the ancient Greek world were also effective repeaters. They could also attract a large group of participants from across a continent, as a portable community. These festivals also often took place outside the city, sometimes in entirely rural settings far removed from the urban centre. Tradition was a fixative point and they were clearly subject to rules or norms of behaviour, often prescribed.[21] Finally, they helped establish a similar social hierarchy, demarcated by civic officials and cult personnel, the delegates, and especially the competitors, athletes, and musicians, as the stars, followed by the local community and the foreigners, those who made up the crowd. Ancient festivals were normative and also served to educate especially the young or foreign.[22] Yet, there were also significant differences that this comparison serves to highlight. Major festivals in the Graeco-Roman world, while often rural, were anything but anti-urban. They were instead densely intertwined with the structuring principles of the *polis*, establishing social roles and hierarchies, as we know from elite profiling or participation via civic, population, and ethnic divisions. Much has been written about *polis* religion,[23] which I will not repeat here, but it is evident that state cult was given a new impulse in the Hellenistic period through festivals, even though it was of utmost importance that it appeared to derive from an age-old tradition.[24] Festivals were at the heart of a sacred urban imaginary, celebrating myths and rituals that were written into the foundation of the city. Highly regulated, festivals provided a venue that at once institutionally galvanised the city while providing a symbolic and emotional focus for its constituents.[25] Ritual participation was seen as producing and reproducing the very structure of society, while at the same time cementing the place of the individual within the community.[26] Major festivals in antiquity were anything but reactionary. Finally, an important difference concerns the manner of engagement of the portable community—knowledge of bluegrass festivals may spread by word of mouth in an informal, or even random, fashion, and this happened in antiquity too; yet, the constituents of festival communities were, as a rule, also deliberately composed through connections framed in ritual and justified by a deep sense of the past.

Time and festival network building

Time is a strong dimension of festivals. The rhythm of the year is marked by festivals, often coinciding with the seasons. Sacrificial calendars inscribed in civic contexts, as at Erchia in Attika, or Mykonos, coordinated festival time for the entire city.[27] Certain politically central festivals moreover serve to distinguish years, particularly by their priesthoods, such as the Heraia at Panamara for Stratonikeia.[28] This sense of festival time in itself was shared by the community and was part of their identity. The Olympic victor lists are a vivid example. First compiled by Hippias of Elis around 400 BCE, chronographies of Olympionikai were collected and passed through the hands of scholars such as Aristotle, Eratosthenes of Cyrene, Diodoros Siculus, Castor of Rhodes, and Dionysius of Halikarnassos among others, and circulated through inscriptions and papyri until late antiquity.[29] The emerging globalising world across the empires lacked a global calendar, yet the Olympic victories served as a universally accepted chronometer.[30] The lists recorded the achievements of individuals and thereby ensured their fame throughout time, while establishing the chronological depth of the festivals and geographic scope of the victors. They intersected the spheres of time and space, but also the scales of individual, *polis*, region, and oikoumene, the Greek world at large. The lists were a common point of reference and a coordinating mechanism across the Mediterranean but also across time, spanning generations for hundreds of years.[31]

Acute attempts at coming to terms with time are particularly visible in the Hellenistic period, as a rise in historicity and almost an obsession with the past seems to take place, at least in the written record.[32] High honours are awarded to local historians for their role in securing fame for all time, as they record (and inscribe) local histories that place the city and especially its principle deity or sanctuary in the foreground.[33] An eloquent example from the second century BCE is found in the praises awarded to Leon of Samos, perhaps dictated by himself, on the base of his statue in the sanctuary of Hera:

> The Assembly of the people of Samos (dedicated this statue of) Leon, the son of Ariston, to Hera.
>
> Because of time, even a stone grows old, holy bronze is scratched by the misty snowflakes and the strength of steel works out. However, the

> unbreakable fame keeps everything alive, the fame which Leon achieved throughout the city, Leon who has brought accomplishments for the homeland into trustworthy history, sang our native Hera, and showed how the sacrificers adorned her sanctuary with many ships and spoils.[34]

This encomium not only demonstrates the importance of the creation of 'intentional history' or the 'social store of knowledge',[35] but also the crucial role of sanctuaries as a vital link in connecting the *polis* to the outer world. The words 'homeland' (*patras*), 'native' (*autochthona*)—Hera was believed to have been born on Samos—and 'sacrificers' (*rexantes*) imply the watchful eye of a non-Samian audience. Leon elevated the city through its famous shrine onto a central stage of the Greek world, with his local 'trustworthy' histories that were not only intentional but also promotional. By this period, the sacred narrative of the past, whether true, elaborated, or wholly fictitious, had itself become a kind of commodity that worked at three scales simultaneously: as it confirmed the social capital of the historian within a *polis*, it also consolidated the identity of that *polis* via the deity while spreading the fame of both cult and *polis* well beyond. Leon could position Samos as a global player, and the success is reflected in the widely acknowledged and privileged status of *asylia*, or inviolability, obtained for the shrine, which the Samians later defended in Rome.[36] Their defence consisted of an Amphictyonic decree supposedly from the time of the foundation of Greek cities in Asia Minor; however unlikely this document may have been, the Romans accepted it and the status was confirmed.[37]

Local histories with a demonstrable connection to the deep past, preferably one that included the Trojan wars, had a highly valued currency and were thus a strategy employed by cities to gain favour and attention in the Greek world.[38] A surge in the demand for consumer-histories and aetiologies led to the high political profile that local historians and mythographers had acquired.[39] The document produced by the Samians may well have been fabricated, as was an earlier decree by the Magnesians in support of their claims of a Cretan origin.[40] Similarly, the Lindian 'chronicle', listing votive gifts from time immemorial to the recent historical past at the sanctuary of Athena Lindia in Lindos, on the island of Rhodes, has been referred to in the context of 'state fiction'.[41] But the real issue was not the authenticity of the documents, but the credibility of the claim.

This credibility took root in a ritual setting, where time itself acquires a new dimension. Assmann designates Festzeit as significantly different from Alltagzeit.[42] Drawing on Halbwachs, he states that festivals are the 'place' of cultural memory; in the flow of everyday life, they form '"islands of time", islands of a completely different temporality, suspended in time [...] memory spaces of "retrospective contemplativeness" [retrospektiven Besonnenheit]'.[43] With all of their trappings and rich sensory triggers—performances with music, sounds, smells, and tastes—and framed by tradition, festivals are the perfect memory alternators that can immediately and effectively substitute the experience of the everyday world with an alternate reality, where other rules apply. Spectacle can drive these impressions deep into the recall system, as with Whitehouse's 'flashbulb' memories in which the shock of an overwhelming (ritual) experience produces heightened acuity with regard to detail, much more so than under normal circumstances.[44] While the regularity of festivals in antiquity may temper the emotional overload, it is nonetheless the case that the element of spectacle was increasingly deployed as a mechanism for joint attention, as well as a platform for elite-showcasing through benefactions, with splendour becoming ever more central. But even less spectacular rituals cemented patterns of communal memory through their frequency and variety of collective rituals.[45] Processions were increasingly staged; they served not only to link critical places in the urban (and rural) topography, but also to establish the reality of social hierarchies.[46] Sacrifices were the primary channels of communication with the divine, but created meanwhile a common, joint focus of attention, and not without an element of competition as it was important for them to be beautiful and impressive in order to be pleasing to the gods.[47] Feasting involved the communal distribution of sacrificial meat, whether this was distributed equally or reinforced hierarchies.[48] But feasting together would have been as definitive for the community as it is in nearly every culture.[49] Finally, athletic, musical, or dramatic competitions were increasingly central to many festivals. Rivalry and connectivity are two sides of the same coin, exemplified by the subjugation to a common set of rules, but especially the shared value system of godlike excellence. Although the religious nature is debated, I do not hesitate to count agonistic events among collective rituals, especially due to the sacred context in which they took place and their emotional effect, discussed further below.[50] Other collective rituals, such as oaths, singing of hymns, and dancing were surely much more common than the sources suggest.[51]

As a locus of collective identity, festivals are flashpoints of cultural memory. They create communities by providing a common centre that is amplified by ritual. Studies on the function of ritual often claim that it is the ritual itself, rather than the deity or the signified, that subconsciously bolsters both individual cognitive structure and group like-mindedness, thereby creating an innate sense of belonging to a larger community.[52] Durkheim described the 'collective effervescence',[53] or emotional energy, brought on by ritual, but this goes even farther. As Assmann states, the festival community transcends not only the experience of the everyday, but even the group of mortals who are present.[54] The past is clearly part of the present at sanctuaries—memories are preserved in inscriptions but also other less official wall writings, such as 'graffiti' and material dedications, and sometimes tombs lining the sacred way.[55] Festivals in these sacred spaces blur the boundaries between the living and the dead but also the mortal and the divine. The growing testimonies of epiphanies in the Hellenistic era, often inscribed on the very walls of the temple of the manifesting god, bears witness to an increasing need to tangibly locate the divine within a community.[56] Meadows speaks of the 'Great Transformation' in the later second century BCE as cities increasingly promote a tutelary deity, and the types of epiphanies recorded are generally those in which the deity actively protects his or her community.[57] All of this points to a multitude of participants, past, present, and supernatural, that were sensed in various ways at festivals.

Therefore, when we look at the social networks that were being forged via sanctuaries, we should consider them against the larger canopy of the perceived festival community, present and past, mortal and divine, but also local and global.[58] The victor lists, as mentioned above, are a small mirror of this: the collection of names represents at some level the continued presence of the contestants themselves at the sanctuary. The same may be said of lists of priests, or the delegates, and of course the great marble host, the throng of honorific statues and inscriptions, reminding visitors of those who deserve to be remembered.[59]

There were many reasons to invest in festivals and to participate in them. The memory culture of festivals and their antiquity continuously expanded with the spread of urbanism. As cities were founded or reorganised, new festivals emerged (see Figure 5.1), yet even with these, tradition and the appearance of an ancient past were critical factors to their success.[60] Another Leon, from Stratonikeia, revived the festival of Zeus during his priesthood at nearby Panamara in the early second century BCE.[61] His efforts began with researching the history of the shrine and discovering old documents, *archaia grammata*,[62] from which he construed ancient honours and rights of *asylia*, which he then used to persuade the communities of the shrine to further invest in the festival.[63] This action shows the weight given to historical claims and a desire to involve others where possible, even at a regional scale. Epiphanies further confirmed the investment of the deity in his or her community. Delphi was the prime authority that could confirm *asylia* for shrines, but involving the kings was also vital in the Hellenistic era.[64] Later this power shifted to Rome, and at Panamara the saving epiphany of Zeus gained it recognition of *asylia* from the Roman senate.[65] Grand festivals celebrating the antiquity of cult, especially those with games, clearly appealed to a shared system of values that encompassed both the weight of the past, the spectacle of ritual, the element of victory and human achievement, and divine and political authority.

The question in the Hellenistic era is exactly which community was being shaped at these festivals. As the city put itself on display, these extravagant urban festivals were increasingly directed towards the Greek world at large, as noted especially by Robert and Parker.[66] Others have moreover observed the general thickening web of cities and inter-urban interaction in this period.[67] While it is no surprise that such festivals could attract people from far and wide, it is important to examine how these connections were forged and the degree of intentionality in order to understand this nascent 'small world' of festival ties.

Activating ritual ties

An important factor in studying the formation of ties in the ancient Greek world is that this world largely considered itself to be connected already. Latent networks, based on kinship, myth, or historical reciprocity, were activated at festivals through ritual. Festivals created their own potent social realities and so it mattered very much who attended them. Depending on their openness, they could attract a wide-ranging crowd, including religious travellers, such as Pausanias, as well as merchants selling food and other goods (festivals often had market functions), and of course the locals themselves. Slater even argues that they drew 'mimes, jugglers, strong men, sideshow artists, story-tellers, animal shows'.[68] But a concerted effort was often made to invite certain other cities or individuals who were considered especially important. These invitations were carefully construed, either by the organisers of the festival, usually the priests and personnel, or by the civic body, and the responses of acceptance were inscribed at the sanctuary. As hubs, festivals thus provided channels of interstate diplomacy. They often overlapped with other forms, such as the public honour of *proxeny*, through which individuals in one city represented the interests of another.[69] Cities used festivals to bond at different scales, starting from within, then outwards to other cities in the region or across the Mediterranean and beyond. The growing inter-urban connectivity evident in the Hellenistic world has been analysed by Ma in terms of 'peer polity interaction'.[70] Drawing largely on epigraphic data, he identifies various indicators of links: *syngeneia*, or kinship; *asylia*, the widespread recognition of the inviolability of a sanctuary; *theoria*, delegations representing various cities at festivals; and arbitrators, or 'foreign' judges who presided at court hearings to ensure a neutral, non-biased outcome. Inscribing these links does not just record the event, but actively reproduces the connections, perpetuating them over time as a mental map of connections.[71] Ma is primarily interested in the institutionalised forms of interaction and stops there. Yet, all of his indicators in fact surface within the context of festivals (with the possible exception of the arbitrators, although they were publicly honoured during a festival). It is worth focusing on especially the factors of declared kinship (*syngeneia*), the commissioning of delegations (*theoria*), as well as the positioning of contestants, which Ma does not include but which we believe is essential in the context of festivals,[72] in order to better understand how these relationships were framed.

Syngeneia

The authority of the past, through relations and traditions, is a strong connective (or divisive) force which is arguably at its height in ties of kinship. As Herodotus states, the Greeks believed themselves to be linked through blood, speech, and religion.[73] In the Hellenistic world, *syngeneia* features prominently among diplomatic relations as cities strive to display their Greekness, but also their interdependencies.[74] With festivals as the primary outlet of civic identity, *syngeneia* increasingly appears as a persuasive argument for communal engagement. In Caria, pre-*polis* sanctuaries often thrived as the common centre of a kinship group that identified itself as a *syngeneia* and that operated as a (semi-)autonomous body. An example is the Pelekos *syngeneia* that regularly met during the festival of the Carian god Sinuri, near Mylasa (Figure 5.1).[75] The community founded its collective identity in the cult and especially through its rituals, which brought them together and which they used to publicly honour their members. Even as the area was absorbed by Mylasa in the Hellenistic era, the shrine remained a focal point for the *syngeneia*, although it operated under a different name.[76] *Syngeneia* was more commonly used as a qualifier in interstate relations. At the sanctuary of Zeus at Panamara, near Stratonikeia, an unnamed priest wrote numerous letters to various surrounding communities, inviting them to join in the festival and mysteries of Zeus and appealing to their shared *syngeneia*.[77] This more common usage implies a distant shared ancestor, whether through blood or myth. Kinship ties were significant bartering chips and came with a sense of obligation, as Ma pointed out in the case of Kytenion in Boiotia and distant Xanthos (Figure 5.1).[78] The fact that they were often based on mythical ancestry should not be taken lightly, as mythical time was considered a deeper dimension of historical time, and just as authentic.[79] The appeal of *syngeneia* demonstrates a real or at least desired sense of brotherhood, whether this concerned a sub-division of a city, extended over an entire region, or spanned the known world. The usage of this term is a firm indication of the perceived network behind the intentional links being reified and commemorated through festivals.

Delegations

Besides general participation, a claim of *syngeneia*, or at least a shared heritage, was commonly used to engage delegations, or *theoroi*. This is clearly the case with the theoric connections established by Magnesia on the Maeander. When the initial announcement, in 221/220 BCE, of the new festival for Artemis Leukophryene failed to draw the crowds they were hoping for, the Magnesians renewed their efforts and launched an impressive diplomatic offensive, in which some 160 cities and kings were approached to secure their acknowledgment of the *asylia* (inviolability) of the sanctuary, but especially their participation in the festival through *theoroi* and athletes.[80] The extant epigraphic record shows the strategies of these missions and the appeal to religious authority but especially to a common heritage in order to ensure a positive response. Rutherford neatly summarises the arguments used by the Magnesians into four types: the epiphany of the goddess; affinity of kinship; reciprocity for services rendered; and peer pressure.[81] In several cases these relations were couched in specific terms of *syngeneia* as well as *philoi* (friendship).[82] Wiemer argues that both the festival and these missions emanated from the underlying goal of the Magnesians to promote their history.[83] He observes how the embassies traversed the Greek world armed with a dossier that included the Delphic oracle mandating the festival, but also documentation demonstrating shared histories with individual poleis, and the deeds performed by Magnesia for the Greeks at large. The 'intentional history' of the Magnesians stated in the official narrative in *I.Magnesia* 17, and in their brief to Epidamnos (*I.Magnesia* 46), but especially in *I.Magnesia* 20, the fabricated document from the federation of the Cretans that presumably substantiated their claims of antiquity and migration to Asia Minor.[84] The text was apparently faked, but no one called their bluff on it as their claims were deemed legitimate.

Wiemer is correct to emphasise the role of the past in their argumentation, but it is equally clear that festival connectivity offered the Magnesians a strategy for securing their position in the present. Like the letters from the Stratonikeian priest discussed above, the Magnesian envoys were not just after any kind of public recognition, but were carefully constructing their place in the Greek world by reactivating old ties to other cities, and investing in their own deep past. Festivals such as the Leukophryena were directed at the Greek world, but inter-urban festivals could also be targeted to a very specific subset, such as federations, for example, the Doric Hexapolis in the Dodecanese.[85] In either case, the deliberation of these embassies makes clear that these were anything but 'random' or weak-link ties. Given the diplomatic nature of the connections, this should not surprise us, yet the semantic intimacy speaks of a deep, intense, and even affective relationship at an inter-urban level, more so than we would normally associate with interstate politics.[86] Relations of personal trust surely developed in the process, and the same individual could fulfil the position multiple times. Moschion, son of Kydimos, was *theoros* on various occasions for Priene, but also an important benefactor who ensured the supply of grain for the city.[87] Delegates were usually drawn from the elite and represented a tight and trusting relationship with the city. They were selected with great care by the council and then confirmed by public election.[88] This was the responsibility of the *polis*, although a person aspiring to the function may have been able to arrange the appointment himself. Rutherford observes this in at least two instances: Demosthenes of Athens was said to have volunteered as *arkhitheoros* (chief delegate) to Olympia in 324 BCE so that he could meet Nikanor of Stagira;[89] Theopropos from Kalynda in Caria, went to Ptolemaic Alexandria in 247 BCE as *theoros*, using the trip to conveniently arranging matters of private business.[90] Both cases illustrate the further complexity of festival networks as actors take on different roles simultaneously.

Festival delegations were a push-pull process, with the hosting city wanting to attract participation, an audience of peers to impress, and cities wanting to impress others as well. The urban counterpart of the travelling *theoros* was the *theorodokos*, the representative within the city who hosted the *theoros* during his stay and ensured the success of his mission. This was also a diplomatic post and 'economic status and participation in foreign affairs and state religion [...] contributed to the selection of individuals to serve as *theorodokoi*', as Perlman surmises.[91] Returning to network theory, we could easily designate *theoros* and *theorodokos* as brokers establishing a weak tie between the strong-tie networks of their respective poleis, were it not for the fact that there was no perceived structural hole between these cities. Instead, they were already imagined as part of a larger interdependent community, and the underlying links were deliberately and carefully crafted. Although the individuals may not have known each other personally, the collective memories of their *poleis* already intersected and it is this overlap of identities that was being reactivated through festival ties.

Competitors

The goal of theoric missions was to establish channels of communication and tighter connectivity between communities, underscoring their agency in festival networks. Yet another kind of network agent may be found in the contestants, those who performed the network through multiple musical and athletic *agones*.[92] Much has been written on this topic and it is my aim here to focus only on the intentionality behind their participation. Contestants were not randomly chosen, but underwent careful training and selection as they represented their hometown.[93] In this regard it should be observed that most poleis across the Hellenistic world had a *gymnasion*, if not theatre and *stadion*, of their own.[94] Human excellence, evidenced through local but especially interstate competitions, was a value shared across the Greek world, and was cultivated from an early age onwards. The abundance of '*nike*' graffiti, especially in late imperial cities in the east, shows the deep permeability of this concept in society.[95] Victory was embedded in the Greek social imaginary and contestants could acquire pop-star status, but this should not overshadow their role as ritual actors too.[96]

Victors were amply rewarded by their own native *poleis* and received high publicity.[97] This was kept in public memory, not just by the archives, but by having the 'sacred victors', the *hieronikai*, or winners at the sacred games (especially the big four panhellenic festivals of Olympia, Delphi, Nemea, and Isthmia) appear in public events time and again, wearing their crowns. In Pergamon, for example, the *hieronikai* take their place next to the chief magistrates such as the archons, the generals, the priests and priestesses of the twelve gods and of the ruler cult, the gymnasiarchs, and the ephebes as they welcomed the return of the king, Attalos III, into the city.[98] In the imperial period, sacred victors could be included among governing bodies, as a decree from Aphrodisias shows, which lists after the *boule*, *demos*, and *gerousia* (council of elders), the *hieronikai*, the *pleistonikai* (winners of multiple games), and the *stephanitei* (crowned victors).[99] At a more organised level, the Dionysiac artists, actors who travelled to theatrical contests, were often considered a para-civic organisation and were called upon to organise and finance festivals. Aneziri has shown that their participation is generally solicited and formalised (even with fines being assigned to performers who fail to show up).[100] Participation was not always tidily organised and Slater is right to argue that the lines between the 'fringe performers' and the more official participants should be blurred when it comes to the spectacle value.[101] While the haphazard connections would certainly increase the value of weak-tie contacts, this does not diminish from the very intentional connections that were being established through the contestants in the main events. In fact, their involvement should be regarded as part of the 'push' factor by cities, but also by gymnasiarchs, *agonothetes*, and of course the individual contestants who wanted to appear, and to win, at the festival in question and before an audience of peers.

Victorious performances could have political resonance. In her study of the middle ground between Phoenician and Greek relations, Corinne Bonnet discussed the case of Diotimos of Sidon who competed and won the chariot race in the Nemean games around 200 BCE.[102] Especially the inscribed epigram of his victory is interesting, as the poet situated Diotimos's victory as a win not just for Sidon, but also for Thebes by framing it through deep ties of mythical kinship between Thebes and Sidon via Agenor, first king of Sidon and son of the king of Argos, and father of Phoenix (founder of Phoenicia), Europa, and Kadmos, founder of Thebes and credited with introducing the Phoenician alphabet to the Greeks. Diotimos' win at Nemea meant a glorious reunion between Sidon and Thebes, calling for a celebration on both sides. The popularity of such panhellenic games was vast, due not only to the antiquity of cult, but also simply, and circularly, to their central place in the common knowledge and shared value system of Hellenic culture. The magnetism of victory was further reinforced by epinikian poetry, particularly Pindar, in his exaltation of athletes and their near godlike qualities. Building on Kurke's argument of the economy of kudos, Aloni recently examined this poetry in connection with cities, concluding that such epic praise surfaces more in cities defined by prominent elite families (that produce winners) than in more egalitarian communities where collective civic identity is foregrounded.[103] Nonetheless, epinikian poetry, along with commemorative statues and ceremonial appearances, would have contributed significantly to the high value placed on victory and the social capital of winners across the Greek world.

The commemorative nature of this kind of mediatisation shows another dimension of connectivity that festivals, and particularly the contestants, could offer—that with the past. Victor lists are, again, especially interesting in this regard.[104] Normally there would normally be little or no overlap between successive victors, except for those competing in circuits within roughly the same generation, and most would not have known each other personally. At the surface, this would problematise placing them within a social network, since the model assumes a transmission of ideas via direct communication or contact. Nonetheless, victors are connected to each other through a community of stone, whether in lists or sculpture; a community that was made very public, and one which rising athletes surely aspired to join. So although they could never have met face to face, Andromedes of Corinth, for example, who won the *stadion* in the 119th Olympiad (304 BCE), may nonetheless have felt a kind of kindred camaraderie with his compatriot Desmon, who won in the same event, but centuries before, in the 14th Olympiad (724 BCE).[105] The inscriptions themselves inspire admiration but also create ties in the mind of the reader, whether athlete or spectator, and that is surely one of their primary intentions. More than an inventory, such inscriptions forged a new community, one of victors, and one that transcended time by interjecting the past into the present.

Agency is thus located both in the actions of individuals functioning as brokers, the *theoroi* and *theorodokoi*, and the competitors. Agency is also found in objects such as inscriptions, statues, but also coins that connect the city, the god, and the festival,[106] and the (for us) more ephemeral objects, such as the crowns won and later dedicated—all of these bore testimony to the expanding festival community over time. Ultimately it is the larger framework of the community, the common heritage, the reactivated *syngeneia*, or brotherhood that generates and is in turn regenerated by these festival connections.[107] These ritualised ties were anything but random. They were deliberately put in place to strengthen the network across boundaries of territory, region, and kingdoms. While there were surely casual and even 'random' visits that would further the knowledge of festivals across different communities,[108] festivals served as a conduit for ties shaped by kinship, heritage, parity, and reciprocity, and served to draw cities together in an increasingly tighter network of inter-urban communities. Of course, none of this would have had any effect without the presence of the crowd, who had a normative role of their own and whose participation was critical in the flow of information across the Greek world. This common knowledge is key to the function of festival networks.

Common knowledge through ritual

If festivals were a prime vehicle for creating inter-urban ties across the Greek East, this is because ritual was the oil that ensured a common foundation. In the first section we observed the central role of festivals in not only retaining but also producing cultural and collective memory. Yet besides this institutional function, it is the subliminal effect of rituals themselves that are especially coercive in both structuring the community and aligning the individual mind within this community. Festivals clearly had a certain logic of their own.[109] Different from the routine of quotidian life, they were highly regulated, as the plethora of ritual norms, or 'sacred laws' demonstrates.[110] While no two festivals were exactly the same, as Slater argues, there was a shared grammar of interaction through which the general rules were understood by all, despite the local variations.[111] This grammar is part of the language of ritual. One of the most detailed examples is the lengthy prescription of behaviour at the Andania festival near Messene.[112] This document contains precise instructions on preparations for the procession and festival, such as abstinence, uniform attire that ensured a measure of equality among the participants, but also the prohibition of certain kinds of behaviour and penalties, reinforced with corporal punishment (by whip). Where competitions were concerned, the rules of the game were even more strictly monitored to ensure fair play. Such regulations enforced a common approach and their inscription was intended to disseminate common knowledge. Yet even without prescriptions, festivals are in themselves normative in their expectations of collective behaviour, as with the unspoken but nonetheless reinforced regulations at bluegrass festivals, discussed above.[113] The combination of rules with rivalry is an especially a powerful accelerator of common knowledge, making the thrill of victory that much more intense.[114] 'We Won!' is not so much a statement of fact as a cry of collective emotion, bringing with it the sense of belonging to something immensely larger than oneself.[115]

Yet at the same time, rituals work at a much more intuitive, subliminal level. In their research on the cognitive science of religion, McCauley and Lawson argue how the human brain is wired for ritual, creating synaptic shortcuts that enhance thought processes.[116] Their approach focuses on how neural pathways are reinforced by the compulsive nature of ritual, in which things must proceed in a particular fashion, with the correct words, gestures, or objects, in order not to fail. This is especially effective in collective ritual, where the element of performing the ritual correctly, but especially of performing it correctly together, triggers an unparalleled sense of unity and belonging. Turner identifies the resulting *communitas* as 'the direct, immediate, and total confrontation of human identities which tends to make those experiencing it think of mankind as a homogeneous, unstructured and free community'.[117] Like Assmann, Turner underscores the alterity of the ritual experience, but positions ritual as antithetical to the structure of everyday.

Key is the performance itself. Drawing on Goffman and Durkheim, Randall Collins' theory of interaction ritual chains (IRC) emphasises the importance of physical assembly in ritual as people engage in a 'homogeneity of movement, synchrony'.[118] Festivals facilitate this kind of spontaneous organisation as bodies naturally react to each other; for example, the synchronous applause after a good performance at a modern festival or concert. An example from the ancient world is found at Panamara (Figure 5.1), where during an attack at the sanctuary the population continuously 'shouted in a great voice, Great is Zeus Panamaros'.[119] Although provoked more by stress than ritual, this example nonetheless shows the kind of collective organisation that can spontaneously take place in a large gathering, as people, that is, bodies respond to each other. As with swarms, Barabási argues for the self-organisation of the 'scale-free' networks that spontaneously turn disorder into order.[120] This is especially relevant to the emotional energy of such situational social interactions charged by ritual, again Durkheim's 'collective effervescence'.[121]

Game theorist Chwe observes the efficacy of 'rational rituals', mass ceremonies, as coordinating mechanisms that create a joint focus, thereby generating common knowledge.[122] Common knowledge, in his view, is the underlying guide in decision-making. This works for both mass ceremonies in place, for the ancient world think of sacrifices, feasting, but also competitions, and those on the move, such as processions. Such public spectacles are a factor in aligning collective thought.[123] Ancient festivals and their spectacles also worked in this way. As with Collins' IRC theory, the innate common knowledge gained by doing the same thing together, in the same way and at the same time, ensures a subliminal kind of bonding that may later find expression through myths and aetiologies. Collins emphasises physical co-presence, while Chwe further stresses the importance of eye-contact in instantaneously conveying emotions and triggering the mutual reciprocity of thought, that is, 'I know that you know that I know… etc.'.[124] Ritual space is thus most powerful in what Chwe calls the 'inward-facing circle', a centralising ceremonial space that facilitates mutual eye-contact and intensifies the collective experience, similar to what Huizenga called the 'magic circle', a space where other rules apply.[125] These are powerful, volatile spaces, and we should remember that besides collective harmony they are also the scenes of mass riots and violence, for exactly the same reasons.[126]

With the power of communication that these spaces possess, and the need to communicate to increasingly larger audiences, it is little wonder that the surge of inter-urban festivals went hand in hand with the appearance of theatres, stadia, and enclosed peristyle sanctuaries, often kilometres away from the urban centre. All of these served to separate the visitor from the surrounding environment and focus public attention inward.[127] As more and more cities across the Greek East developed their prime sanctuaries along similar lines and for this exact purpose, we should consider these spaces as network repeaters of common knowledge, disseminating this sameness, albeit with couleur locale, via the wide-ranging audience that convened at the shrines.[128]

These places and their festivals reproduced common knowledge that reinforced the strong ties linking communities across the Mediterranean and well beyond. Knowledge of the festival, the rituals, the winners and the myths, all helped to create a festival culture that spanned across the Greek world. Yet more than as rational ties, the kind of common knowledge produced through ritual was profoundly subliminal and cemented these relations in a visceral way, probably fostering the same kind of emotional energy felt among modern festival-goers. In this sense they engendered a portable community, much like the bluegrass festivals discussed by Gardner, even though individuals could come and go, the sense of community remained constant.[129] The material act of leaving something of one's self behind in these places, for example, through votives as gifts for the gods, testimonies of miraculous healing, or informal writings or markings on architecture ('graffiti') testify to a need to belong, 'I too was here', and a desire to perpetuate one's presence over time.

The portable community that emerged from these festivals lent their participants the role of broker, in communicating the common knowledge from the festivals to their homelands and beyond. Besides the athletes, musicians, and official delegations (*theoroi*), others who were less formally involved, the story-tellers, street artists, soothsayers, merchants and peddlers, tourists and the public-at-large, were also ritual agents in this sense, spreading the word whether intentionally or out of excitement or for other reasons. Furthermore there is a sense of spectacle rivalry, culminating in the imperial period, wherein the burden generally lay with the priests to make their festival a most memorable event for all attending, as we already saw with Leon and Panamara.[130] Whether the spectacles were intensive enough to qualify as one of Whitehouse's 'flashbulb memories' may be debated, but they surely led to an altered sense of time as Assmann argued, situating the memory of festival in a different level of cognisance than quotidian experience.[131]

Rituals are undeniably central to the formation of community in every age, but ancient festivals had the additional dimension of divine sanction and (perceived) weight of the past. The spectacles, the games, the sacrifices, the processions and hymns, all took place under the watchful gaze of the gods and heroes. Religious authority and deep tradition were the main drivers behind the rituals, and were also what made them such strong shapers of society, even with seemingly new elective cults.[132] Again this had to do with the shared value system that was cyclically reinforced through festivals. Festivals were not solely dependent on strong tie networks, but they certainly helped to create them, fostering even stronger bonds among the participants. The common knowledge that proceeded from this was deeply entrenched, and surely one of the reasons that a traveller could readily feel at home in festivals across the ancient world. While there were certainly the obvious local differences, they shared the same underlying festival grammar; like bluegrass festivals, they generated a common ground for the increasingly portable communities.

Conclusion

This consideration of the context and strength of festival ties in antiquity has brought us to three main points pertaining to time, intentional relations, and common knowledge. In the first place, the festival was a prime locus of collective memory, and was central to urban identity. Festivals are time machines in the sense that they transcend the quotidian experience of life. They provide an alternative experience through ritual, often in a spatial setting separated or even far removed from the everyday world, presumably even for much of the rural population living in the countryside of the city. Despite their alterity, major festivals laid the foundations of social and urban structure, with political events marked by festival cycles. At the same time, sanctuaries were urban memory banks, harbouring dedications to the gods, but also state archives, public and honorific decrees, lists of winners of contests that had been hosted, but also multiple layers of dedications, sometimes for several generations or over centuries. The sense of deep time increased the authority of the festival as well as its fame and the scope of attention that it was likely to attract.

In the second place, the position of the city in the larger world of cities was shaped through the dynamics of festivals contacts. Strategies, sometimes highly complex as with Magnesia, were deliberately crafted to appeal to other communities. Ties were intentionally forged through kinship, overlapping collective memories and myth. Both push and pull factors were at work as hosting cities sought to ensure the acknowledgment and participation of key players and royal powers, and as invited cities saw opportunity to put themselves on display before their peers by sending their best delegations as well as their best contestants. Much was at stake on every side. Yet in the end host, delegation and contestant, who may well have met for the first time personally during the festival, would have envisaged themselves as part of the same 'extended' family, in a strong-tie network. Meanwhile, the roles of the agents of this network shifted and overlapped, for example, *theoroi* may have had other business matters to attend to while friendships and other relationships surely spontaneously formed as business combined with pleasure.

Finally, through their rituals, these festivals served as highly effective conduits of common knowledge. Cognitive science has underscored ways that ritual aligns thought processes.[133] Collective ritual especially enables common knowledge to be transmitted almost instantaneously. In successful rituals, situational body interaction and synchrony trigger a subliminal feeling of belonging. Essential here is not the meaning, symbolism or knowledge of the ritual, but the ritual itself and the very fact that it is performed collectively. The memory of the event is often retained through inscriptions, imagery and objects, architecture and the very topography, which in themselves become transmitters, or agents, of common knowledge over time. Common knowledge goes much further than a shared experience as it is normative and implies a measure of group predictability. This, combined with the necessary mobility, is what makes turns participants of these expanding festival networks into a portable community, as the underlying grammar of the individual rituals became more and more aligned with the larger festival culture.

This festival culture was a circular production system, as it was the product of perceived strong ties while forging them at the same time. Most interesting is the need to actually come together. It was not enough to simply acknowledge a festival, but delegations were expected and rituals were to be performed jointly. These networks collapsed both the physical and cultural distance between the urban centres of the new Greek world as local communities sought a connection to what they perceived as a shared Greek past. This emerging festival network was purposefully crafted, and the brokers carefully selected before being brought together. This solicited a propinquity, especially within the context of ties of kinship and friendship, that would have fostered the sort of homophily, that is, 'birds of a feather', that leads to strong-tie networks and especially group-think, even though these groups spanned much of the eastern Mediterranean and well beyond.[134]

No ancient historian will be surprised at the deliberation with which contacts were forged, given the abundance of documentation in this regard. The essential problem lays in the application of network theory as defined in the small-world model, where innovation is presumed to come from outside the strong-tie network, seen as static, via brokers as random links, that is, weak ties. The primary aim of this chapter has been to establish that the broker role at sanctuaries was not randomly filled but, on the contrary, was a process carefully construed by all parties involved and built on what was believed to be pre-existing relationships based on kinship, myth, or reciprocity. While the locus of innovation in this network remains for a future study, this chapter offers some important considerations. In the first place, festival networks allowed for an extremely rapid transfer of information via ritual as coordinating mechanism and the strength of the festival ties, especially as these networks gathered momentum and grew increasingly dense. In the second place, before turning to external factors, usually applied to interpret these expansion and changes in ritual, the internal factors within a festival community need to be examined much more closely. The assumption that strong-tie networks are by nature static is surely false, considering group dynamics and internal asymmetries. Social and political imbalances in themselves lead to change.[135] Finally, external factors need not always be negative, such as anxiety, impending irrelevance or a crisis in faith. What if cities found new opportunities in reaching out beyond their borders, and securing not just alliances but also new avenues for trade, or sometimes even a good solution to an internal crisis?[136]

In sum, the binary classification of ties as either weak or strong is not particularly informative regarding festival networks. In contrast with modern urban anonymity, the ancient Greek world was generally framed within a known reference system, in which everyone, and every community or city, and every sanctuary had a place. Moreover, actors in festival networks fulfilled a variety of roles simultaneously, some of them official and deliberately crafted, while other encounters will have been more spontaneous. Broker functions took place at many levels, and second-degree ties, that is, ties emerging between two communities connected for differing reasons to a third common community, need to be examined more closely, as does the emergence of clusters and asymmetries in these networks. Of course, numerous other connections must have randomly developed side-by-side with these intentional networks. People of all kinds were drawn to festivals and their paths crossed, but they also crossed city streets and open countryside in the same physical environment. A next step would be to investigate the variety and overlap of interests, levels of social connections and portable communities that evolved from these intentional ritual networks.

Abbreviations

BNJ Worthington, I. (2006–) *Brill's New Jacoby* (online revised edition of Jacoby's *Die Fragmnete der Griechischen Historiker* I–III), https://referenceworks.brillonline.com/browse/brill-s-new-jacoby (accessed 2021).

CGRN Carbon, J.M., S. Peels and V. Pirenne-Delforge (2016–) *A Collection of Greek Ritual Norms* (CGRN), Liège http://cgrn.ulg.ac.be (accessed 2021).

FD **III** Colin, G., R. Flacelière, A. Plassart and J. Pouilloux (1930–1976) *Fouilles de Delphes. Tome III, Épigraphie. Fasc. IV, Inscriptions de la terrasse du temple et de la région nord du sanctuaire, 4 vols.* Paris: de Boccard.

iAph Reynolds, J., C. Roueché and G. Bodard (2007) *Inscriptions of Aphrodisias*, http://insaph.kcl.ac.uk/iaph2007/index.html (accessed 2021).

I.Ephesos Engelmann, H. and D. Knibbe (1979–1984) *Die Inschriften von Ephesos. 8 vols. in 9 parts, with a Supplement*, Inschriften griechischer Städte aus Kleinasien, 11,1–17,4, Bonn: Rudolf Habelt.

I.Iasos Blümel, W. (1985) *Die Inschriften von Iasos*, Inschriften griechischer Städte aus Kleinasien, Band 28, Bonn: Rudolf Habelt.

I.Lindos Blinkenberg, C.S. and K.F. Kinch (1941) *Lindos. Fouilles et recherches, 1902–1914. II: Inscriptions publiées en grande partie d'après les copies de K. F. Kinch, avec un appendice contenant diverses autres inscriptions rhodiennes*, Berlin: de Gruyter.

I.Magnesia Kern, O. (1900) *Die Inschriften von Magnesia am Maeander*, Berlin: Spemann.

I.Priene Hiller von Gaertringen, F. (1906) *Inschriften von Priene*, Berlin: Reimer.

IvP *Fränkel, M. (1890) VIII. Die Inschriften von Pergamon. Teil 1. Bis zum Ende der Königzeit, Altertümer von Pergamon*, Berlin: Spemann.

I.Stratonikeia Şahin, M.Ç. (1981–2010) *Die Inschriften von Stratonikeia, 3 vols., Inschriften griechischer Städte aus Kleinasien, Bd. 21–23*, Bonn: Habelt.

MAMA VIII Calder, W.M. and J.M.R. Cormack (1962) *Monumenta Asiae Minoris Antiqua (MAMA) VIII: Monuments from Lycaonia, the Pisido-Phrygian borderland, Aphrodisias, Publications of the American society for archaeological research in Asia Minor, 8*, Manchester: University Pressin assoc. with the Society.

Syll.3 Dittenberger, W. (1915–1924) *Sylloge inscriptionum graecarum, 3rd ed.*, Leipzig: Hirzel.

Notes

1. This chapter is part of a triad (also Williamson in press a and b) written during a research fellowship at the Max-Weber-Kolleg at the University of Erfurt, in the context of the project 'Religion and Urbanity: reciprocal formations'. The research also stems from the projects 'Deep-mapping sanctuaries' and

'Connecting the Greeks' at the University of Groningen, sponsored by the Netherlands Organisation for Scientific Research (NWO), and which I co-direct with Onno van Nijf: see http://www.connectingthegreeks.com and http://www.deepmappingsanctuaries.org. My gratitude extends to the fellows at the Max-Weber-Kolleg and to the project team members in Groningen, including Tom Britton, Robin van Vliet, Adam Wiznura, and Sjoukje Kamphorst, for their helpful comments. My gratitude extends to the editor and reviewer for their helpful suggestions. Any remaining errors are entirely my own.

2. Smith 2019, 67–87, and 69, with the metaphor 'prehistoric Woodstock'. On Göbekli Tepe: Notroff et al. 2014, with references.
3. E.g., Morgan 1990; Malkin 2011. See also the contributions in this volume by Mooring and Daniels.
4. *SEG* 35, 1476; Petropoulou 2006, for a pre-urban context of this foundation as part of a *synoikismos*; Roueché and Sherwin-White 1985.
5. Chaniotis 1995, the Database of Hellenistic Athletes (University of Mannheim) http://www.athletes.geschichte.uni-mannheim.de, and the Connected Contests database (University of Groningen) http://www.connectedcontests.org.
6. Duffy and Waitt 2011, 44.
7. *Syll.*[3] 390, l. 22, *FD* III 4:357, l. 15–17. On the surge in festivals and crowned games: Robert 1984; Chaniotis 1995; Parker 2004a; Slater and Summa 2006; Mann 2018.
8. Esp. Chaniotis 1991; discussions in Beck and Wiemer 2009a.
9. Van Nijf and Williamson 2016, and now the project in Groningen: 'Connecting the Greeks. Multi-scalar festival networks in the Graeco-Roman period', see *connectingthegreeks.com*.
10. See introduction, also among many others: Kadushin 2012; Barabási and Pósfai 2016; and on objects: Knappett 2008; Brughmans et al. 2016.
11. Granovetter 1973.
12. Described in Malkin 2011, 36–39, with references.
13. Watts and Strogatz 1998; also Malkin 2011, regarding the development of the Archaic Mediterranean into a 'Small Greek world' via colonisation.
14. Barabási and Pósfai 2016, 72–111 discusses this problem in detail.
15. McPherson et al. 2001; White 2008; Kadushin 2012, 35.
16. These are being incorporated in the online database http://www.connectedcontests.org.
17. Examples abound. The Lykaia festival, for example, was believed to be among the oldest games, while being steeped in local myth (e.g. Paus. 8.2.1–2); the Lindian 'chronicle' blurs any boundaries between mythological and historical time, measured through votive objects and their donors, discussed further below, along with other examples.
18. Gardner 2004.
19. Gardner 2004, 170, and 156, citing Bellah et al. 1985 on the difference between segmented lifestyle enclaves and communities of memory. See also Turner's *communitas*, in which festivals provide a necessary anti-structure to the structure of everyday life, creating alternative communities that are distinct from the city or the state, Turner 1974a. Also, Duffy and Waitt 2011, 55: 'the festival is, in fact, a paradoxical thing; festival events function as a form of social integration and cohesion, while simultaneously they are sites of subversion, protest or exclusion and alienation. It is precisely this paradoxical nature that creates the festival's socio-spatial and political significance for notions of community and belonging'. cited in Frost 2016, 571.
20. Gardner 2004, 163, cites 'norms of inclusivity, intimacy, and simple and natural living' as the three key motives that participants list for travelling thousands of miles to attend such festivals.

21. Boyer 1990.
22. E.g. Bruit Zaidman and Pantel 1992, 63–79; Wijma 2014. See also the Collection of Greek Ritual Norms (CGRN), http://www.cgrn.ulg.ac.be, for further discussion.
23. Sourvinou-Inwood 1990, 304–305, 'The Greek *polis* articulated religion and was itself further articulated by it…'; for a critical view, see Kindt 2012.
24. Discussed in Beck and Wiemer 2009b; Funke and Haake 2013.
25. Bluegrass festivals are, by contrast, 'institutional lite', Gardner 2004, 174.
26. Durkheim 1912; Rüpke 2015 on structure and agency; Williamson in press a discusses the inclusiveness of festival space.
27. *CGRN* 52 (Erchia); *CGRN* 156 (Mykonos); see Carbon 2015, 539–543.
28. Williamson 2020, 149.
29. Christesen 2007, Tables 3–5, and 45–160 on Hippias of Elis, his context, motives, and methods.
30. Polybios *Hist.* 1.3, for example, dates his narrative to the 140th Olympiad (220–216 BCE) and *Hist.* 12.10.4, on Timaeus' construction of chronologies based on comparative lists, discussed in Christesen 2007, 277–289.
31. See further Christesen 2007 on the accessibility of the texts.
32. On the rise in historicity, see esp. Herrmann 1984; Chaniotis 1988; Gehrke 2001; Dillery 2005; Beck and Wiemer 2009a; Chiai 2013.
33. Prominent examples include Syriskos of Chersonesos and the authors of the Lindian 'chronicle'; see among others Rostowzew 1920; Higbie 2003; Dillery 2005; Platt 2011.
34. *IG* 12.6.1.285; *BNJ* 540 T1 (transl. A. d'Hautcourt, 2007): ὁ δῆμος ὁ Σαμίων | Λέοντα Ἀρίστωνος |χαλκὸς ἀπ' ἠερίας δρυπτόμενος |νιφάδος, (5) | καὶ τὸ σιδάρειον κάμνει σθένος· ἁ δ' ἀπὸ δόξας | ἄθραυστος φάμα πάντα μένει βίοτον. | τᾶς δὲ Λέων ἐκύρησε κατὰ πτόλιν, ὃς περὶ πάτρας | πράξιας εἱς πινυτὰς ἄγαγεν ἱστορίας, | ὑμνήσας Ἥραν αὐτόχθονα καὶ πόσα ναυσὶν (10) | ῥέξαντες σκύλοις ἱερὸν ἀγλάισαν. Also discussed and translated in Dillery 2005; Chaniotis 1988.
35. Gehrke 2001.
36. Due to the abundance of claims to *asylia* in the Greek world, the Roman emperor Tiberius held a council in 23 BCE, investigating each case, Tacitus, *Ann.* 4.14. Belloni 1984; Rigsby 1996, 580–586.
37. Rigsby 1996, 394–398, and no. 2, discussing Herrmann 1960, 90–93, no. 5. The inscription confirms the *asylia* of Samos together with that of Kos, for the Asklepieion.
38. Especially connections to Homeric heroes, see also Scheer 1993.
39. Ma 2003; Beck and Wiemer 2009a.
40. *I.Magnesia* 20, Wiemer 2009, 88 discusses the Cretan formulas; also below.
41. *I.Lindos* 2; Kosmetatou 2012. Other interpretation of the inventory of votives: Gehrke 2001; Shaya 2005; Platt 2011, 161–169.
42. Assmann 1991.
43. Halbwachs 1925; Assmann 1995, 129, drawing on Aby Warburg's Mnemosyne Atlas project with the term *retrospektiven Besonnenheit.* On sanctuaries and collective memory, see also Alcock 2002; Beck and Wiemer 2009a; Cusumano 2013 among many others.
44. Whitehouse 2004.
45. McCauley and Lawson 2002.
46. Chaniotis 1995; Chaniotis 2013.
47. Bremmer 2007; Mylonopoulos 2006, esp. 71–84.
48. Schmitt Pantel 1990; Schmitt Pantel 1992; Strootman 2018.
49. Nielsen and Nielsen 2001; Mylonopoulos 2006, 77–79.
50. Winkler and Zeitlin 1990; Kyle 2007; Mikalson 2007; van Nijf 2013a; Mann 2017.

51. Chaniotis 1988; Lonsdale 1993; Naerebout 1997; Kowalzig 2007b.
52. Geertz 1980; Chwe 2001; McCauley and Lawson 2002.
53. Durkheim 1912.
54. Assmann 1991, 26: 'Die Gemeinschaft, die im Fest zusammenfindet, geht aber über die Gruppe der real Anwesenden hinaus. Zu ihr gehören typischerweise die Toten und – damit zunächst aufs engste zusammenhängend – das Heilige oder Kosmische'.
55. Esp. in Asia Minor at outlying sanctuaries, as at the Asklepieion in Pergamon, Labraunda, Panamara. This topic is understudied.
56. A prominent example is the epiphany of Zeus at Panamara, near Stratonikeia in Caria, *I.Stratonikeia* 10, discussed in Rivault 2018. On epiphanies: Rostowzew 1920; Wheeler 2004; Platt 2011; Petridou 2016. On epiphanies and temples, Williamson 2018.
57. Meadows 2018, on coinage. On epiphanies: Rostowzew 1920; Wheeler 2004; Platt 2011; Petridou 2016. On epiphanies and temples, Williamson 2018.
58. A similar host of witnesses—past and present, tangible, and intangible—may be observed at Ephesos, in the Salutaris Procession (Rogers 1991) or the inclusion of statues of tribes, personifications, deities, and local heroes at meetings of the popular assembly in the theatre of Ephesos (*I.Ephesos* 28–36), as discussed by Chaniotis 2007. The inclusion of human and non-human ritual agents is further discussed in Rüpke 2015.
59. Traveller Pausanias is careful to elaborate on monuments and local stories at sanctuaries, in his *Periegesis*, written in the second century CE.
60. Chaniotis 1995.
61. van Bremen 2004, who discusses this from a political perspective.
62. *I.Stratonikeia* 7, line 4.
63. *I.Stratonikeia* 7, lines 4–9, 'from the [letter]s? and from the old documents, and having reconstructed that the above-mentioned honours and (grants of) *asylia* adhered to Zeus and to the Panamareis, he persuaded the entire people to make the sacrifices more splendid and better and going to certain *demoi* he persuaded them, too, to participate in sacrificing'. transl. van Bremen 2004, 240–241. This was issued by the Panamareis, the community at the shrine. Inscriptions from other communities include Kallipolis (*I.Stratonikeia* 1401) and the Laodikeis (*I.Stratonikeia* 1401), both in translation in van Bremen 2004, 241–244.
64. As at Magnesia on the Maeander, *I.Magnesia* 16, or Bargylia *I.Iasos* 613, discussed in Wiemer 2009, 117.
65. *I.Stratonikeia* 10 records the epiphany of Zeus; Rivault 2018, with references. Panamara put the *polis* of Stratonikeia in the unique situation of defending (successfully) the rights of *asylia* at both shrines before the council of Tiberius, Tacitus, *Ann.* 4.14, mentioned above.
66. Robert 1984; Parker 2004a.
67. Rostovtzeff 1941, 1056–1057; Millar 2004, 131; Beck and Wiemer 2009a; Meadows 2018.
68. Slater 2007, 44.
69. Festival delegates were often awarded *proxeny*, see Kowalzig 2007a and especially Mack 2015.
70. Ma 2003. Peer polity interaction was developed by Renfrew and Cherry to interpret the Bronze Age networks of the Aegean, Renfrew and Cherry 1986. See also Chiai 2013, although he does not discuss connections in these terms.
71. Ma 2003, 21: 'All these symbolic maps do more than just reflect or participate in peer polity interaction: to a great extent, they *are* peer polity interaction'.
72. See the website *connectedcontests.org*.

73. Hdt. 8.144.2 'the kinship of all Greeks in blood and speech, and the shrines of gods and the sacrifices that we have in common, and the likeness of our way of life' as one reason for never surrendering to the Persians.
74. On *syngeneia* in interstate contexts see esp. Bresson and Debord 1985; Jones 1999; Curty 2001; Patterson 2010.
75. Wilhelm 1947; Williamson 2016 and 2021.
76. The *syngeneia* of Pormounos appears at the end of the fourth or early third century, Robert 1945.
77. *I.Stratonikeia* 25 (Alinda), 27 (Nysa), 33–35 (unnamed), 37 (Miletos), probably all from the imperial period; see also Williamson 2013.
78. *SEG* 38.1476, discussed in Ma 2003, 9–12; also Chiai 2013, 86. Kytenion appealed to Xanthos for aid in building their city wall, elaborately spelling out their common ancestry and arguing that by supporting their endeavour the Xanthians would not only be doing them a favour, but also to all the Aetolians, Dorians, and even King Ptolemaios. Although Xanthos ultimately gave them very little, the inscription, inscribed in the temple of Leto at Xanthos, bears witness to their acknowledgment of the obligation.
79. As the fluidity in the Lindian 'chronicle' or the Parian marble shows; Dillery 2005; also Patterson 2010.
80. *I.Magnesia* 16, and *I.Magnesia* 17–87, nearly 100 cities appeared only in lists. Rigsby 1996, 179–279, nos. 66–131. On the crowned games: Slater and Summa 2006; Thonemann 2007.
81. Rutherford 2013, 271. He also stresses the use of primary sources and the 'mini-library' that the Magnesians carried with them as evident in *I.Magnesia* 17 and 20, 'perhaps they even performed the poetic texts for the benefit of their audience'.
82. *Syngeneia* appears in at least thirteen cases. Besides Antioch in Persis and Antioch in Pisidia, both of which were partly colonised by Magnesians, the relations with Knidos, Gonnoi, the Phokian League, Megalopolis, Messene, Epidamnos, Chalkis, Mytilene, Syracuse, and two unknown cities, are framed as *syngeneia*, and in many cases *philoi* as well.
83. Wiemer 2009, 87.
84. Chaniotis 1999; Carless Unwin 2017, 169–188. On 'intentional history', Gehrke 2001.
85. Mann 2018, 306–307.
86. Sjoukje Kamphorst addresses the language of connectivity in her doctoral thesis, 'Carving communities in stone. Inscriptions as a medium of Hellenistic interactions', at the University of Groningen. On the use of kinship and *philia* to express relations within the city, see van Nijf 2013b.
87. *I.Priene* 108.
88. On theoric appointments see Rutherford 2013, 156–173, and Perlman 2000, 37–62 on the duties and appointment of the *theorodokoi.*
89. Dem. 82, cited in Rutherford 2013, 164, n. 53. In the same note, Rutherford refers to an unpublished oracle at Dodona, in which the inquirer asks whether he should act as *theoros.*
90. Rutherford 2013, 164, and 257–258, n. 35, referring to Zenon Papyrus 59341(a), with further references. The business matter pertained to the reimbursement of 250 drachmas that he had paid for wine at a festival at nearby Kypranda.
91. Perlman 2000, 45.
92. van Nijf and Williamson 2016; Mann 2018.
93. van Nijf 2010, 2011.
94. Jones 1964, 10 argues that if a place like Jerusalem had a *gymnasion* (II Maccabees 4:12–15), then most other cities would have had one too.

95. Remarked by Ziebarth 1909, 104. My thanks to Onno van Nijf for relaying this reference.
96. Mann 2017 gives a good discussion of the dismissal of sport as a religious event in scholarly tradition; also Murray 2014, van Nijf 2013a.
97. Slater and Summa 2006; Slater 2015; Mann 2018.
98. *IvP* 1.246, found near the harbour town Elaia.
99. *iAph* 13.124; *MAMA* VIII.495.
100. Aneziri 2007.
101. Slater 2007, 45.
102. Bonnet 2013, 50–52, with references.
103. Kurke 1993; Aloni 2012.
104. The online database of performers, http://www.connectedcontests.org, is largely based on such lists.
105. As recorded in Eusebius' *Chronographia*, listed in Christesen 2007, 30–31, 388–389.
106. Festival coinage: Nollé and Nollé 1994; Meadows 2018.
107. Rüpke 2015, 351: 'Structure and agency constitute each other…'
108. E.g. Slater's 'fringe performers', or tourists like Pausanias, or the two women in Herondas' fourth mime who visit the Asklepieion on Kos, but also merchants, and the crowd at large.
109. Assmann 1991, 22–23.
110. E.g. Parker 2004b; Carbon and Pirenne-Delforge 2012.
111. Slater 2007. See also Parker 2018, who finds enough consistency in ritual norms across the Greek world to be suggestive of 'something we can call Greek religion'.
112. Deshours 2006; Wiemer 2009, 96–101; Gawlinski 2012.
113. Gardner 2004.
114. Appadurai 1996, on cricket in India as a 'hard' form of British culture that was entirely absorbed locally.
115. Caillois 1961; Serres 2011.
116. McCauley and Lawson 2002; McCauley and Lawson 2007. Rüpke 2015, 358 discusses the problems of such cognitive approaches in explaining ritual diversity.
117. Turner 1974b, also Boissevain 2016, 623 on dynamic festivals: '*Communitas* is an ideal status that can be achieved, albeit temporarily, by celebrating together'. On *communitas* and ancient festivals, Kowalzig 2007a, Iddeng 2012.
118. Collins 2004, 33–35; Durkheim 1912; Goffman 1974.
119. *I.Stratonikeia* 10, line 13: ἔτι δὲ ἀναβοών[των] μεγάλῃ τῇ φωνῇ Μέγαν εἶναι Δία Πανάμαρον.
120. Barabási and Pósfai 2016, 110.
121. On emotion and ritual, see esp. Chaniotis 2010.
122. Chwe 2001.
123. Chwe 2001, 22–23, on the role of ceremony as coordinating mechanism in the French Revolution, with references. Also McCauley and Lawson 2002; Collins 2004, 31–36; Whitehouse 2004, on cognitive focus through emotional intensity of ritual.
124. Chwe 2001, 30–33, also noting the obsession with round communal spaces in the French Revolution; also Tomasello 2009.
125. Chwe 2001, 30–35; Huizinga 1949; Williamson in press a.
126. Collins 2004, 42 regards the transformative role of ritual mobilisation in broader terms, also in connection with revolutions.
127. Discussed at length in Williamson in press a.
128. Discussed at length, with examples, in Williamson in press a, also b.

129. Gardner 2004, 174: 'Devoted members of the RMBS attend festivals faithfully and arrange their work and vacation schedules to ensure consistent attendance from festival to festival and from year to year'.
130. Discussed above. See also Chaniotis 1997; Koch Piettre 2018; also Rüpke 2013 and the 'logic of escalation'; Rüpke 2015.
131. Whitehouse 2004; Assmann 1991.
132. Martin 2006; Price 2012.
133. E.g. Whitehouse 2004; Chwe 2001; Collins 2004.
134. On homophily and group-think: McPherson et al. 2001; White 2008. See also Mooring (this volume).
135. White 2008; Mack 2015 on asymmetric proxeny relations.
136. Mackil 2004 on cities in the face of dissolution.

Bibliography

Alcock, S.E. (2002) 'Archaeologies of memory', in *Archaeologies of the Greek Past. Landscape, Monuments, and Memories*, Cambridge: Cambridge University Press, 1–35.

Aloni, A. (2012) 'Epinician and the "*polis*"', *Bulletin of the Institute of Classical Studies of the University of London* 55, 21–37.

Aneziri, S. (2007) 'The organisation of music contests in the Hellenistic period and artists' participation. An attempt at classification', in P.J. Wilson, ed., *The Greek Theatre and Festivals. Documentary Studies*, Oxford: Oxford University Press, 67–84.

Appadurai, A. (1996) 'Playing with modernity. The decolonization of Indian cricket', in A. Appadurai, ed., *Modernity at Large. Cultural Dimensions of Globalization*, Minneapolis: University of Minnesota Press, 89–113.

Assmann, J. (1991) 'Der zweidimensionale Mensch. Das Fest als Medium des kollektiven Gedächtnis', in J. Assmann, ed., *Das Fest und das Heilige. Religöse Kontrapunkte zur Alltagswelt*, Gütersloh: Gütersloher Verlagshaus G. Mohn, 13–30.

——— (1995) 'Collective memory and cultural identity', *New German Critique* 65, 126–133.

Barabási, A.-L. and M. Pósfai (2016) *Network Science*, Cambridge: Cambridge University Press.

Beck, H. and H.-U. Wiemer (2009a) 'Feiern und Erinnern. Eine Einleitung', in *Feiern und Erinnern. Geschichtsbilder im Spiegel antiker Feste*, Studien zur Alten Geschichte, Berlin: Verlag Antike, 9–54.

——— eds. (2009b) *Feiern und Erinnern. Geschichtsbilder im Spiegel antiker Feste*, Berlin: Verlag Antike.

Bellah, R., R. Madsen, W. Sullivan, A. Swidler and S. Tipton (1985) *Habits of the Heart. Individualism and Commitment in American Life*, New York: Harper & Row.

Belloni, G.G. (1984) ''Asylia' en santuari greci dell'Asia Minore al tempo di Tiberio', *Contributi dell'Istituto di Storia Antica dell'Università del Sacro Cuoro* 10, 164–180.

Boissevain, J. (2016) 'The dynamic festival. Ritual, regulation and play in changing times', *Ethnos. Journal of Anthropology* 81, 617–630.

Bonanno, D. (2013) 'Memory lost, memory regained. Considerations on the recovery of sacred texts in Messenia and in Biblical Israel. A comparison', in N. Cusumano, V. Gasparini, A. Mastrocinque and J. Rüpke, eds., *Memory and Religious Experience in the Greco-Roman World*, Stuttgart: Franz Steiner Verlag, 63–80.

Bonnet, C. (2013) 'The religious life in Hellenistic Phoenicia: 'Middle Ground' and new agencies', in J. Rüpke, ed., *The Individual in the Religions of the Ancient Mediterranean*, Oxford: Oxford University Press, 41–57.

Boyer, P. (1990) *Tradition as Truth and Communication. A Cognitive Description of Traditional Discourse*, Cambridge: Cambridge University Press.

Bremmer, J. (2007) 'Greek normative animal sacrifice', in D. Ogden, ed., *A Companion to Greek Religion*, Malden, MA: Blackwell Pub., 132–144.

Bresson, A. and P. Debord (1985) 'Syngeneia', *REA* 87, 191–211.

Brughmans, T., A. Collar and F. Coward (2016) *The Connected Past. Challenges to Network Studies in Archaeology and History*, Oxford: Oxford University Press.

Bruit Zaidman, L. and P. Schmitt Pantel (1992) *Religion in the Ancient Greek City*, Cambridge: Cambridge University Press.

Caillois, R. (1961) *Man, Play, and Games*, New York: The Free Press of Glencoe.

Carbon, J.-M. (2015) 'Ritual cycles. Calendars and festivals', in E. Eidinow and J. Kindt, eds., *The Oxford Handbook of Ancient Greek Religion*, Oxford: Oxford University Press, 537–550.

Carbon, J.-M. and V. Pirenne-Delforge (2012) 'Beyond Greek "sacred laws"', *Kernos* 25, 163–182.

Carless Unwin, N. (2017) *Caria and Crete in Antiquity. Cultural Interaction between Anatolia and the Aegean*, Cambridge: Cambridge University Press.

Chaniotis, A. (1988) 'Als die Diplomaten noch tanzten und sangen. Zu zwei Dekreten kretischer Städte in Mylasa', *Zeitschrift für Papyrologie und Epigraphik* 71, 154–156.

_______ (1991) 'Gedenktage der Griechen. Ihre Bedeutung für das Geschichtsbewußtsein griechischer Poleis', in J. Assmann, ed., *Das Fest und das Heilige. Religiöse Kontrapunkte zur Alltagswelt*, Gütersloh: Gütersloher Verlagshaus G. Mohn, 123–145.

_______ (1995) 'Sich selbst feiern? Städtische Feste des Hellenismus im Spannungsfeld von Religion und Politik', in M. Wörrle and P. Zanker, eds., *Stadtbild und Bürgerbild im Hellenismus*. München: Beck, 147–172.

_______ (1997) 'Theatricality beyond the theater. Staging public life in the Hellenistic world', in B. Le Guen, ed., *De la scène aux gradins. Théâtre et représentations dramatiques après Alexandre le Grand*, Toulouse: Press Universitaires du Mirail, 219–259.

_______ (1999) 'Empfängerformular und Urkundenfälschung. Bemerkungen zur Urkundendossier von Magnesia am Mäander', in R.G. Khoury, ed., *Urkunden und Urkundenformular in klassischen Altertum und in den orientalischen Kulturen*, Heidelberg: Universitätsverlage Winter Heidelberg, 51–69.

_______ (2007) 'Theatre rituals', in P.J. Wilson, ed., *The Greek Theatre and Festivals. Documentary Studies*, Oxford: Oxford University Press, 48–66.

_______ (2010) 'Dynamic of emotions and dynamic of rituals. Do emotions change ritual norms?', in C. Brosius and U. Hüsken, eds., *Ritual Matters. Dynamic Dimensions in Practice*, London: Routledge, 208–233.

_______ (2013) 'Processions in Hellenistic cities. Contemporary discourses and ritual dynamics', in R. Alston, O.M. van Nijf and C.G. Williamson, eds., *Cults, Creeds and Contests in the Greek City after the Classical Age*, Leuven: Peeters, 21–47.

Chiai, G.F. (2013) 'The origins and deeds of our gods. Inscriptions and local historical-religious memories in the Hellenistic and Roman world', in N. Cusumano, V. Gasparini, A. Mastrocinque and J. Rüpke, eds., *Memory and Religious Experience in the Greco-Roman World*, Stuttgart: Franz Steiner Verlag, 81–114.

Christesen, P. (2007) *Olympic Victor Lists and Ancient Greek History*, Cambridge: Cambridge University Press.

Chwe, M.S.-Y. (2001) *Rational Ritual. Culture, Coordination, and Common Knowledge*, Princeton: Princeton University Press.

Collins, R. (2004) *Interaction Ritual Chains*, Princeton: Princeton University Press.

Curty, O. (2001) 'Les parentés entre cités chez Polybe, Strabon, Plutarque et Pausanias', in V. Fromentin and S. Gotteland, eds., *Origines gentium*, Paris: de Boccard, 49–56.

Cusumano, N. (2013) 'Memory and religion in the Greek world', in N. Cusumano, V. Gasparini, A. Mastrocinque and J. Rüpke, eds., *Memory and Religious Experience in the Greco-Roman World*, Stuttgart: Franz Steiner Verlag, 17–20.

Deshours, N. (2006) *Les mystères d'Andania. Étude d'Épigraphie et d'histoire religieuse*, Scripta antiqua 16, Pessac: Ausonius.

Dillery, J. (2005) 'Greek sacred history', *The American Journal of Philology* 126, 505–526.

Duffy, M. and G. Waitt (2011) 'Rural festivals and processes of belonging', in C. Gibson and J. Connell, eds., *Festival Places. Revitalising Rural Australia*, Bristol: Channel View, 44–60.

Durkheim, É. (1912) *Les formes Élémentaires de la vie religieuse. Le système totémique en Australie*, Paris: Alcan.

Frost, N. (2016) 'Anthropology and festivals. Festival ecologies', *Ethnos. Journal of Anthropology* 81, 569–583.

Funke, P. and M. Haake, eds. (2013) *Greek Federal States and their Sanctuaries. Identity and Integration*, Stuttgart: Franz Steiner Verlag.

Gardner, R.O. (2004) 'The portable community. Mobility and modernization in bluegrass festival life', *Symbolic Interaction* 27, 155–178.

Gawlinski, L. (2012) *The Sacred Law of Andania. A New Text with Commentary*, Berlin: De Gruyter.

Geertz, C. (1980) *Negara. The Theatre State in Nineteenth-century Bali*, Princeton, NJ: Princeton University Press.

Gehrke, H.-J. (2001) 'Myth, history and collective identity. Uses of the past in ancient Greece and beyond', in N. Luraghi, ed., *The Historian's Craft in the Age of Herodotus*, Oxford: Oxford University Press, 236–313.

Goffman, E. (1974) *Frame Analysis. An Essay on the Organization of Experience*, New York: Harper & Row.

Granovetter, M. (1973) 'The strength of weak ties', *American Journal of Sociology* 78, 1360–1380.

Halbwachs, M. (1925) *Les cadres sociaux de la mémoire*, Paris: Librairie Félix Alcan.

Herrmann, P. (1960) 'Die Inschriften römischer Zeit aus dem Heraion von Samos', *Athenische Mitteilungen* 75, 68–183.

——— (1984) 'Die Selbstdarstellung der Hellenistischen Stadt in den Inschriften. Ideal und Wirklichkeit', in A.G. Kalogeropoulou, ed., *Praktika tou E' Diethnous Synedriou Hellenikes kai Latinikes Epigraphikes, Athena, 3–9 Oktovriou 1982*, Athens: Hypourgeio Politismou kai Epistēmōn; Musée Épigraphique, 109–119.

Higbie, C. (2003) *The Lindian Chronicle and the Greek Creation of their Past*, Oxford: Oxford University Press.

Huizinga, J. (1949) *Homo Ludens. A Study of the Play-Element in Culture*, London: Kegan Paul, Trench, Trubner.

Iddeng, J.W. (2012) 'What is a Graeco-Roman festival? A polythetic approach', in J. Rasmus Brandt and J.W. Iddeng, eds., *Greek and Roman Festivals. Content, Meaning, and Practice*, Oxford: Oxford University Press, 11–37.

Jones, A.H.M. (1964) 'The Hellenistic age', *Past & Present*, 3–22.

Jones, C.P. (1999) *Kinship Diplomacy in the Ancient World, Revealing Antiquity* 12, Cambridge, MA: Harvard University Press.

Kadushin, C. (2012) *Understanding Social Networks. Theories, Concepts, and Findings*, New York: Oxford University Press.

Kindt, J. (2012) *Rethinking Greek Religion*, Cambridge: Cambridge University Press.

Koch Piettre, R. (2018) 'Anthropomorphism, Theatre, Epiphany. From Herodotus to Hellenistic Historians', *Archiv für Religionsgeschichte* 20, 189–209.

Kosmetatou, E. (2012) 'Inventories', in R.S. Bagnall, K. Brodersen, C.B. Champion, A. Erskine and S.R. Huebner, eds., *The Encyclopedia of Ancient History*, Wiley-Blackwell, Wiley Online Library, doi:10.1002/9781444338386.wbeah9781444309130

Knappett, C. (2008) 'The neglected networks of material agency. Artefacts, pictures and texts', in C. Knappett and L. Malafouris, eds., *Material Agency. Towards a Non-anthropocentric Approach*, New York: Springer, 139–156.

Kowalzig, B. (2007a) 'Mapping out communitas. Performances of theoria in their sacred and political context', in J. Elsner and I. Rutherford, eds., *Pilgrimage in Graeco-Roman and Early Christian Antiquity*, Oxford: Oxford University Press, 41–72.

——— (2007b) *Singing for the Gods. Performances of Myth and Ritual in Archaic and Classical Greece*. Oxford Classical Monographs, Oxford: Oxford University Press.

Kurke, L. (1993) 'The economy of *kudos*', in C. Dougherty and L. Kurke, eds., *Cultural Poetics in Archaic Greece. Cult, Performance, Politics*, Cambridge: Cambridge University Press, 131–163.

Kyle, D.G. (2007) *Sport and Spectacle in the Ancient World*, Malden, MA; Oxford: Blackwell Pub.

Lonsdale, S.H. (1993) *Dance and Ritual Play in Greek Religion*, Baltimore: Johns Hopkins University Press.

Ma, J. (2003) 'Peer polity and interaction in the Hellenistic age', *Past and Present* 180, 9–39.

Mack, W. (2015) *Proxeny and Polis. Institutional Networks in the Ancient Greek World*, Oxford: Oxford University Press.

Mackil, E. (2004) 'Wandering cities. Alternatives to catastrophe in the Greek *polis*', *AJA* 108, 493–516.

Malkin, I. (2011) *A Small Greek World. Networks in the Ancient Mediterranean*, New York: Oxford University Press.

Mann, C. (2017) 'Losverfahren in der antiken Agonistik. Überlegungen zum Verhältnis von Religion und Sport', *Gymnasium* 124, 429–448.

——— (2018) 'Cash and crowns. A network approach to Greek athletic prizes', in M. Canevaro, A. Erskine, B. Gray and J. Ober, eds., *Ancient Greek History and Contemporary Social Science*, Edinburgh: Edinburgh University Press, 293–312.

Martin, L.H. (2006) 'Cognitive science, ritual, and the Hellenistic mystery religions', *Religion and Theology* 13, 383–395.

McCauley, R.N. and E.T. Lawson (2002) *Bringing Ritual to Mind. Psychological Foundations of Cultural Forms*, Cambridge: Cambridge University Press.

——— (2007) 'Cognition, religious ritual, and archaeology', in E. Kyriakidis, ed., *The Archaeology of Ritual*, Los Angeles: Cotsen Institute of Archaeology, University of California, 209–254.

McPherson, M., L. Smith-Lovin and J.M. Cook (2001) 'Birds of a feather. Homophily in social networks', *Annual Review of Sociology* 27, 415–444.

Meadows, A.R. (2018) 'The Great Transformation. Civic coin design in the second century BC', in P.P. Iossif, F. de Callatäy and R. Veymiers, eds., *TYPOI. Greek and Roman Coins Seen through their Images. Noble Issuers, Humble Users?*, Liège: Presses Universitaires de Liège, 297–318.

Mikalson, J.D. (2007) 'Gods and athletic games', in O. Palagia and A. Choremi-Spetsieri, eds., *The Panathenaic Games*, Oxford: Oxbow Books, 33–40.

Millar, F. (2004) *Rome, the Greek World, and the East. Vol. 2. Government, Society, and Culture in the Roman Empire*, Chapel Hill: University of North Carolina Press.

Morgan, C. (1990) *Athletes and Oracles. The Transformation of Olympia and Delphi in the Eighth Century B.C.*, Cambridge: Cambridge University Press.

Murray, S.C. (2014) 'The role of religion in Greek sport', in P. Christesen and D.G. Kyle, eds., *A Companion to Sport and Spectacle*, Chichester: Wiley-Blackwell, 309–319.

Mylonopoulos, J. (2006) 'Greek sanctuaries as places of communication through rituals. An archaeological perspective', in E. Stavrianopoulou, ed., *Ritual and Communication in the Graeco-Roman World*, Liège: Centre International d'Étude de la Religion Grecque Antique, 69–110.

Naerebout, F.G. (1997) *Attractive Performances. Ancient Greek Dance: Three Preliminary Studies*, Leiden: Brill.

Nielsen, I. and H. Nielsen (2001) *Meals in a Social Context. Aspects of the Communal Meal in the Hellenistic and Roman World*, Aarhus: Aarhus University Press.

Nollé, J. and M.K. Nollé (1994) *Götter, Städte, Feste. Kleinasiatische Münzen der römischen Kaiserzeit*. Munich: Staatliche Münzsammlung.

Notroff, J., O. Dietrich and K. Schmidt (2014) 'Building monuments, creating communities. Early monumental architecture and pre-pottery Neolithic Göbekli Tepe', in J.F. Osborne, ed., *Approaching Monumentality in Archaeology*, Albany: SUNY Press, 83–105.

Parker, R. (2004a) 'New 'panhellenic' festivals in Hellenistic Greece', in R. Schlesier and U. Zellmann, eds., *Mobility and Travel in the Mediterranean from Antiquity to the Middle Ages*, Münster: LIT, 9–22.

——— (2004b) *'What are Sacred Laws?', in E.M. Harris and L. Rubenstein, The Law and the Courts in Ancient Greece*, London: Duckworth, 57–70.

——— (2018) 'Regionality and Greek ritual norms', *Kernos* 31, 73–81.

Patterson, L.E. (2010) *Kinship Myth in Ancient Greece*, Austin, TX: University of Texas Press.

Perlman, P.J. (2000) *City and Sanctuary in Ancient Greece. The Theorodokia in the Peloponnese*, Göttingen: Vandenhoeck & Ruprecht.

Petridou, G. (2016) *Divine Epiphany in Greek Literature and Culture*, Oxford: Oxford University Press.

Petropoulou, M.-Z. (2006) 'A seleucid settlement on Failaka', *Epigraphica Anatolica* 39, 139–147.

Platt, V. (2011) 'Epiphany and authority in Hellenistic Greece', Chapter 3 in *Facing the Gods. Epiphany and Representation in Graeco-Roman Art, Literature and Religion*, Greek Culture in the Roman World, Cambridge: Cambridge University Press, 124–169.

Price, S. (2012) 'Religious mobility in the Roman Empire', *Journal of Roman Studies* 102, 1–19.

Renfrew, C. and J.F. Cherry (1986) *Peer Polity Interaction and Socio-political Change*, Cambridge: Cambridge University Press.

Rigsby, K.J. (1996) *Asylia. Territorial Inviolability in the Hellenistic World*, Berkeley: University of California Press.

Rivault, J. (2018) 'Les mises en scène de l'épiphanie du nouveau dieu de Stratonicée, Zeus Panamaros', *Pallas. Revue d'Études antiques* 107, 95–116.

Robert, L. (1945) *Le sanctuaire de Sinuri près de Mylasa. Première partie. Les inscriptions grecques*, Paris: De Boccard.

——— (1984) 'Discours d'ouverture', in C. Pelekides, D. Peppa-Delmouzou and B.C. Petrakos, eds., *Πρακτικά του Η' Διεθνούς Συνεδρίου Ελληνικής και Λατινικής Επιγ ραφικής, Αθήνα, 3–9 Οκτωβρίου 1982, τόμος Α*, Athens: Hypourgeio Politismou kai Epistēmōn, 35–45.

Rogers, G.M. (1991) *The Sacred Identity of Ephesos. The Foundation Myths of a Roman City*, London; New York: Routledge.

Rostovtzeff, M.I. (1941) *The Social and Economic History of the Hellenistic World*, Oxford: Clarendon Press.

Rostowzew, M. (1920) ''Επιφάνειαι', *Klio* 16, 203–206.

Roueché, C. and S.M. Sherwin-White (1985) 'Some aspects of the Seleucid Empire. The Greek Inscriptions from Failaka in the Persian Gulf', *Chiron* 15, 13–39.

Rüpke, J. (2013) 'On religious experience that should not happen in sanctuaries', in N. Cusumano, V. Gasparini, A. Mastrocinque and J. Rüpke, eds., *Memory and Religious Experience in the Greco-Roman World*, Stuttgart: Franz Steiner Verlag, 137–143.

——— (2015) 'Religious agency, identity, and communication: Reflections on history and theory of religion', *Religion* 45, 344–366.

Rutherford, I. (2013) *State Pilgrims and Sacred Observers in Ancient Greece. A Study of Theōriā and Theōroi*, Cambridge: Cambridge University Press.

Scheer, T.S. (1993) *Mythische Vorväter. Zur Bedeutung griechischer Heroenmythen im Selbstverständnis kleinasiatischer Städte*, Munich: Maris.

Schmitt Pantel, P. (1990) 'Collective activities and the political in the Greek city', in O. Murray and S. Price, eds., *The Greek City from Homer to Alexander*, Oxford: Clarendon Press, 199–213.

——— (1992) *La cité au banquet. Histoire des repas publics dans les cités grecques*, Rome: École Française de Rome.

Serres, M. (2011) 'Vertigo', in M. Serres and R. Burks, eds., *Variations on the Body*, Minneapolis: University of Minnesota Press, 111–151.

Shaya, J. (2005) 'The Greek temple as museum. The case of the legendary treasure of Athena from Lindos', *AJA* 109, 423–442.

Slater, W.J. (2007) 'Deconstructing festivals', in P.J. Wilson, ed., *The Greek Theatre and Festivals. Documentary Studies*, Oxford: Oxford University Press, 21–47.

——— (2015) 'Victory and bureaucracy. The process of agonistic rewards', *Phoenix* 69, 147–169.

Slater, W.J. and D. Summa (2006) 'Crowns at Magnesia', *GRBS* 46, 275–299.

Smith, M.L. (2019) *Cities. The First 6000 Years*, New York: Viking Press.

Sourvinou-Inwood, C. (1990) 'What is *polis* religion?', in O. Murray and S. Price, eds., *The Greek City from Homer to Alexander*, Oxford: Clarendon Press, 295–322.

Strootman, R. (2018) 'The return of the king. Civic feasting and the entanglement of city and empire in Hellenistic Greece', in J.H. Blok, R. Strootman and F. van den Eijnde, eds., *Feasting and Polis Institutions*, Leiden: Brill, 273–296.

Thonemann, P.J. (2007) 'Magnesia and the Greeks of Asia (*I.Magnesia* 16.16)', *GRBS* 47, 151–160.

Tomasello, M. (2009) *Why We Cooperate*, Cambridge, MA: MIT Press.

Turner, V. (1974a) *Dramas, Fields, and Metaphors. Symbolic Action in Human Society*, Ithaca NY: Cornell University Press.

——— (1974b) 'Pilgrimages as social processes', in V. Turner, ed., *Dramas, Fields, and Metaphors. Symbolic Action in Human Society*, Ithaca, NY: Cornell University Press, 166–230.

van Bremen, R. (2004) 'Leon son of Chrysaor and the religious identity of Stratonikeia in Caria', in S. Colvin, ed., *The Greco-Roman East*, Cambridge: Cambridge University Press, 207–244.

van Nijf, O.M. (2010) 'Athletics, festivals and Greek identity in the Roman East', in J. König, ed., *Greek Athletics*, Edinburgh: Edinburgh University Press, 175–197.

——— (2011) 'Les athlètes et les artistes comme médiateurs culturels dans l'empire romaine', in A. Gangloff, ed., *Médiateurs culturels et politiques dans l'Empire romain: voyages, conflits, identités*, Paris: de Boccard, 71–82.

——— (2013a) 'Ceremonies, athletics and the city. Some remarks on the social imaginary of the Greek city of the Hellenistic period', in E. Stavrianopoulou, ed., *Shifting Social Imaginaries in the Hellenistic Period. Narrations, Practices, and Images*, Leiden: Brill, 311–338.

——— (2013b) 'Affective politics. The emotional regime in the imperial Greek city', in A. Chaniotis and P. Ducrey, eds., *Unveiling Emotions II. Emotions in Greece and Rome. Texts, Images, Material Culture*, Stuttgart: Franz Steiner Verlag, 351–368.

van Nijf, O. M., & C. G. Williamson (2016) 'Connecting the Greeks. Festival networks in the Hellenistic world', in C. Mann, S. Remijsen, & S. Scharff, eds., *Athletics in the Hellenistic World*, Stuttgart: Franz Steiner Verlag, 43–71.

Watts, D.J. and S.H. Strogatz (1998) 'Collective dynamics of 'small-world' networks', *Nature* 393, 440–442.

Wheeler, G. (2004) 'Battlefield epiphanies in ancient Greece. A survey', *Digressus* 4, 1–14.

White, H.C. (2008) *Identity and Control. How Social Formations Emerge*, Princeton: Princeton University Press.

Whitehouse, H. (2004) 'Rites of terror. Emotion, metaphor, and memory in Melanesian initiation cults', in J. Corrigan, ed., *Religion and Emotions*, Oxford: Oxford University Press, 133–148.

Wiemer, H.-U. (2009) 'Neue Feste—neue Geschichtsbilder? Zur Erinnerungsfunktion städtischer Feste im Hellenismus', in H. Beck and H.-U. Wiemer, eds., *Feiern und Erinnern. Geschichtsbilder im Spiegel antiker Feste*, Berlin: Verlag Antike, 83–108.

Wijma, S.M. (2014) *Embracing the Immigrant. The Participation of Metics in Athenian Polis Religion (5th–4th c. BC)*, Stuttgart: Franz Steiner Verlag.

Wilhelm, A. (1947) *Zu den griechischen Inschriften aus dem Heiligtum des karischen Gottes* ΣΙΝΥΡΙ*; Zu einem Beschluss der Athener aus dem Jahre 128 v. Chr; Epigramme aus Side*, Vienna: R.M. Rohrer.

Williamson, C.G. (2013) 'Civic producers at Stratonikeia. The priesthoods of Hekate at Lagina and Zeus at Panamara', in M. Horster and A. Klöckner, eds., *Cities and Priests. Cult Personnel in Asia Minor and the Aegean Islands from the Hellenistic to the Imperial Period*, Religionsgeschichtliche Versuche und Vorarbeiten 64, Berlin: De Gruyter, 209–245.

Williamson, C.G. (2013) 'As god is my witness. Civic oaths in ritual space as a means towards rational cooperation in the Hellenistic *polis*', in R. Alston, O.M. van Nijf and C.G. Williamson, eds., *Cults, Creeds and Identities in the Greek City after the Classical Age*, Leuven: Peeters, 119–174.

Williamson, C.G. (2016) 'A Carian shrine in a Hellenizing world. The sanctuary of Sinuri, near Mylasa', in M. Paz de Hoz Garcia-Bellido, J.P. Sánchez Hernández and C. Molina Valero, eds., *Between Tarhuntas and Zeus Polieus. Cultural Crossroads in Temples and Cults of Graeco-Roman Anatolia*, Leuven: Peeters, 75–100.

——— (2018) 'Filters of light. Greek temple doorways as portals of epiphanies', in S. de Blaauw and E. van Opstall, eds., *Sacred Thresholds. The Door to the Sanctuary in Late Antiquity*, Leiden: Brill, 309–340.

——— (2020) 'Hera on the mountain. Complexities of the cult of Zeus at Panamara under Stratonikeia', in H. Bumke, J. Breder and I. Kaiser, eds., *Kulte im Kult. Sakrale Strukturen extraurbaner Heiligtümer*, Rahden: Verlag Marie Leidorf, 143–168.

——— (2021) *Urban Rituals in Sacred Landscapes in Hellenistic Asia Minor*, Leiden: Brill.

——— (in press a) 'Sacred circles. Enclosed sanctuaries as urban repeaters in festival networks', in Z. Newby, ed., *The Material Dynamics of Festivals in the Graeco-RomanEast*, Oxford: Oxford University Press.

——— (in press b) 'Crowned heights. Sacred mountains and developing political landscapes in Asia Minor', in B. Vergnaud and N. Carless Unwin, eds., *Anatolian Landscapes. Inhabiting Western Anatolia in Antiquity*, Istanbul: Institut Français d'Études Anatoliennes.

Winkler, J.J. and F.I. Zeitlin (1990) *Nothing to do with Dionysos? Athenian Drama in its Social Context*, Princeton: Princeton University Press.

Ziebarth, E. (1909) *Aus dem griechischen Schulwesen*, Leipzig: B.G. Teubner.

6 A network analysis of the Mithraic tauroctony: local innovation and diversity in Roman Mithras-worship

Kevin Stoba

Introduction

In this chapter, I use network analysis paired with a qualitative approach to assess iconographical differences within the tauroctony (see Figure 6.1), the cult image of Roman Mithras-worship in which the deity stabs a bull surrounded by several other elements. Echoing recent trends in Mithraic scholarship, I emphasise local variation within the cult and argue that the iconography produced within local variants can be used to trace their development. This is because these variants emerged through interaction between Mithraic groups, through established sociocultural and professional networks, and with other local (religious) forms. Analysis of the multiple, overlapping processes which resulted in these interactions is key to understanding the Mithraic communities in which they occurred, and there are some broader patterns which can be observed. I argue that future scholarship needs to foreground these processes and patterns, and firmly move away from attempts to universalise Mithraic evidence, which I contend cannot be justified.

Mithras was worshipped in the Roman world from the late 1st-early 2nd centuries CE with early activity in Rome and along the Rhine and Danube rivers.[1] Mithraic activity spread profusely around the empire in the following two centuries, to the extent that there are now over 600 surviving tauroctonies (many of which are fragmentary) with findspots ranging from Egypt to Britain and from Spain to Syria. Contemporary literary evidence of the cult is rare, and often the product of heavily biased Christian authors.[2] This cannot be taken at face value. The rich iconography, some of which is explored in this chapter, is abundant with ambiguous (and often demonstrably multivalent) meanings.

Archaeology reveals that there were some broad consistencies across Mithraic sites. Temples (termed in modern literature 'mithraea') often had several features designed to mimic a cave.[3] They usually featured two parallel benches which were used for communal dining, where it was very common across the empire for Mithras-worshippers to eat chicken.[4] Temples were also usually focused on a tauroctony, placed at the end of a central aisle clearly highlighting its ritual significance. Tauroctonies could come in a range of

DOI: 10.4324/9780429429217-8

shapes and sizes, and some were evidently for personal use.[5] The presence of a tauroctony seems to have been essential for Mithraic cult activity.

Despite these and other consistencies, Mithraic activity was not uniform: there is much variation within the surviving evidence. This chapter is devoted to exploring iconographic variation, which occurs alongside diversity within the archaeological and epigraphic evidence of Mithraic cult practices around the Roman world. Variations include differences in ritual practice, the social composition of memberships, leadership structures, inclusiveness, interactivity between discrete Mithraic groups, and motivations for participating (cultic vs professional). These differences, which are indeed often reflected in Mithraic iconography, attest to the fundamental diversity of Roman Mithras-worship over space and time.

In this chapter, I analyse a database of 528 tauroctonies, 444 of which are recorded in Vermaseren's *CIMRM*, which remains an extremely useful resource more than 60 years after its publication, despite its flaws. The remainder were either unknown to Vermaseren or discovered more recently.[6] The collection includes all types, sizes, shapes, and materials of tauroctony, from tiny personal gems to large relief sculptures on stone. The tauroctonies were recorded in the rows of the database, while the columns constituted 105 features of tauroctonies (iconographic and physical). Each tauroctony has a score of 0 or 1 in each column depending on the presence or absence of the feature recorded therein.[7] These 105 features were selected qualitatively. As well as distinct iconographic features (e.g. the inclusion of a zodiac, a lizard, or a swastika) and physical features (e.g. an aedicular relief or a reversible tauroctony), the columns mostly represent the multiple possible variants of each of the major iconographic details (which are outlined below). For example, a raven may be depicted on Mithras' cloak, on the border of the scene, on a rock perch, on a plant perch, or flying next to Mithras' head; so too might the torchbearers be depicted in multiple ways, and so on. Ties between tauroctonies in the networks in this chapter are based on shared patterns of affiliation to these features, although some tweaking was required to represent the significance of rarer features (see below). Two tauroctonies which share a strong tie are therefore very similar in terms of these 105 features.

Crucial to the first study in this chapter ('Tauroctony Network 1') is the separate attributes database. This records the dates, materials, and findspots of each tauroctony, along with confidence measures for dates and findspots.[8] I ask: how likely is it that tauroctonies share ties with other tauroctonies from the same region, province, or local area? I use the geographical information in the attributes database to measure geographical homophily in the network ties. The evidence suggests that patterns at provincial or regional levels are not consistent, which means that regional/provincial typologies are problematic.

Following this, I shift to a microscopic analysis of one feature of the tauroctony depicted in Figure 6.1: Luna, the personified moon, is shown looking away from the bull-stabbing scene, as she is in a small number of other

Figure 6.1 A tauroctony from the Roman settlement of Aequiculi (CIMRM 650), now on display at the Museum of the Baths of Diocletian in Rome; late 2nd century CE.

Source: Image taken from https://commons.wikimedia.org/wiki/File:Tauroctony_Apronianus_Terme.jpg.

tauroctonies. The relationship between this and other features is examined statistically to elucidate how, where, and why this motif spread. Finally, we look at visualisations of 'Tauroctony Network 2', which only represents the strongest ties between tauroctonies (i.e. those most similar to each other). These visualisations confirm the impact of local innovation, but also its potential to spread, and the multiple interactions between different influences which underpin and/or co-occur with local innovation.

The tauroctony: the apex of Mithras-worship

More than 600 tauroctonies survived antiquity, and they were in some way fundamental to Mithras-worship. In all tauroctonies, Mithras restrains a flailing bull and stabs a dagger into its shoulder or neck.[9] They are therefore identifiably representative of the same essential narrative and visual idiom, the recognition of which was possibly a kind of 'secret code' which legitimised membership in a Mithraic group.[10] Nevertheless, they are all distinct from one another in numerous ways.

This study begins with the 2nd century CE tauroctony in white marble seen in Figure 6.1. It is from the settlement of Aequiculi, a small town in the Abruzzo mountains c. 100 km from Rome, and below the tauroctony is a dedicatory inscription:

> *Apronianus rei p(ublicae) ark(arius) sua pecunia fecit.*
>
> ('Apronianus, municipal treasurer, set this up at his own expense'.)[11]

The same municipal treasurer also dedicated two other inscriptions to Mithras at Aequiculi, probably at the same mithraeum. The first indicates that Apronianus paid for the mithraeum to be restored, calling the temple 'the cave of the Unconquered Sun Mithras' (*spelaeum Solis Invicti Mithrae*). The renovation was done 'for the health of the Mithraic community and the people' (*pro salute ordinis et populi*).[12] Apparently, the mithraeum had 'collapsed of old age' (*vetustate [conlap]sum*) and Apronianus restored it 'with the permission of the Mithraic community' (*permissu ordinis*).

Apronianus' third inscription allows more insight:

> *Invicto Mithrae/Apronianus arkar(ius)/rei p(ublicae) d(onum) d(edit)./ Dedicatum VII kal(endas) Iul(ias)/Maximo et Orfito co(n)s(ulibus)/per C. Arennium Rea/tinum patrem.*
>
> ('Apronianus, municipal treasurer, gave a gift to the Unconquered Mithras. Dedicated on the 7th day before the kalends of Iulius [25th June] on the year Maximus and Orfitus were consuls [CE 172] through the father C. Arennius Reatinus'.)[13]

The dating coincides with the summer solstice, which was celebrated on 24th June in this period.[14] It is possible that Apronianus' tauroctony and two further inscriptions were contemporary (in CE 172), and thus that a new tauroctony and renovation of the temple were completed to co-occur with the solstice. A direct parallel is found on an inscription from a 3rd century CE Mithraic community in Virunum in Noricum.[15] Here, the mithraeum was also renovated after the previous one had collapsed due to old age, and this renovation also coincided with the summer solstice. For comparison, another Mithraic group at Tienen (modern Belgium) also held a large feast to coincide with the summer solstice in the late 3rd century CE.[16] It seems likely that the Mithraic communities at Aequiculi, Tienen, and Virunum adhered to a broader Mithraic trend of associating the solstice with Mithras-worship.

Apronianus himself, meanwhile, also dedicated to Isis and Serapis at Aequiculi, while he was a public slave (*Aequiculorum servus*),[17] which he was no longer in the Mithraic inscriptions. It is well-established that it was not necessary to maintain exclusivity between different cultic memberships,[18] and Apronianus was doing nothing controversial in dedicating to Mithras as well as Isis and Serapis. Indeed, there is some evidence from 4th century CE Ostia of even deeper confluence between the cults of Mithras, Isis, and Cybele, including shared sacred spaces.[19]

From these inscriptions, we see that Mithras-worship at Aequiculi adhered to a principle also evident at Virunum and Tienen and shared some interaction with other local cults. The character of Mithras-worship at Aequiculi would have reflected these influences and other factors, such as the (e.g. professional and social) backgrounds of cult participants. It is the central thesis of this chapter that analysis of the variable iconography produced by Mithraic groups can often help us discern these factors. What, then, can we learn from Apronianus' tauroctony itself?

The stabbing of the bull in Figure 6.1 unfolds before celestial observers: Sol holding a whip, and Luna with her crescent, in the top left and right corners, respectively; the personified sun and moon. Attending either side of Mithras are Cautes and Cautopates, the deity's two aides, who usually carry torches pointing upwards or downwards, respectively. They are generally interpreted as mirroring the sunrise and sunset, the summer and winter solstices, or the vernal and autumnal equinoxes: scholarship broadly agrees that they had a solar significance rooted in their oppositionality.[20] In Apronianus' tauroctony, Cautes is paired with Sol (torch up: sunrise, summer solstice, vernal equinox, etc.): both are on the left side, and both look at Mithras. Cautopates is paired with Luna (torch down: sunset, winter solstice, autumnal equinox, etc.): both are on the right side, and both avert their eyes from the scene.

Around the bull in Figure 6.1 are some animals.[21] A snake and a dog lap the blood from the bull's wound, while a scorpion pinches its testicles. A raven is perched on Mithras' cape, keenly observing the bull's suffering, but blocking eye contact between Mithras and Sol. The bull's tail transforms into grain, signifying that the bull's suffering in some way sustains life. Peripheral to the central scene are also six smaller scenes depicting Mithras engaged in a number of interactions with the bull and with Sol, which collectively attest to some kind of mythic narrative.[22] The presence of a narrative is, according to the cognitive scientific approach, key to the transmission of the scene around the Roman world.[23]

In the absence of any ancient source describing the function or context of the tauroctony, scholars have speculated on its significance for centuries. Interpretations have ranged from attempts to connect the elements of the tauroctony to Persian precedents,[24] or astrological symbols,[25] and readings that the bull-slaying act released creative forces.[26] My aim in this chapter, however, is not to interpret the tauroctony. Instead, it is to offer a strategic approach to understanding its variability, and to explore the significance of doing so. Although all tauroctonies conform to the crucial feature of Mithras stabbing (or about to stab) a bull, every detail (including the bull-stabbing itself) is prone to considerable variation. The features of the tauroctony vary in presence/absence, but also in their position, design, or interaction with other features.

Tauroctonies appear as reliefs, statues, altars, paintings, vases, engravings on gems, and images on coins. Of the reliefs, there are marble, limestone, and sandstone examples, along with much larger rock-cut tauroctonies. Reliefs may be circular, rectangular, or aedicular (shrine-shaped). There are grand marble statues and small bronze statuettes. The smallest gems are <3 cm^2 and the largest rock-cut reliefs can be >150 cm^2. This variety attests to differences in function: smaller tauroctonies were portable and may have served as items of personal worship,[27] while the rock-cut relief at Močići was visible even to passers-by, in contrast to most other reliefs, which were positioned internally within mithraea.[28] Apronianus' tauroctony is quite large (height = 81 cm, width = 104 cm, depth = 8 cm), and would have been visible to a wider group from its probable position at the end of the central aisle inside a mithraeum.

We only have the space to explore one variation in Mithraic iconography: depictions of Luna. In Apronianus' tauroctony, the bull-stabbing scene is framed by Sol and Luna. Luna appears on the right and averts her gaze from the scene. Apronianus' relief is one of just 42 tauroctonies (of the 528 in the database) featuring Luna's averted gaze, 23 of which are from Italy. Meanwhile, 110 tauroctonies (mostly concentrated along the Danube) have Sol and Luna both gazing out at the viewer. Twenty-nine more (again mostly concentrated along the Danube) have Luna gazing at the viewer while Sol looks at Mithras. Luna instead looks directly at the tauroctony in a further 31 examples, with no discernible geographical tendency. Sol rides a *quadriga* in 25 tauroctonies, and Luna accompanies him on a *biga* in the majority of these. Most unusually, Luna is on the left while Sol is on the right in 14 depictions from all around the empire.

Modern scholarship on diversity within Mithras-worship

Through most of the 20th century, scholars often downplayed diversity within tauroctonies, or even dismissed it as superficial. Iconographic variations were considered primarily the result of inexperienced craftsmen failing to reproduce a 'standard' tauroctony or ignorance of this standard among (usually provincial) worshippers. Into this bracket, we can put Martin Vermaseren's frequent description in his CIMRM of 'primitive' monuments which do not conform to classical norms. So too Roger Beck's dismissal of worshippers who produced scenes featuring two downward-pointing torchbearers as 'ignorant' of Mithras-worship and having 'little or no company among the worshippers of Mithras'.[29] Ernest Will also described regional variations of Mithraic iconography as 'barbaric work' with 'awkward proportions'.[30]

In recent decades, however, scholarship has begun to foreground Mithraic diversity. Studies have found people worshipping Mithras from markedly different backgrounds at different sites,[31] performing different rituals,[32] eating different communal meals,[33] writing different inscriptions,[34] worshipping in different environments,[35] building different types of temples,[36] and, of course, producing different iconography.[37] Mithras-worshippers thus had different ambitions for, different experiences of, and perhaps even different understandings of the salient features of their cults. These differences were fundamental to the nature of Mithras-worship as it manifested in each place: no two Mithraic cults were exactly the same (even if some were obviously similar), and it is important to reflect on the suitability of the aggregation of evidence between them on a case-by-case basis.

The tide has now turned convincingly away from the universalisation of Mithraic evidence. Clauss claims that 'often only the language and imagery remain constant' while everything else is subject to variation.[38] Beck now rejects the view that Mithras-worship had 'anything like a systematic and coherent body of teaching',[39] despite adhering to that perspective in the past.[40] Martin writes that it is 'unlikely that any standardized accounts of Mithraic myth or teachings circulated widely'.[41] Szabó states that 'the interior and local dynamic of the cult [had] specific characteristics [which probably] differed from community to community'.[42]

Richard Gordon argues that leaders of small religious groups were personally critical in defining Mithras-worship in their own locality, building their own norms upon loose traditions around a heroic myth (often using local tropes for the sake of simplicity). He declares that it was 'every [small-group religious leader] for himself' in the case of Mithras-worship; all Mithraic cults would have been distinct.[43] Local innovation was certainly key in Mithraic diversity, but local versions of the cult could also emerge through distinct patterns of adherence to established trends (see below).

For Jaime Alvar, diversity occurs as minor interpretative deviations from a standardised form of Mithras-worship. These deviations develop through peripheral adaptation to the local environment. In this view, 'universalist tendencies were compatible with local particularities' within the worship of Mithras, Isis, and Cybele.[44] Mithraic variation simply shows the creative thinking of Mithraic leaders and the flexibility of the central core of Mithras-worship to mould its presentation to the (spiritual or social) needs of different communities. Dészpa,[45] writing on the worship of Jupiter Dolichenus, also suggests that the tendency to look for the particular (rather than the universal) is postmodernist, while the truth lies in their combination. In terms of iconography, a recent approach has been to emphasise provincial artistic regionalisation as the personalisation of classical norms to 'make them more potent, hybrid signifiers for peoples with their own imagistic traditions'.[46]

Several studies have attempted to make typologies of Mithraic iconographic variation, usually at regional or provincial levels. Such typologies usually group together depictions of the tauroctony from a defined area, observe which iconographic features are most common among them, and log these features as indicative of 'types' from that particular area with varying degrees of inter- or intra-regional factors. Leroy Campbell's typology was the earliest and most influential. His *Typology of Mithraic Tauroctonies*

(1954–1955) split images of the tauroctony into eight 'types' according to the varying shape/layout of the overall scene. A further eight 'subtypes' (independent of his types) were split according to divergences in Mithras' interaction with the bull. These types/subtypes were largely geographically based, but Campbell acknowledged anomalies and 'conflations' of multiple types/subtypes. His *Mithraic Iconography and Ideology* (1968) then applied his original typology to the massive variation of other features of the tauroctony, often leading to impractical lists of extant variants *within* his types/subtypes.[47]

Campbell admitted that he 'blazed a path' through variation in Mithraic iconography,[48] hoping that someone in the future would take the reins and perform more nuanced analysis. Few have attempted this. Campbell's work was criticised for its unsuitability for the data and its dated reliance on attempts to link most iconographic features to Persian precedents.[49] Despite this, Campbell's approach underlies many discussions of Mithraic iconographic regionalisation.[50] Indeed, its methodology is mirrored in the three other major typologies of Mithraic iconography.[51] More specific regional or provincial typologies of Mithraic iconography have also been produced for Dalmatia,[52] the Near East,[53] southeast Europe,[54] Dacia,[55] and Pannonia.[56]

In practice, however, we find that adherence to any typology (regional or otherwise) is limited. Sicoe observed, for example, that a feature of the Dacian 'type' of tauroctony is a depiction of the torchbearers holding two torches each.[57] Indeed, of the 20 cases of this feature, 16 are from Dacia (and a large group from Sarmizegetusa in particular, where it probably originated). In contrast, there are another 84 tauroctonies from Dacia which *do not* have this feature. In the following section, I use network analysis to demonstrate that this is not a unique occurrence but instead a much more fundamental pattern in our evidence. It follows that we should focus on the iconographic strains themselves; geographical data should inform our analysis, not define it.

A network approach: reasons and methods

One of the major strengths of network analysis is that it can treat all evidence equally. Within this study, there is no assumption of a standardised tauroctony or a deviation from it. Each case enters the database and analysis on equal footing, generating its own unique profile of influences and informing the bigger picture of the overall distribution of tauroctonies. What emerges are indications that tauroctonies generally do not consistently tend towards regional or provincial types and are instead much more significantly likely to be similar to their 'neighbours' from the same city or settlement. This likelihood, which I term 'localisation at site level', attests to a complex range of multilateral influences and interactions (Mithraic and other), tendencies towards and against interactivity, and levels of conservatism and innovation.

'Localisation at site level' may involve adherence to a particular set of trends which are unique and not found at any other site (i.e. local innovation). However, it is much more likely to involve adherence to a *unique pattern* of trends (i.e. its tauroctonies conform in an idiosyncratic way to trends which may be empire-wide, region-specific, or even site-specific). When regional trends are abundant in any given case, evidence is still often 'localised at site level' (and not regional level) because the pattern of particular regional trends at any one site is not matched by comparable patterns at other sites. It may also involve adherence to supraregional or extraregional trends. Usually, we find combinations of these influences. In this study, I have created 'Tauroctony Network 1' to demonstrate 'localisation at site level'. The process of its creation is outlined below.

Within our database of 528 tauroctonies logging 105 features, some features are obviously more common than others. For example, Mithras usually wears superficially 'Persian' clothes,[58] including a tunic, *anaxyrides* trousers, Phrygian cap, and flying cloak, and so these four features occur frequently in the dataset. As such, the co-occurrence of these features in two tauroctonies is not especially significant: these two tauroctonies are not necessarily linked in any meaningful way, because almost every tauroctony also ticks those boxes.

Conversely, depictions of Mithras nude are extremely rare,[59] so it is much more significant for two tauroctonies to share this feature. Aiming to evenly weight the impact of (variation within) each feature, all 105 features were grouped into 16 broader categories and then hierarchically clustered.[60] This means, for example, that all the features relating to Mithras' clothes were grouped together into one such category and tauroctonies were assigned to clusters based on their patterns within these features. All the tauroctonies which featured Mithras' tunic, *anaxyrides*, Phrygian cap, and flying cloak would belong to one cluster, while all tauroctonies featuring a nude Mithras—and hence no other clothes—would belong to another. Tauroctonies featuring Mithras' regular clothes but with a crown instead of a cap would belong to a further cluster. A tauroctony would then also be assigned to other clusters for its patterns of features relating to the torchbearers, to the animals, and so on.[61] A single tauroctony would thus belong to 16 different clusters (one for each category). We can confidently call two tauroctonies which shared membership in multiple clusters very similar.

Within the attributes database, each of the 528 tauroctonies was given a score for 'Completeness' from 1 to 3, with 3 representing a relatively intact tauroctony and 1 representing a tauroctony which is either extremely fragmentary or whose image I was unable to source. In a tauroctony with a score of 3 for 'Completeness', we can be confident that the absence of a feature within the database probably corresponds to a genuine absence of the feature in the original tauroctony—that the relevant part of the tauroctony has not simply perished, as may be the case with a score of 1. In Tauroctony Network 1, I focused only on the 176 tauroctonies with a score of 3 for 'Completeness'. The geographical distribution of 165 of these tauroctonies can be seen in Figure 6.2, and the remaining 11 are of unknown provenance. This widespread distribution is a reasonably accurate reflection of the overall distribution of Mithraic finds.

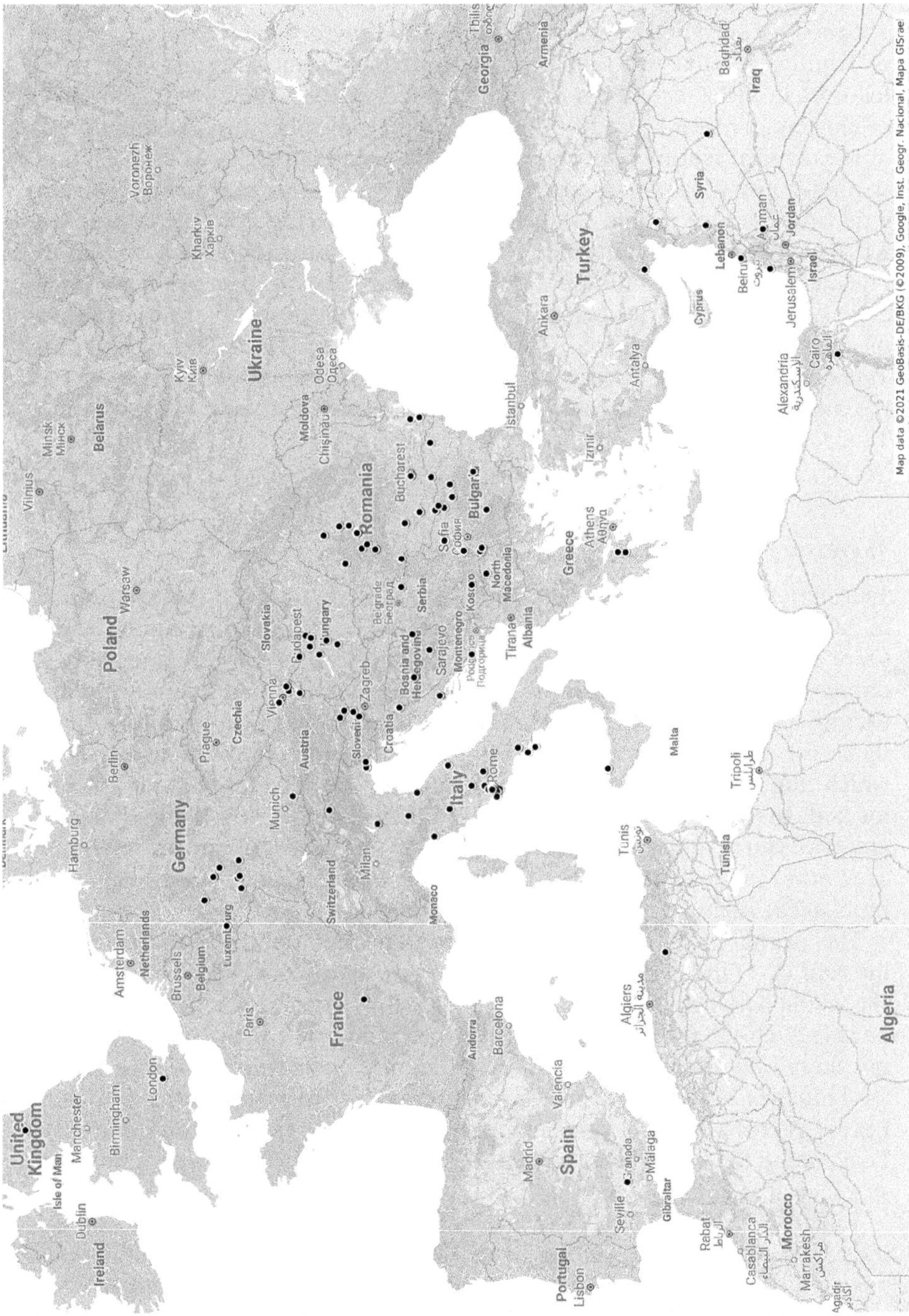

Figure 6.2 Distribution of tauroctonies with a 'Completeness' score of 3.[62]

I then built a network of tauroctonies, where ties between two tauroctonies depended on how many cluster memberships they shared.[63] Ties were assigned a continuous value between 0 and 1, where, for example, a score of 1 indicated that two tauroctonies shared co-membership of clusters of features in all categories, a score of 0.5 indicated co-membership in 50% of clusters, while a score of 0 indicated no co-membership. As mentioned above, several features are extremely widespread across tauroctonies, and so almost every tauroctony shared a tie with every other tauroctony (i.e. had a tie valued above 0), and most shared several ties. The average tie strength was 0.484. To focus on tauroctonies which were more significantly similar to one another, these results were dichotomised. Dichotomisation is the process of replacing all ties below a certain value with 0, and all ties equal to or above that value with 1. I dichotomised the network twice: firstly, into Tauroctony Network 1, which recorded all relationships equal to or above the average tie score in the original network (≥0.484), and secondly into Tauroctony Network 2, recording only the strongest relationships (≥0.85). We return to Tauroctony Network 2 later.

Tauroctony Network 1: localisation

The analysis of Tauroctony Network 1 is based on homophily. Homophily, in network terms, is the tendency of nodes to be tied to other nodes which share some significant *attribute*. As detailed above, the separate attributes database for this study contains geographical information about each tauroctony (among other details). I recorded three levels of geographical specificity for each tauroctony: the site at which it was found, the province in which that site is located, and the broader region in which that province is located. In three separate calculations, I then explored how likely it was for any two tauroctonies from the same region, province, or site to be tied to each other in the network. The results show that tendencies towards homophily at the regional or provincial level are low.

The primary calculation used to measure homophily is the external-internal (E-I) index.[64] This calculation gives a value between –1 and 1 for the whole network based on the tendency for ties to be formed with other nodes which share the same attribute (internal ties) or with nodes which do not share that attribute (external ties). The closer the value to –1, the higher the tendency towards internal ties; a value closer to 1 indicates a tendency towards external ties. A value of 0 indicates that the ties are divided evenly. I calculated the E-I index three times for this network for each geographical specificity: region, province, and site.

Table 6.1 shows that, as the attributes become more localised, the E-I index moves significantly closer to –1. The E-I index of at the level of regions has a value of –0.164, which is not indicative of a significant tendency for tauroctonies from the same region to be similar to one another. Although there is already weak to moderate tendency towards localisation at the provincial level (–0.291), there is a very large step up at the individual site level

Table 6.1 The E-I index for attributes of geographical localisation

	E-I Index (rescaled)	*p-value*
Regions	−0.164	0.023
Provinces	−0.291	<0.001
Individual sites	−0.641	<0.001

(–0.641). This suggests that this is the level that it is more likely to consistently find similarities between tauroctonies. It is possible thus to confidently describe the features of tauroctonies as localised at the level of individual sites, and, more significantly, it is not possible to do so at the level of regions or provinces.

This does not mean that every site was innovative or that tauroctonies from one site tend to be completely divorced from those from other sites (within its province or beyond). Instead, it means that the patterns of influences seen in each tauroctony were unique, and they were more likely to share lots of patterns with others from the same site rather than further afield. These patterns could include local innovation, but they could also include regional artistic trends (i.e. the kind identified in regional typologies), broader global trends, influence from specific Mithraic networks (e.g. evidence of communication between Mithraic groups), extra-cultic communication (such as military or trade networks), and more.

As detailed above, it is common in modern scholarship to use regional or provincial typologies to represent variation in Mithraic iconography. The results in this section suggest that this model is not concordant with the data, which is distributed in more complex patterns. I argue that these patterns are better understood (and more useful for understanding the wider phenomena of Mithras-worship) if we explore how groups of iconographic features tended to spread together, indicative of distinct strains of Mithras-worship, rather than focusing on the geographic preponderances of individual features. In the following section, I return to Apronianus' tauroctony and Luna's averted gaze to demonstrate this point.

Luna's averted gaze: distinct translocal variants of Mithras-worship

In Figure 6.1, we saw that Sol and Luna appeared in the top left and right corners, respectively, of Apronianus' tauroctony. This positioning is common among tauroctonies, but the motif was generally common in antiquity: it was not a Mithraic invention. It appeared on public buildings, including the east pediment of the Parthenon in Athens and the temple of Jupiter Capitolinus in Rome.[65] It also appeared on personal objects and on some Roman coinage.[66] Especially when it appears alongside a narrative scene, such as the Mithraic tauroctony, the motif is often interpreted as a framing device signifying 'the inherently fluctuating nature of eternal stability'.[67] The sun and the moon are thus detached from the narrative itself, instead representing the importance of the scene in cosmic terms. Nowhere is this more apparent than on a 3rd century CE sarcophagus from Ostia,[68] on which the story of Endymion is depicted and the scene is framed by Sol and Luna. Luna, of course, appears twice in two distinct roles: as a framing device paired with Sol, and as a narrative character with Endymion. Viewers of a tauroctony in the ancient world would likely have been familiar with the motif.

In a Mithraic twist, however, Apronianus' tauroctony plays with the lines between symbol and narrative. Mithras looks back over his shoulder, seemingly to make eye contact with Sol and/or the raven. Sol *engages*, looking back. Luna is not passively framing the scene, inert and eternal. Instead, she is repelled somehow by the stabbing of the bull; she has to look away. By responding to the scene, Sol and Luna become interactive participants within the narrative, but maintain their non-narrative framing role.

We have seen that Luna's averted gaze is shown in 42 tauroctonies. They are not all identical. In Figure 6.3, a white marble relief from Aquileia in northeast Italy, the pairing which occurred in Apronianus' tauroctony between Luna and Cautopates is absent (Cautopates is on Sol's side instead), and Mithras does not appear to be looking at Sol, but Luna is still very clearly avoiding the bull-stabbing scene.

Figure 6.3 A white marble tauroctony from Aquileia showing Luna's averted gaze, CIMRM 736

Source: Image taken from Carole Raddato on Flickr, https://www.flickr.com/photos/carolemage/21314852873/

Multiple modern interpretations of Luna's averted gaze exist. Several scholars have contended that Luna and the bull are fundamentally linked, [69] following Porphyry,[70] in which the moon is a bull and Taurus is the exaltation of the moon. Thus, Luna looks away because the bull's suffering is her own. Faraone instead saw Luna's look away as protective, comparing the tauroctony scene to depictions of the evil eye, which cannot be looked at without incurring harm.[71] Campbell, meanwhile, simply thought that the bull-stabbing was 'in some important aspect unfavourable' to Luna and Cautopates.[72] Conversely, it has also been proposed that Luna is not averting her eyes from the tauroctony itself, but is heading to the right because she is escaping the pursuit of Sol: a separate celestial drama reflecting the waxing and waning moon independent of Mithras' own narrative.[73]

This section does not offer an alternative to or confirmation of any of these readings, although we shall see that there are some suggestions that Sol and Luna are integrated into the narrative of the tauroctony rather than their own independent narrative. This analysis looks at correlations between depictions of Luna and other significant features based on geographical distribution. What emerges are patterns of features which tend to occur together. For the producers (and viewers) of these tauroctonies, these combined features evidently contributed to the definition of what a tauroctony 'should' look like. Usually, any individual tauroctony will be an amalgamation of several such patterns. Network analysis allows us to trace the movements of these patterns, some of which tend to be associated with regions or provinces. I argue that these patterns are the product of distinct strains of Mithras-worship, characterised by unique associations with particular socioeconomic or professional groups, influential and/or innovative people, motivations for participation in the cult, ritual practices, versions of sacred myths, and several other factors.

We have seen that 23 (54.8%) of the 42 tauroctonies featuring Luna's averted gaze are from Italy (including 16, or 38.1%, from Rome). These 23 make up just 20% of the 128 tauroctonies from Italy in the entire database. However, as in Tauroctony Network 1, it is necessary to discount extremely fragmentary tauroctonies from the analysis (because they may count as false negatives: they could lack a particular feature simply because that part of the scene has not survived). Here, I have only analysed tauroctonies with a 'Completeness' score of ≥2.[74] There are 279 tauroctonies which meet this requirement, and 36 (12.9%) of them have Luna's averted gaze. 90 are from Italy, 22 (24.4%) of which have Luna's averted gaze. The other 189 tauroctonies are from outside Italy, and 14 (7.4%) of them have Luna's averted gaze. The feature is very likely to have originated in Italy, based on the high proportion of cases from there and further evidence.[75]

In Table 6.2, we see a cross-tabulation of tauroctonies featuring Luna's averted gaze with another detail of Apronianus' tauroctony, the raven on Mithras' cloak. We see that 21 (23.3%) of the 90 tauroctonies from Italy

Table 6.2 Cross-tabulation of tauroctonies from Italy featuring a raven on Mithras' cloak and Luna's averted gaze

	Luna's gaze not averted	*Luna's gaze averted*	*Total*
No raven on Mithras' cloak	60	9	69
Raven on Mithras' cloak	8	13	21
Total	68	22	90

feature the raven in this position. Because 24.4% of the dataset has Luna's averted gaze, if the distribution were random, we would expect 24.4% of the 21 with the raven on Mithras' cloak to also have Luna's averted gaze (about five cases). Instead, there are 13 cases (61.9% of the 21). The two features are thus significantly linked, and tend to occur together: the value of the phi correlation coefficient shows that their relationship is relatively strong (phi = 0.481 with p <0.001).[76]

Conversely, Table 6.3 shows the same cross-tabulation for tauroctonies found *outside* Italy. 59 (31.2%) feature a raven on Mithras' cloak, and only 14 (7.4%) feature Luna's averted gaze. Just seven (3.7%) have both. In contrast to the data from Italy, the phi correlation coefficient confirms that there is no significant relationship between these two features. We therefore have a feature (Luna's averted gaze) which appears to be slightly more common in Italy than elsewhere, but which, at least in terms of this particular association, has a different pattern inside Italy than outside. The raven is often interpreted as an intermediary between Mithras and Sol,[77] and this is most pointedly the case when the bird is positioned on Mithras' cape, physically perched between Mithras and Sol. Its depiction on the cape is consistent with the involvement of Sol and Luna in the narrative of the tauroctony through their interactions with the scene—but, of course, we only get this confirmation in Italy.

Of the seven tauroctonies from outside Italy which include both features, I draw attention to two from Salona in Dalmatia. The first, a marble tondo possibly dating to the second half of the 3rd century CE (CIMRM 1861),[78] is emblematic of a distinctive Mithraic iconography which developed in Salona,[79] with unique aquatic animals (lobster, crocodile, and dolphin)

Table 6.3 Cross-tabulation of tauroctonies from outside Italy featuring a raven on Mithras' cloak and Luna's averted gaze

	Luna's gaze not averted	*Luna's gaze averted*	*Total*
No raven on Mithras' cloak	123	7	130
Raven on Mithras' cloak	52	7	59
TOTAL	175	14	189

and a circular frame (Figure 6.4). Miletić observed a distinct Italian influence in early Mithras-worship in Salona, observable in CIMRM 1860, 1868, and 1869, brought through the imperial administration.[80] However, greater Pannonian influence subsequently came through ties to the Illyrian customs office starting in the 3rd century. The Mithraic iconography seen in CIMRM 1861 was, according to Silnovic, a product of this Pannonian influence mixed with local Salonitan innovation.[81] The incorporation of the snake/vessel motif into the scene, most other examples of which are from along the Rhine, brings a possible influence from Germania, perhaps via Pannonia.[82] Following the evidence in this chapter, I argue that the depiction of Luna's averted gaze and the raven on Mithras' cloak is a product of further Italian influence with different characteristics than that seen in earlier Salonitan iconography.

Figure 6.4 A sketch of a tondo from Salona with Luna and the raven circled (*CIMRM* 1861).

Source: Author.

The second case from Salona is a portable miniature relief. [83] Because of its small size, it does not include much embellishment, but it was clearly considered necessary to include Luna looking away and the raven on Mithras' cloak, highlighting the significance of these features. The production of miniature tauroctonies for private, portable use appears to have been an invention of the lower Danubian region, and presumably arrived in Salona from there, possibly in the 3rd century CE.[84] This again leaves us with a mix of regional influences in a single tauroctony.

The exact process of transmission is impossible to identify with certainty. Speculatively, it may have been linked to financial roles among imperial slaves. An inscription dedicated to Mithras, discovered in a vineyard at Salona, refers to two imperial slaves named Pamphilius and Fortunatus.[85] This is one of the only sources we have to deduce which social groups worshipped Mithras at Salona, although it is not specifically linked to the two tauroctonies discussed above. Pamphilius worked as a *dispensator* and Fortunatus worked as an *arcarius*, two financial roles within the imperial bureaucracy. We have already seen that Apronianus was also an imperial slave who also worked as an *arcarius*. Atimetus, the dedicator of a marble relief in Rome (CIMRM 546/547), was again an imperial slave who worked as an *actuarius*, another imperial financial position. Clauss counted 23 imperial slaves associated with Mithras-worship,[86] but it is beyond the limits of this study to consider whether it is just coincidence that several of them were linked to this rare way of depicting Luna.

In Table 6.4, the significant correlation between Luna's averted gaze and a range of other features is documented, again split between tauroctonies

Table 6.4 Correlation between Luna's averted gaze and several other features in tauroctonies from Italy and elsewhere

Feature appearing with Luna's averted gaze	*Phi correlation inside Italy*	*p-value*	*Phi correlation outside Italy*	*p-value*
Arched border	0.265	0.012	No correlation	–
Cautes on the left	0.34	0.001	No correlation	–
Cautes cross-legged	0.378	<0.001	No correlation	–
Cautopates cross-legged	0.437	<0.001	No correlation	–
Cautes with a rooster	0.254	0.016	No cases	–
Cautopates with an owl	0.265	0.012	No cases	–
Mithras looking over his shoulder	0.258	0.014	0.223	0.002
Scorpion at the bull's genitals	0.208	0.048	No correlation	–
Snake drinking blood from the bull's wound	0.22	0.037	No correlation	–

from inside and outside Italy. Values are recorded as 'No correlation' when the p value was >0.05 in one of the geographical areas. The pattern is much more clearly established inside Italy than outside. In Italy, if one finds a tauroctony with Luna averting her gaze, it is very likely that some combination of these other features will also be represented, while outside Italy that claim cannot be sustained.

Luna's averted gaze correlates with two features which support a solar/lunar reading of the two torchbearers: depictions of a rooster with Cautes and an owl with Cautopates. The rooster probably signifies Cautes' relationship with the sun, while the owl ties Cautopates to Luna.[87] This pairing is absent in many other tauroctonies—and it was even absent in Figure 6.3, where the torchbearers appeared on the opposite sides. The correlation between Luna's averted gaze and Mithras looking over his shoulder—the only correlation which also seems to occur outside Italy—is, as we have already seen, indicative of the involvement of Sol in the mythical narrative. For individuals in Italy who produced and used tauroctonies featuring Luna's averted gaze, the 'right' image of a tauroctony included a confluence of some of these features.

We can see that this pattern fragmented somewhat when it spread outside Italy. Although we do not have space to demonstrate it here, there was similar patterning and fragmentation at work in the transmission of Mithraic iconography empire-wide. Individual tauroctonies are the products of complex interactions of multiple strains of Mithras-worship, which manifested in (among other ways) distinct combinations of preferred iconographic features. In the final section, we return to the bigger picture of Mithras-worship with some snapshots of the character of the entire dataset.

Tauroctony Network 2: the bigger picture

Tauroctony Network 2 was built from the same data as Tauroctony Network 1: all tauroctonies which have a 'Completeness' score of 3. However, the network was dichotomised at ≥ 0.85, meaning it only shows the strongest ties based on the most features in common (i.e. the tauroctonies which are most similar to one another). This network is, of course, a proxy for a network of Mithras-worshipping individuals or Mithraic communities which we cannot piece together because the evidence is insufficient. The iconography they produced represents them instead. Ties should be read as potential evidence of (indirect or direct) communication, influence, fellowship, or some degree of shared values, the analysis of which permits insight into the transmission of Mithras-worship.

Tauroctony Network 2 has 10 components which contain 89 of the 176 tauroctonies. The remaining 87 tauroctonies were isolates within the network, meaning that they did not share a single tie at this high cut-off point

(≥0.85). This is to be expected: many tauroctonies are too distinct to be regarded as extremely similar to any other. The discussion begins with analysis of the largest component, before turning to the smaller ones.

Figure 6.5 shows the largest component of Tauroctony Network 2, visualising 61 tauroctonies which share at least one very strong tie with another tauroctony within the component. The nodes (each of which is a single tauroctony) are labelled according to their findspots, and each symbol represents the physical type of tauroctony (relief, statue, terracotta, gem, and coin). Most ties with such a high strength are made between tauroctonies of the same medium: statues are more likely to share iconographic features with other statues, reliefs with other reliefs, and so on.

The central group is almost entirely made up of statues from Rome. It is extremely dense and loosely connected to everywhere else in the network. The density of the group suggests a high level of conservatism and conformity among Mithraic statues in Rome, at least in contrast to other tauroctonies. A smaller, less dense group on the left of the network entirely comprises reliefs from Italy and Eastern Europe. In the bottom right corner, some ties emerge between tauroctonies in terracotta or clay.

The number of statues in this component in is predictable, despite the preponderance of reliefs in the database. Of the 528 (fragmentary and complete) tauroctonies in the total database, 432 (81.8%) are reliefs. In this component of 33 tauroctonies, only 14 are reliefs (42.4%). 12 are statues (36.4%), four are terracotta/clay (12.1%), two are gems (6.1%), and one is a coin (3%). The large representation of statues, and particularly those from Rome, is partially a result of the design of the study. Judging from a qualitative perspective, statues of tauroctonies are indeed more likely than other media to be similar to one another. However, this is partly due to the limits of the medium. Many of the 105 features recorded in the database are only practically feasible in the form of flat representations (including but not limited to reliefs), such as Sol and Luna framing the tauroctony, or peripheral scenes added above and beside the main scene. This means that many statues have a lot of zeros in common with one another because they all lack those features, making their tie strength somewhat of a false positive.

We should not, however, dismiss the strong ties between statues simply because they are rooted in their essential lack of certain embellishments: the decision to represent a tauroctony in this way was still a creative and active choice. Our oldest known tauroctony is a statue from Rome (CIMRM 593) dating to the end of the 1st century CE.[88] It is not shown within Figure 6.5, as it does not share any very strong ties with any other tauroctony. However, many of the Roman statues in Figure 6.5 have been dated to the early to mid-2nd century CE, while collectively the dates of other tauroctonies in the component tend to be later. Their conservatism and early dating may indicate they were produced before the iconographic innovations and idiosyncrasies seen in other media. The high interconnectivity between so many other statues from Rome within Figure 6.5 (and, further, their looser

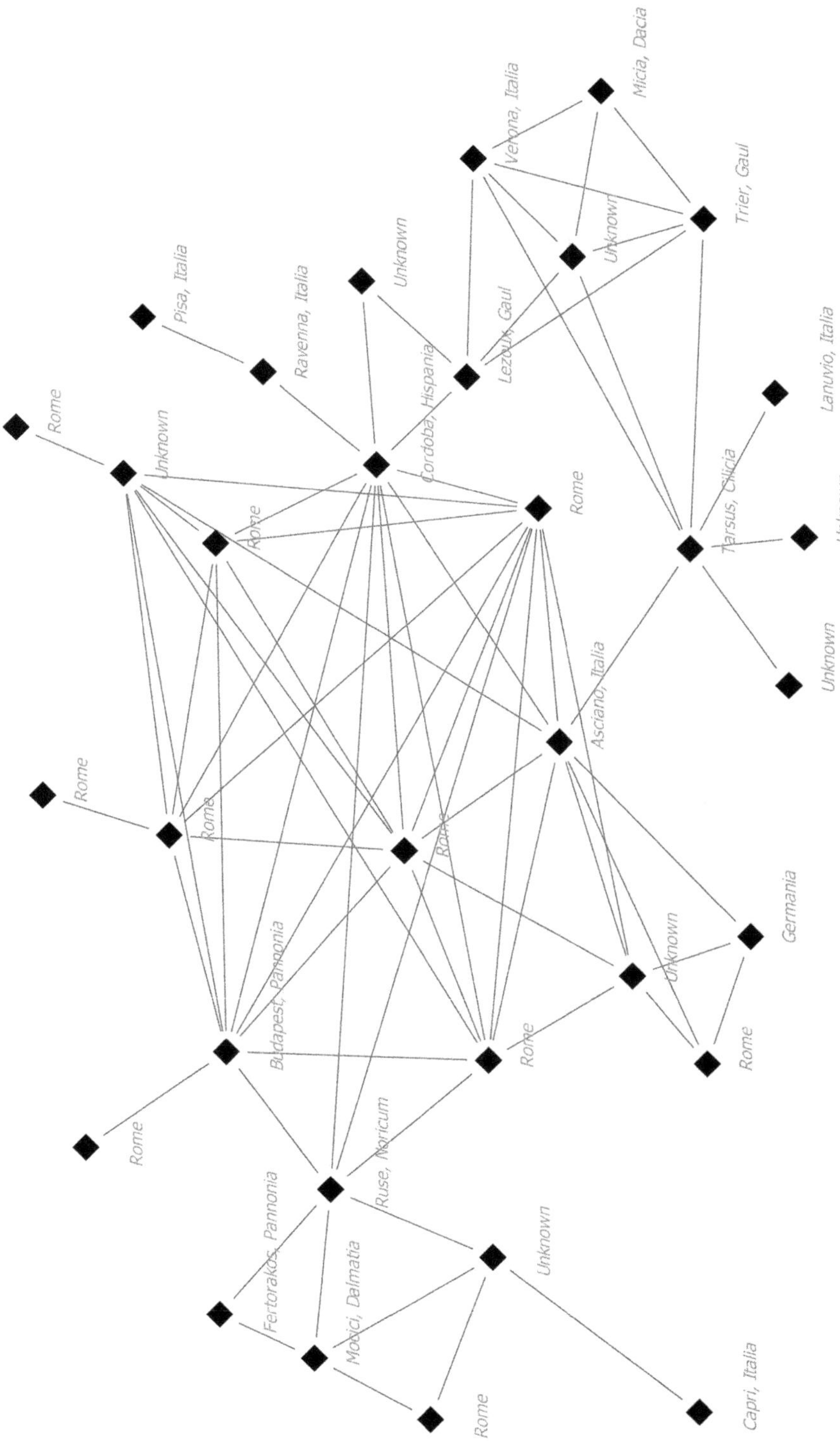

Figure 6.5 The large component of a network showing very strong ties (≥0.85) between tauroctonies (diamond = relief, square = statue, triangle = terracotta/clay, circle = gem, and star = coin).

connectivity with tauroctonies in other media from across the empire) is consistent with the theory that Mithras-worship itself originated in Rome and spread from there.[89]

Figure 6.6 visualises the other nine components, which all contain only reliefs. Echoing the E-I index results of Tauroctony Network 1, many of the ties at this extremely high strength occur between tauroctonies from the same site, with connections emerging in Rome, Carnuntum, Alba Iulia, and Sarmizegetusa. There are also ties between tauroctonies from sites which are very nearby. For example, the dyad marked 1 in Figure 6.6 is between reliefs from the villages of Kreta and Pleven in ancient Moesia Inferior (modern Bulgaria, CIMRM 2255 and 2257) which are just 30 km apart.

The dyad (marked 2) from Alba Iulia (ancient Apulum in Dacia, CIMRM 1974 and 2000) are almost identical. They are part of a set of highly idiosyncratic tauroctonies which, despite all being displayed at different mithraea, all shared some key features.[90] The multiple Mithraic communities at Alba Iulia were 'tightly bound by a common visual idiom', although there were separate socioeconomic and professional groups active in the different mithraea at the site.[91] Nevertheless, despite their similarities—and despite the identicality of CIMRM 1974 and 2000—the tauroctonies of the Alba Iulia type still have a lot of variation (for example, in the positioning of the lion, or of the objects carried by the torchbearers). They are thus a particularly pointed case of local innovation within the cult, but also of artistic diversity which may be the innovation of individual craftsmen or may be the result of alternative influences. There is evidence of influence from Italy in Mithraic iconography at Alba Iulia,[92] and of connections to Mithraic iconography

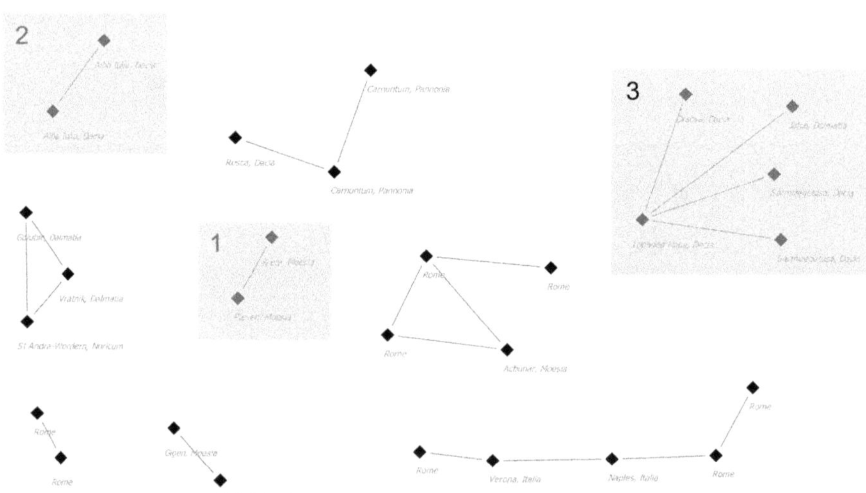

Figure 6.6 The other nine components in the network of very strong ties (≥0.85 similarity).

elsewhere in Dacia.[93] These patterns of influence and innovation are, again, multiple and overlapping.

In the component marked 3 in Figure 6.6, four nodes are tied to one node from Lopadea Nouă,[94] a small town about 40 km from Alba Iulia. This tauroctony was dedicated by Marcus Aurelius Maximus, who declares himself a *flamen* (priest) from one of the two settlements at Alba Iulia. It was possibly produced between 193 and 197 CE, and likely moved to Lopadea Nouă from Alba Iulia.[95] It does not conform to the Alba Iulia type discussed above, instead fitting the style of several reliefs from Sarmizegetusa, the ancient capital city of Dacia: each torchbearer holds two torches and the relief is quite small. Indeed, two of the four tauroctonies tied to it in Figure 6.6 are from Sarmizegetusa: (*CIMRM* 2073 and 2135). The other two (*CIMRM* 2154 from Orsova in Dacia and *CIMRM* 1906 from Jajce in Dalmatia) also share these salient features.

Dirven and McCarty have asserted that these smaller tauroctonies were probably dedicated in large numbers by individual worshippers, an innovation suggested by the many tauroctonies found at Sarmizegetusa.[96] The style was evidently influential around Dacia, as attested by the ties in this component, but could also travel much further afield: Jajce is almost 600 km away from Sarmizegetusa. Once again, our evidence points towards local innovation—but innovation which could spread widely through (Mithraic) networks. The distinctive form of Mithras-worship at Sarmizegetusa probably spread elsewhere through Mithraic association among employees of the customs office, which was key to the diffusion of the cult across the entire Danubian region, and among employees of the salt-mining industry, many of whom were tied to the Mithraic cult in Dacia.[97]

Conclusions

Three ideas have tied this chapter together. Firstly, that Mithras-worship around the Roman Empire had diverse characteristics. This diversity depended largely on the unique ways in which different Mithraic communities adhered to the forms and systems of the cult which they encountered. These forms and systems could be globally familiar, fundamental building blocks of the cult, but their transmission may have varied. They may have been received in different ways, filtered through different sociocultural lenses by different people at different times in different places. They could look similar, but be understood differently. They could be understood similarly, but represented differently. In turn, these subtly different receptions could have been established into new norms, passed with consistency among groups along professional, social, or even purely Mithraic networks—until they too were filtered through an alternative lens; until they clashed with other norms; until somebody worked out that something new made more sense or more true. Other forms and systems were not global: they were local innovations, crafted by individuals. They could attain a range of levels of popularity at a local level—but they could also travel, become provincial and regional, and stretch even further afield. These complex and numerous processes cannot be summed up in (regional or provincial) typologies, and certainly cannot be universalised.

Secondly, the analysis of Mithraic iconography produces results which are relevant to the movement of Mithras-worship around the empire. We have seen that the Mithraic tauroctony is almost ubiquitous as a central motif, but that there is great variability in its many features. I do not contend that all its variation is meaningful from a narrative or interpretative perspective: many of the choices made by artisans and dedicators may have been superficial or arbitrary, although this point should not be overplayed, as our reading of the tauroctony remains uncertain. I do not believe that we can always use Mithraic iconography to discover meaningful details about the praxis or principles of the Mithraic group which produced it, although we sometimes can. I do think, however, that all iconographic variation is meaningful in terms of *transmission*. Iconographic choices are either innovations or echoes of another version which has been seen or described. If they are innovations, the alternative has either never been encountered, or has been discarded. In all cases, they contain valuable information about the movement of forms which can potentially be unlocked.

Finally, statistical analysis, and network analysis especially, is a useful way to explore this data. 528 tauroctonies are too many to weigh against one another qualitatively. 105 features are too many variables to juggle simultaneously in one's head. A statistical approach generally elicits patterns which are not immediately obvious, and which are fundamental and valuable, so long as they are vetted with an appropriate level of qualitative research. The network approach specifically is ideally suited to a task where one is interested in the movement of information, and where the focus needs to be simultaneously macroscopic (looking at overall trends) and microscopic (taking into account individual tauroctonies, sites, or features).

In this study, a network approach has shown us that regional/provincial trends are not common enough in individual tauroctonies from those regions/provinces to be used as the foundation of any analysis. Instead, the character of one tauroctony is impacted by several such regional/provincial trends, and other trends which may be extraregional, supraregional, or very local. Statistical analysis of one such trend, the depiction of Luna's gaze averted from the scene, revealed that the motif travelled with an observably high frequency with a series of other features (a raven on Mithras' cloak, Mithras looking back over his shoulder, a rooster with Cautes, etc.), but not with universal consistency. In this case, patterns occurred particularly within one geographical area (Italy), but this is not always the case. We also saw that the strongest ties between tauroctonies tend to be among those which conform to highly idiosyncratic local variants, including statues from Rome and reliefs from Alba Iulia and Sarmizegetusa. In the latter two cases, however, we saw that these local forms could nonetheless travel far, and qualitative analysis showed that they overlapped with other trends. The characteristics of the cult, then, depended on how communities in each place adopted, changed, or merged the forms with which they had encountered—and this occurred in a wide variety of ways.

Notes

1. See Chalupa 2016 on the earliest archaeological and literary evidence for the worship of Mithras in the Roman world.
2. E.g. Tertullian *De Corona* 15, Origen *Contra Celsum* 6.21–22, Justin Martyr *Dialogue with Trypho* 70.
3. Some had artificial vaulted ceilings, such as the mithraeum at the Baths of Caracalla (CIMRM 457), mirroring the vaulted cave seen in several tauroctonies. Wall decoration sometimes featured materials to create a cave-like experience, such as seashells and pumice (e.g. CIMRM 216, 389, 434). Other mithraea were set slightly below ground (e.g. CIMRM 829, 852). Some were also constructed within real caves, such as the mithraeum at Duino (Stacul 1976).
4. Lentacker et al. 2004.
5. Gordon 2004.
6. Conversely, 65 of the tauroctonies recorded by Vermaseren have not been included in the database, usually because they are not clearly Mithras (e.g. CIMRM 11 and 12), are inaccurately described (e.g. CIMRM 1938), are too incomplete to analyse (e.g. CIMRM 737) or lack images or detailed descriptions (e.g. CIMRM 176).
7. See Wasserman and Faust 1994: 291–344.
8. For example, a tauroctony with a strongly corroborated date would receive a Date Confidence score of 3, while a tauroctony with a speculative date would receive a Date Confidence of 1. This allows cases with low confidence scores to be discounted when necessary.
9. It is not clear whether Mithras is killing or simply wounding the bull (see Faraone 2013 and Palmer 2009).
10. Panagiotidou 2018: 672–674.
11. CIMRM 651. Author's own translation.
12. CIL IX 4109. For the reading of *ordo* as 'Mithraic community', see Vermaseren 1982: 18. Another Mithraic inscription probably from Rome also refers to an *ordo* (CIL VI 2151): here, the *ordo sacerdotum* dedicates to the *pater patrum*.
13. CIL IX 4110. Author's own translation. See Kovács 2009: 272 for the dating of the consulship of Maximus and Orfitus to AD172, with reference to *Hist. Aug.* Comm. 11.14.
14. AE 2003: 1279.
15. CIL III 4800.
16. Martens (2012: 270–273).
17. CIL IX 4112.
18. See e.g. Beard et al. 1998: 307.
19. David 2018: 127–129.
20. See e.g. Beck 2017: 143–144, Mastrocinque 2017: 139–144, Beck 2006: 212, Clauss 2000: 95–97, Beck 1977, and Schwartz 1975.
21. See Clauss 2000: 98–101 for a summary of these and other animals in tauroctonies.
22. The birth of Mithras from a rock, a water miracle, images of Mithras hunting and subduing the bull, and a series of interactions with Sol. See Clauss 2000: 62–77 and 146–154 for an overview of these scenes. They also appear in CIMRM 42, 390, 966, 1083, 1292, 1400 and a painted tauroctony from the mithraeum at Marino (Vermaseren 1982).

23. See Martin 2014: 103–117 for several cognitive theories on potential readings of the tauroctony. See Griffith 2014 for the necessity of narrativity in transmitting a visual idiom.
24. The name 'Mithras' is etymologically linked to a number of eastern deities, including the Persian Mithra. Some groups of Roman Mithras-worshippers seem to have highlighted a connection to Persia, which would have been tentative at most (Gordon 2017b).
25. Most prominently the work of Roger Beck (see especially Beck 2006).
26. See the summary and footnotes in Faraone 2013: 96–99.
27. See Gordon (2004) on the range of sizes.
28. See Adrych et al. 2017: 61–80.
29. Beck 1982: 140.
30. Will 1950: 265 and 268.
31. See e.g. Dirven and McCarty 2014: 134 on the different social profiles of worshippers at the multiple mithraea in Alba Iulia.
32. Such as the reduced emphasis on hierarchy at the Carrawburgh and London mithraea compared to Santa Prisca (Norberg 2016), or the ritualised deposition of small tauroctonies at Sarmizegetusa.
33. We have seen that chickens were usually eaten. The birds were often exclusively male, but at Orbe-Boscéaz 60% were female (Olive 2004: 149). A rubbish pit outside the Wiesloch mithraeum in Germany produced only cattle, pig, sheep/goat, and horse bones (Hensen 1994), and no chickens.
34. Such as the preference for the term '*Invicto Deo Omnipotenti*' in inscriptions from Noricum, Pannonia, Dacia, and Dalmatia (Misic 2013: 79), or the tendency to make dedications '*pro bono communi*' in Apulum (CIMRM 1017 and 1783).
35. Mithraea in the northwest provinces tended to be built outside or towards the periphery of settlements (e.g. at Güglingen and Heidelberg), while those in Rome were in more accessible areas and were probably known more widely in the community (Hensen 2017: 400–401).
36. Mithraea at Güglingen and Wiesloch had timbered first phases. At Krefeld-Gellep, Künzing, Heidelberg, and Tienen, a wooden part was kept throughout. At Wiesloch, Künzing, and Szombathely, mudbrick was used, a material associated with military architecture. See Hensen 2017: 396.
37. In addition to the tauroctony, variation is also prevalent in peripheral Mithraic iconographic scenes, such as the rock-birth. See Schofield (1995) on the regionalisation of this scene.
38. Clauss 2000: 17, although I question his statement that Mithraic imagery was constant!
39. Beck 2004: xxii.
40. E.g. Beck 2000: 158.
41. Martin 2006: 141.
42. Szabó 2013b: 44.
43. Gordon 2017a: 295.
44. Alvar 2017: 30 and 35.
45. Dészpa 2017: 168.
46. Dirven and McCarty 2014: 129.
47. This is especially apparent in his chapter on the peripheral animals of the scene (Campbell 1968: 12–28).
48. Campbell 1968: 5.
49. On the unsuitability of Campbell's method for the data, see Silnovic 2014: 7, Vollkommer 1991, Gordon 1980, and Will 1957. On Campbell's dated tendency to look for Persian precedence, see Chalupa 2016: 69, Beck 2006: 29, 49 and Gordon 1980.

50. See e.g. Fusco 2015: 527, Misic 2013: 238, Belayche 2006, Vrkljan 2005, von Kiernan 2002: 238, Vrkljan 2002 and Tóth 1999.
51. Schofield 1995, Turcan 1993: 45–61 and Vollkommer 1991.
52. Vrkljan 2005 and Gabričević 1951.
53. Belayche 2009: 5–7.
54. Miletić 2005.
55. Sicoe 2014 and Veliciu 2004.
56. Misic 2013.
57. Sicoe 2014: 59–60.
58. See García Sánchez 2012 on the superficiality of Mithras' 'eastern' dress.
59. There are only four: CIMRM 201, 1275, 2196, and 2327.
60. Using Ward's method (Ward 1963). This was performed using R (R Core Team 2013).
61. The 16 categories were: the position/presence of the torchbearers, peripheral details of the torchbearers, Mithras' interaction with the bull, other details associated with the bull, the position of Sol and Luna, peripheral details of Sol and Luna, the direction of Mithras' gaze, Mithras' attire, Mithras' peripheral details, the position of the raven, the positions of other animals, the presence of distinct astrological symbols (zodiacs, stars, etc.), the presence of peripheral scenes such as Mithras' rock birth, the depiction of other deities such as Oceanus or members of the Olympian pantheon, the shape of the tauroctony, and any other details not covered by the above.
62. Produced using Google Maps (2018, http://maps.google.com).
63. This was performed using UCINET (Borgatti et al. 2002).
64. I also calculated other standard measures of correlation: Yule's Q and Pearson's phi. These are not recorded here for the sake of clarity, but their values were all concordant with the values of the E-I index.
65. See Hijmans 2009: 141 on the likelihood of this motif appearing on the temple of Jupiter Capitolinus.
66. See, for example, a lamp produced in North Africa but found in Naples (https://www.britishmuseum.org/collection/object/G_1856-1226-401). See examples of coins from the reigns of Titus, Domitian, Trajan, and Hadrian in Hijmans 2009: 417–419.
67. Hijmans 2009: 563.
68. See images and further information at https://www.metmuseum.org/art/collection/search/254590.
69. Including Walsh 2018: 5, Clauss 2012: 77–88, and Beck 2006: 198.
70. *De antro nympharum* 8.
71. Faraone 2013: 15.
72. Campbell 1968: 35.
73. This view was expressed by Beck 2006: 37 and Lentz 1975: 375.
74. See above on the 'Completeness' score.
75. In a network analysis of all tauroctonies which contain this motif (which I do not have the space to include here), a core/periphery test showed that 100% of the core nodes in the network were from Italy (with 75% from Rome itself). The fit of the core/periphery model was 0.537, indicating that it describes the dataset well. In this context, 'core' nodes are more likely to be the influencers while 'periphery' nodes are more likely to be influenced.
76. On the phi correlation coefficient, see McHugh 2018. On the interpretation of this value as 'relatively strong', see Kotrlik et al. 2011: Table 4.
77. Clauss 2000: 162. See also the raven as a messenger in Virgil (*Eclogues* 9) or Aristophanes (*Birds* 1).
78. Silnovic 2014: 66–67.

79. See Miletić 1996: 13.
80. Miletić 1996: 17–18, 28.
81. Silnovic 2014: 61.
82. Of the nine cases including the Salonitan tondo, five are from Germania (CIMRM 1014, 1137, 1275, 1330, and 1359) and one is from Raetia (Gordon 2004: Figure 19).
83. See Silnovic 2014: 57–59.
84. Silnovic 2014: 59.
85. CIL III 1955, see also CIMRM 1875.
86. Clauss 1992: 270.
87. See e.g. Alvar 2008: 85 and Lentacker et al. 2004: 73.
88. See Gordon 1978 on the dating of CIMRM 593.
89. See Nilsson 1961: 675–676, Merkelbach 1984: 75–77 and 160–161 and Clauss 2000: 7.
90. Outlined in Dirven and McCarty 2014:134–135.
91. Ibid.
92. Depictions of a radiate Sol (CIMRM 1938, 1944, and 1973) were likely influenced by Italian Mithras-worship (Szabó et al. 2014: 152–153).
93. CIMRM 1935 has a parallel from Sarmizegetusa (Sicoe 2014: 126).
94. See Szabó 2007: 60 and Szabó 2013b: 58 for details and http://lupa.at/19291 for images.
95. Szabó 2018: 170.
96. Dirven and McCarty 2014: 135–137.
97. See Szabó 2018: 99–100.

Bibliography

Adrych, P., Bracey, R., Dalglish, D., Lenk, S. and Wood, R., (2017), *Images of Mithra*, Oxford University Press, Oxford.

Alvar, J., (2008), *Romanising Oriental Gods: Myth, Salvation and Ethics in the Cults of Cybele, Isis and Mithras*, tr. R. Gordon, Brill, Leiden.

Alvar, J., (2017), "The 'Romanization' of 'Oriental Cults'", in S. Nagel, J.F. Quack and C. Witschel (eds.), *Entangled Worlds: Religious Confluences between East and West in the Roman Empire*, Mohr Siebeck, Tübingen: 23–46.

Beard, M., North, J. and Price, S., (1998), *Religions of Rome Volume 1: A History*, Cambridge University Press, Cambridge.

Beck, R., (1977), "Cautes and Cautopates: Some Astronomical Considerations", *Journal of Mithraic Studies* 2: 1–17.

Beck, R., (1982), "The Mithraic Torchbearers and 'Absence of Opposition'", *Classical Views* 26 (1): 126–140.

Beck, R., (2000), "Ritual, Myth, Doctrine and Initiation in the Mysteries of Mithras: New Evidence from a Cult Vessel", *Journal of Roman Studies* 90: 145–180.

Beck, R., (2004), *Beck on Mithraism: Collected Works with New Essays*, Ashgate, Aldershot.

Beck, R., (2006), *The Religion of the Mithras Cult in the Roman Empire: Mysteries of the Unconquered Sun*, Oxford University Press, Oxford.

Beck, R., (2017), "If So, How? Representing 'Coming Back to Life' in the Mysteries of Mithras", in F.S. Tappenden and C. Daniel-Hughes (eds.), *Coming Back to Life: The Permeability of Past and Present, Mortality and Immortality, Death and Life in the Ancient Mediterranean*, McGill University Library and Archives, Montreal: 129–151.

Belayche, N., (2006), "Note sur l'Imagerie des Divinités »Orientales« dans le Proche-Orient Romain", in C. Bonnet, J. Rüpke and P. Scarpi (eds.), *Religions Orientales – Culti Misterici. Neue Perspektiven – Nouvelles Perspectives – Prospettive Nuove*, Steiner, Stuttgart: 123–133.
Belayche, N., (2009), "Beobachtungen zur bildlichen Darstellung der »orientalischen« Gottheiten im Vorderen Orient zu römischer Zeit", *Trivium* 4: 1–15.
Borgatti, S.P., (2002), *NetDraw Software for Network Visualization*, Analytic Technologies, Lexington.
Borgatti, S.P., Everett, M.G. and Freeman, L.C., (2002), *UCINET for Windows: Software for Social Network Analysis*, Analytic Technologies, Harvard.
Campbell, L.A., (1954–1955), "Typology of Mithraic Tauroctonies", *Berytus* 11: 1–60.
Campbell, L.A., (1968), *Mithraic Iconography and Ideology*, Brill, Leiden.
Chalupa, A., (2016), "The Origins of the Roman Cult of Mithras in the Light of New Evidence and Interpretations: The Current State of Affairs", *Religio* 24 (1): 65–96.
Clauss, M., (1992), *Cultores Mithrae: die Anhängerschaft des Mithras-Kultes*, Steiner, Stuttgart.
Clauss, M., (2000), *The Roman Cult of Mithras: The God and his Mysteries*, tr. R. L. Gordon, Edinburgh University Press, Edinburgh.
Clauss, M., (2012), *Mithras: Kult und Mysterium*, Philipp von Zabern, Darmstadt.
David, M., (2018), "A Newly Discovered Mithraeum at Ostia", *Acta Antiqua Hungaricae* 58: 117–131.
Dészpa, M. L., (2017), "Jupiter Optimus Maximus Dolichenus and the Re-Imagination of the Empire: Religious Dynamics, Social Integration, and Imperial Narratives", in S. Nagel, J.F. Quack and C. Witschel (eds.), *Entangled Worlds: Religious Confluences between East and West in the Roman Empire*, Mohr Siebeck, Tübingen: 113–181.
Dirven, L. and McCarty, M.M., (2014), "Local Idioms and Global Meanings: Mithraism and Roman Provincial Art", in L.R. Brody and G.L. Hoffman (eds.), *Roman in the Provinces: Art on the Periphery of Empire*, Chestnut Hill, Chicago: 127–143.
Faraone, C., (2013), "The Amuletic Design of the Mithraic Bull-Wounding Scene", *Journal of Roman Studies* 103: 96–116.
Fusco, U., (2015), "A New Mithraic Relief from Veii", *Archeologia Classica* 66 (2.5): 519–546.
Gabričević, B., (1951), *Mitrin kult na području rimske Dalmacije*, PhD dissertation, University of Zagreb.
García Sánchez, M., (2012), "The Dress and Colour of Mithraism: Roman or Iranian Garments?", in S. Schrenk and K. Vössing (eds.), *Kleidung und Identität in Religiösen Kontexten der Kaiserzeit*, Schnell & Steiner (Verlag), Regensburg: 123–134.
Gordon, R., (1975), "Franz Cumont and the Doctrines of Mithraism", in J. Hinnells (ed.), *Mithraic Studies Volume I*, Manchester University Press, Manchester: 215–248.
Gordon, R., (1978), "The Date and Significance of CIMRM 593 (British Museum, Townley Collection)", *Journal of Mithraic Studies* 2: 148–174.
Gordon, R., (1980), "Panelled Complications", *Journal of Mithraic Studies* 3: 220–227.
Gordon, R., (2004), "Small and Miniature Reproductions of the Mithraic Icon: Reliefs, Pottery, Ornaments and Gems", in M. Martens and G. de Boe (eds.), *Roman Mithraism: The Evidence of the Small Finds*, Instituut voor het Archeologisch Patrimonium, Brussels: 259–284.

Gordon, R., (2008), "Institutionalized Religious Options: Mithraism", in J. Rüpke (ed.), *A Companion to Roman Religion*, Blackwell, Oxford: 392–405.

Gordon, R., (2017a), "Projects, Performance and Charisma: Managing Small Religious Groups in the Roman Empire", in R.L. Gordon, J. Rüpke and G. Petridou (eds.), *Beyond Priesthood: Religious Innovators in the Roman Empire*, de Gruyter, Berlin: 277–316.

Gordon, R., (2017b), "From East to West: Staging Religious Experience in the Mithraic Temple", in S. Nagel, J.F. Quack and C. Witschel (eds.), *Entangled Worlds: Religious Confluences between East and West in the Roman Empire*, Mohr Siebeck, Tübingen: 413–442.

Griffith, A.B., (2014), "Dead Religion, Live Minds: Memory and Recall of the Mithraic Bull-slaying Scene", *Journal of Cognitive Historiography* 1 (1): 72–89.

Hensen, A., (1994), "Das Mithräum im Vicus von Wiesloch", *Archäologische Nachrichten aus Baden 51/52*: 30–37.

Hensen, A., (2017), "Templa et spelaea Mithrae. Unity and Diversity in the Topography, Architecture and Design of Sanctuaries in the Cult of Mithras", in S. Nagel, J.F. Quack and C. Witschel (eds.), *Entangled Worlds: Religious Confluences between East and West in the Roman Empire*, Mohr Siebeck, Tübingen: 384–412.

Hijmans, S.E., (2009), *Sol: The Sun in the Art and Religions of Rome*, PhD thesis, University of Groningen.

Kotrlik, J.W., Williams, H.A. and Jabor, M.K., (2011), "Reporting and Interpreting Effect Size in Quantitative Agricultural Education Research", *Journal of Agricultural Education* 52 (1): 132–142.

Kovács, P., (2009), *Marcus Aurelius' Rain Miracle and the Macromannic Wars*, Brill, Leiden.

Lentacker, A., Ervynck, A. and van Neer, W., (2004), "The Symbolic Meaning of the Cock: The Animal Remains from the Mithraeum at Tienen (Belgium)", in M. Martens and G. de Boe (eds.), *Roman Mithraism: The Evidence of the Small Finds*, Instituut voor het Archeologisch Patrimonium, Brussels: 57–80.

Lentz, W., (1975), "Some Peculiarities not Hitherto Full Understood of 'Roman' Mithraic Sanctuaries and Representations", in J.R. Hinnells (ed.), *Mithraic Studies: Proceedings of the First International Congress of Mithraic Studies*, Volume II, Manchester University Press, Manchester: 358–377.

Martens, M., (2012), *Life and Culture in the Roman Small Town of Tienen: Transformations of Cultural Behaviour by Comparative Analysis of Material Culture Assemblages*, PhD thesis, Vrije Universiteit Amsterdam.

Martin, L.H., (2006), "The Roman Cult of Mithras: A Cognitive Perspective", *Religio* 14 (2): 131–146.

Martin, L.H., (2014), *The Mind of Mithraists: Historical and Cognitive Studies in the Roman Cult of Mithras*, Bloomsbury, London.

Mastrocinque, A., (2017), *The Mysteries of Mithras: A Different Account*, Mohr Siebeck, Tübingen.

McHugh, M.L., (2018), "Phi Correlation Coefficient", in B.B. Frey (ed.), *The SAGE Encyclopedia of Educational Research, Measurement, and Evaluation*, SAGE, Thousand Oaks: 1252–1253.

Merkelbach, R., (1984), *Mithras*, Hain, Königstein im Taunus.

Miletić, Z., (1996), *Mitraizam u rimskoj provinciji Dalmaciji*, PhD thesis, University of Zadar.

Miletić, Ž., (2005), "Typology of Mithraic Cult Reliefs from Southeastern Europe", in M. Sanader and A. Rendić-Miočević (eds.), *Religion und Mythos als Anregung für die provinzialrömische Plastik. Akten des VIII. Internationalen Kolloquiums über Probleme des provinzialrömischen Kunstschaffens*, Golden marketing-Tehnička knjiga, Zagreb: 269–275.

Misic, B., (2013), *Cults and Religious Integration in the Roman Cities of the Drava Valley (Southern Pannonia)*, PhD thesis, University of London.

Nilsson, M.P., (1961), *Geschichte der griechischen Religion II*, C.H. Beck, Munich.

Norberg, A., (2016), *In the Cave of Mysteries: Analyzing Ritual Space within the Roman Cult of Mithras through the Examples of Santa Prisca, Walbrook, and Carrawburgh*, MA thesis, University of Stockholm.

Olive, C., (2004), "La faune exhume des mithraea de Martignny (Valais) et d'Orbe-Boscéaz (Vaud) en Suisse", in M. Martens and G. de Boe (eds.), *Roman Mithraism: The Evidence of the Small Finds*, Instituut voor het Archeologisch Patrimonium, Brussels: 147–156.

Palmer, G., (2009), "Why the Shoulder? A Study of the Placement of the Wound in the Mithraic Tauroctony", in G. Casadio and P.A. Johnston (eds.), *Mystic Cults in Magna Graecia*, University of Texas Press, Austin: 314–323.

Panagiotidou, O., (2018), "Secrecy in the Mithras Cult: Concealment, Cognition and Social Cohesion", *Acta Antiqua Hungaricae* 58: 667–679.

R Core Team, (2013), *R: A Language and Environment for Statistical Computing*, R Foundation for Statistical Computing, Vienna, Austria. Available online at https://www.R-project.org/.

Schofield, A., (1995), "The Search for Iconographic Variation in Roman Mithrasim", *Religion* 25: 51–66.

Schwartz, M., (1975), "Cautes and Cautopates", in J.R. Hinnells (ed.), *Mithraic Studies*, Vol. 2, Manchester University Press, Manchester: 406–423.

Sicoe, G., (2014), *Die Mithraischen Steindenkm aler aus Dakien*, Mega Publishing House, Cluj-Napoca.

Silnovic, N., (2014), *Soli Deo Stellam et Fructiferam: The Art of the Mithraic Cult in Salona*, MA thesis, Central European University, Budapest.

Stacul, G., (1976), "The Cave of the Mitreo near San Giovanni di Duino", *Antichira Altoadriatiche* 10: 29–38.

Szabó, A., (2007), *Dáciai papság*, Opitz, Budapest.

Szabó, C., (2013a), "The Mithraic Statue of Secundinus from Apulum", *Revistei Doctoranzilor în Istorie Veche şi Arheologie* 1: 45–64.

Szabó, C., (2013b), "Microregional Manifestation of a Private Cult: The Mithraic Community of Apulum", in I. Moga (ed.), *Angels, Demons and Representations of Afterlife within the Jewish, Pagan and Christian Imagery*, Editura universitatii Alexandru Ioan Cuza, Iaşi: 43–72.

Szabó, C., (2018), *Sanctuaries in Roman Dacia: Materiality and Religious Experience*, Archaeopress, Oxford.

Szabó, C., Boda, I., Bunoiu, V. and Timoc, C., (2016), "Notes on a new Mithraic Inscription from Dacia", in R. Ardevan and E. Beu-Dachin (eds.), *Mensa Rotunda Epigraphica Napocensis*, Mega Publishing House, Cluj-Napoca: 91–104.

Szabó, C., Bounegru, G. V. and Sava, V., (2014), "Mithras Rediscovered. Notes on CIMRM 1938", *Ziridava Studia Archaeologica* 28: 149–156.

Tóth, I., (1999), "The Composition of the Mithras relief from Intercisa", *Antaeus* 24: 535–537.

Turcan, R., (1993), *Mithra et le mithriacisme*, Les Belles Lettres, Paris.
Veliciu, O., (2004), "Reprezentările lui Cautes şi Cautopates în Dacia", *Revista Studenţească de Studii de Istorie şi Arheologie* 2: 1–9.
Vermaseren, M.J., (1956–1960), *Corpus Inscriptionum et Monumentorum Religionis Mithriacae (CIMRM)*, Martinus Nijhoff, The Hague.
Vermaseren, M.J., (1982), *The Mithraeum at Marino*, Brill, Leiden.
Vollkommer, R., (1991), "Studien zu einer Typologie der Stieropferszene auf mithras bildwerken", *Mélanges de l'Ecole française de Rome: Antiquité* 103: 265–281.
von Kiernan, P., (2002), "A New Mithras Relief in the Römisch-Germanisches Museum Cologne", *Kölner Jahrbuch* 35: 237–240.
Vrkljan, G.K., (2002), "Prilog Čitanju Sadržaja I Kompozicije Donjih Polja Konjičke Tauroktonije", *Prilozi Instituta za arheologiju u Zagrebu* 19: 131–141.
Vrkljan, G.K., (2005), "Some Examples of Local Production of Mithraic Reliefs from Roman Dalmatia", in M. Sanader and A. Rendić-Miočević (eds.), *Religion und Mythos als Anregung für die provinzialrömische Plastik. Akten des VIII. Internationalen Kolloquiums über Probleme des provinzialrömischen Kunstschaffens*, Golden marketing-Tehnička knjiga, Zagreb,: 249–258.
Walsh, D., (2018), *The Cult of Mithras in Late Antiquity: Development, Decline and Demise ca. AD270-430*, Brill, Leiden.
Ward, J.H., (1963), "Hierarchical Grouping to Optimize an Objective Function", *Journal of the American Statistical Association* 58 (301): 236–244.
Wasserman, S. and Faust, K., (1994), *Social Network Analysis: Methods and Application* Cambridge University Press, Cambridge.
Will, E., (1950), "La Date du Mithréum de Sidon", *Syria* 27: 261–269.
Will, E., (1957), "Review: Leroy Campbell – Typology of Mithraic Tauroctonies", *Syria* 34: 388–393.

7 Networks, apostolic itineraries, and Mediterranean witnesses to the oral traditions of south India

Nathanael Andrade

By the sixth century, reports about the preaching, martyrdom, and death of Thomas the Apostle had circulated throughout much of ancient Afro-Eurasia. We hear of them from Christian and Manichaean authors situated in Upper Mesopotamia, Syria, Egypt, Libya, north Africa, Italy, Spain, and Gaul.[1] We also learn from many of these same authors that local communities in Upper Mesopotamia and India boasted of sacred sites devoted to Thomas' life and death. A remarkable feature about these reports is their general consistency. Despite local variations, these reports, it seems, were based on a single general narrative that circulated in the landmass between the Atlantic Ocean and the Indian subcontinent and that underwent various transformations over time. This chapter examines the networks that enabled the circulation of this general narrative and thus its local derivations.

To begin our inquiry, we can invoke the example of Gregory of Tours, writing from late sixth-century northern Gaul. Gregory reports that Thomas the Apostle had travelled to India to preach and had died there. His original tomb could be found in India, where locals maintained a church and a monastery in his honour. A contact named Theodore had visited it and reported that a lamp at Thomas' tomb burned in perpetuity despite never being fed with fuel. He also points out that this tomb of Thomas was empty; after his death, Thomas' remains had been brought to the Syriac-speaking city of Edessa in Upper Mesopotamia, where Mediterranean Christians overwhelmingly travelled to interact with his relics.[2] Unfortunately, Gregory does not elaborate on who Theodore was, but this most probably was the contemporary bishop of Marseilles of that name.[3] As we will see, Theodore, being bishop of a port city in southern France, was potentially linked to the see of Alexandria, from where Roman Egyptians maintained direct contact with India in the sixth century.

Significantly, Gregory's testimony synthesises various sources of information. His general reference to Thomas' preaching in India is clearly derived from the apocryphal *Acts of Thomas*, a narrative to which Gregory or contemporaries apparently had access in a Latin variation.[4] The *Acts of Thomas* (in my view) was written in Syriac in Roman Upper Mesopotamia (probably Edessa) during the third century and translated into Greek

DOI: 10.4324/9780429429217-9

almost immediately. It celebrated how Thomas the Apostle travelled to India and evangelised its people, and it enjoyed a wide circulation in the Mediterranean and Middle East in a vast array of languages thereafter.[5] After its initial circulation, various parties produced oral or written traditions through which they 'updated' the basic narrative contained in the original text. As we will see, these newer traditions often identified Thomas' original burial site or the current location of his relics, which were never specified in the *Acts of Thomas* itself.

It is on this score that we can invoke a second origin point for Gregory's material. His information on the relic cult of Thomas at Edessa ultimately originated from local traditions that had taken shape at Edessa after the initial late third-century circulation of the *Acts of Thomas*. As early as the fourth century, the Edessenes were rigorously promoting their claim to have acquired his relics. The Latin-speaking pilgrim Egeria thus visited a church reportedly housing Thomas' body there in the 380s, and Edessa's reputation as Thomas' resting place was generally known to aspiring Christian pilgrims throughout late antiquity.[6] Gregory's information appears to be based on contemporary pilgrim reports or the information of Syrians resident in Gaul.[7] These were presumably responsible for the 'updates' regarding Thomas' tomb in Edessa that appear in Latin iterations of the *Acts of Thomas* composed during the fourth to sixth centuries (the original Syriac and Greek texts do not specify a burial at Edessa). The Edessenes had in effect augmented the original Thomas narrative by claiming that Thomas' relics had been transferred to Edessa from India. The premise was almost universally accepted in the Mediterranean world and thus appears in Latin texts of the period.

Finally, Gregory's information on the tomb for Thomas in India, by all appearances, was based on a local tradition at home in south India, even if it is tempting to dismiss Gregory's testimony as a creative invention. No doubt, excavating this tradition raises various epistemological difficulties. Around 1600 CE, a priest on the Keralan coast of India transcribed in Malayalam an oral hymn that celebrated how Thomas the Apostle arrived in south India, preached in the region, and was then martyred and buried at the site of Mylapore, where St. Thomas Christians locate Thomas' tomb today.[8] Scholars have long noted the similarities between south Indian traditions like this and the narrative of the apocryphal *Acts of Thomas*. They have sometimes even deemed them corroborating evidence for the early arrival of Thomas or Christianity in India.[9]

In my recent book, I have pursued a different line of interpretation on whose implications I intend to expand in this chapter. Sasanian Persian Christians from Mesene and subsequently coastal Fars brought and anchored Christianity, and the Thomas narrative, in south India during the fifth century. Their activity occurred at a time of accelerated Persian settlement in south India and the extension of Sasanian commercial links to south-east Asia and beyond. Subsequently, Christians in south India created oral traditions on the basis of the Thomas narrative, and this explains why various episodes from the hymns of St. Thomas Christians bear resemblance to episodes contained in the *Acts of Thomas*, but with much greater cultural and topographical specificity to the Kerala and Coromandel coasts. These Christians also established a tomb for Thomas the Apostle, and news of it travelled to the Middle East and the Mediterranean, thus leaving various traces there.[10] Assuming that this interpretation is more or less correct, it would mean that Christianity, local variations on the Thomas narrative, and a tomb for Thomas were active in south India during the sixth century (but probably not long before). Gregory's information on Thomas' tomb in India thus seems to have its origins in actual local traditions in south India that pilgrims, like his informant Theodore, were now bringing to the sixth-century Mediterranean.

By invoking Gregory of Tours as a starting point, this chapter aims to explore whether reflections on the networks that connected the worlds of the Mediterranean and the Indian Ocean enable us to address the seemingly intractable question of how the south Indian oral traditions on Thomas the Apostle took shape. As mentioned, these were only transcribed in Malayalam in the seventeenth and eighteenth centuries, after the arrival of Portuguese Christians in the subcontinent. While some scholars treat the oral traditions as quite ancient in origin, some hallmarks of Portuguese contact with India after 1500 are evident,[11] and it is difficult to discern when exactly oral traditions about Thomas began to take root in south India or what their features were. This chapter probes how an understanding of late ancient and early medieval networks and their ability to bring traces of oral traditions to the Mediterranean can enable us to outline a theory for how certain key aspects of the oral traditions took shape in the sixth century. These networks were determined by a variety of intersecting organisational dispositions that were commercial, ecclesiastical, diplomatic, and pilgrimage in nature. In combination, they enabled Christians to forge across vast geographical spaces a host of ties that were to varying degrees 'strong' (defined by multifaceted, dynamic, and frequent contact) or 'weak' (defined by one-dimensional or infrequent contact).[12] As this chapter posits, Christians in south India maintained various social connections to the Middle East and the Mediterranean that enabled them to shape their local cult for Thomas. In turn, these connections enabled reports about their cult to circulate throughout much of Afro-Eurasia.

The Thomas narrative in late antique Mesopotamia and India

The *Acts of Thomas* is a text of immense complexity, and interpretations of the date, original language, provenance, and historical value of its narrative have varied.[13] I will proceed according to the premise that it was composed at mid-to-late third-century Edessa, originally in Syriac but also in Greek shortly after.[14] Also at variance are how scholars have understood the relationships between the *Acts of Thomas* and the oral traditions of Thomas Christians. Do they constitute separate witnesses for the same historical phenomenon, or are the oral traditions derived from the original narrative of the *Acts of Thomas*? Simply put, the *Acts of Thomas* depicts the preaching of Thomas the Apostle throughout India, particularly at the courts of a king named Gudnaphar and then a king named Mazdai.[15] While Gondophares was certainly the name of Indo-Parthian kings of first-century north India, the character is a literary construct attached to a name that the Palmyrene commercial network probably brought from north India to Syria in the late first century.[16] Every other character in the narrative (and even Gudnaphar) appears to be entirely contrived, and India is depicted in an entirely general way. No toponyms, distinctive topographies, or unique cultural practices are mentioned at all. Eventually, Thomas is martyred by the soldiers of king Mazdai for converting members of his court and buried. By contrast, the oral traditions of Thomas Christians on the Keralan coast have a much higher degree of cultural specificity. These include Thomas' converting the members of various caste groups and Thomas' death at the hands of Brahmins in Mylapore for miraculously destroying a shrine for Kali. They also feature an episode in which Thomas builds a palace in heaven for a king, just as he does for Gondophares in the *Acts of Thomas*. In other scholarship, I have asserted that the oral traditions are ultimately derived from the general narrative of the *Acts of Thomas* (with its very general India) but reflected various layers of regional, cultural, and religious specificity embedded in south India.[17] Sometime after Persian Christian settlers in south India brought Christianity to south India, and with it the Thomas narrative, during the fifth-sixth centuries, local Christian communities there situated various elements of the narrative in their own unique physical and cultural topographies.

But if the south Indian oral traditions are ultimately inspired by the original narrative of the *Acts of Thomas*, this raises many outstanding questions. Since the surviving *Acts of Thomas* and the transcribed versions of the oral traditions are roughly 1300 years apart, one wonders whether it is possible to identify intermediary stages, especially in the absence of explicit south Indian witnesses between 500 and 1500. By reflecting on ancient networks, we may be able to address this pressing question and identify traces of the oral traditions that made their way into late antique and early medieval source material in the Mediterranean, particularly in the sixth century.

The Acts of Thomas, Edessa, and India

During the fourth-sixth centuries, the narrative of the *Acts of Thomas* circulated throughout the Mediterranean world. So did reports of Thomas' burial site at Edessa, a claim promoted by the Edessenes themselves. The complicated web of commercial and ecclesiastical networks that enabled the circulation of these traditions are beyond the scope of this chapter. May it suffice to say that Christian pilgrims from various parts of the Mediterranean were in direct contact with Edessa during this time. Likewise, we have seen how Syrians were active in the western Mediterranean. Eventually, Edessene local traditions began to appear in 'updated' versions of the *Acts of Thomas* that circulated in the Mediterranean after the fourth century. For example, the Latin *Passion of the Holy Apostle Thomas* (the late-fourth century) reports the apocryphal premise that the Edessenes acquired Thomas' relics in the 230s when the emperor Alexander Severus negotiated with the Persian king 'Xerxes' (apparently Ardashir I) to acquire them.[18] Although some scholars treat this report as referring to an actual event,[19] more probably it reflects how some late fourth-century Edessene Christians had historicised their own claims of possessing Thomas' relics. As they surmised, Thomas had preached and died in India during the apostolic age, but they had acquired his relics a couple of centuries later (but not so recently that their claim might appear bogus). What is most pertinent to the present chapter, however, is that narratives based on the Edessenes' promotion of their cult for Thomas apparently circulated throughout the Mediterranean world during the fourth and fifth centuries. In fact, Edessa was generally recognised as the home of Thomas' body for the next 1000 years, whether due to literary traditions or engagement with an actual burial site.[20]

News about Thomas' resting place at Edessa also moved eastward and reached India. As a result, when Christians in south India promoted their own tomb for Thomas, it was an empty tomb. They claimed that Thomas' remains had once been there, but were there no more. When the Portuguese first established direct contact with south India and St. Thomas Christians c. 1500, they allegedly encountered Christians of the Keralan coast and residents of the Coromandel coast (including non-Christians) who boasted of a tomb for Thomas at Mylapore. When they investigated the tomb in 1524, it only harboured trace skeletal remains. Some locals and Portuguese at the time claimed that these were the burial remains of Thomas' body, but most eventually conceded that at least some of his remains had to have been transferred in Edessa.[21] While any relationship between the ancient and modern traditions regarding Thomas' burial site(s) remains a tortuous issue, it is nonetheless intriguing that when the Portuguese arrived in south India, a tradition that Thomas' body was not in his initial tomb in India but had instead been transferred to Edessa had long circulated. As we have seen, Gregory of Tours, writing in sixth-century Gaul, noted an empty tomb for Thomas in India and a resting place at Edessa. During the one millennium interval between Gregory's lifetime and the arrival of the Portuguese in India, certain Christians in India were apparently aware that Edessenes were widely recognised to have Thomas' corpse and had fashioned their traditions accordingly. Such activity was not consistent; the testimony of various late medieval Islamic and European authors (principally 1300–1500 CE) indicates that some contemporary Christians in south India did eventually claim that Thomas' body was housed locally, and this may be due to their relative eventual isolation from the traditions of Edessa at that time.[22] Still, the situations as they existed c. 500–600 and c. 1500–1600 seem to be characterised by a key parallel. An awareness that the Edessenes were recognised to have Thomas' body apparently encouraged locals and visitors to south India to posit that Thomas' tomb there was empty. In fact, some scholars have even regarded the various commentaries on Thomas' empty tomb, separated by vast time and distance, to be evidence that Thomas himself must have died and been buried in India before his relics were relocated.[23] But in my view, Christians in India during late antiquity, and eventually during the early modern period, were coordinating their own narrative with that of the Edessenes. While conceding that the Edessenes now had Thomas' corpse, they boasted of the tomb in which he was originally buried.

It is significantly in the sixth century that both the Edessene and Indian claims began to find their home in Mediterranean texts. This marked an important shift in the discourse on Thomas the Apostle's remains. Previously, Edessa in Upper Mesopotamia had been overwhelmingly treated as his resting place.[24] But in the sixth century, authors in the Mediterranean and Europe began to describe a pilgrimage site in India that held an empty tomb where Thomas was reportedly buried first. As we have seen, Gregory of Tours is an important witness. He notes a pilgrimage site in India that his informant Theodore had visited. Theodore had reported that a monastery and church adorned the site of Thomas' tomb and that a lamp or flame burned perpetually without anyone tending to it.[25]

But Gregory is not alone in providing such a testimony. A number of Greek and Latin apostolic lists and itineraries that clearly became popular in the Mediterranean between c. 400 and c. 1000 support that this dynamic was widely accepted.[26] A couple of generations or so before the activity of Gregory of Tours, these itineraries began to impart another intriguing article of information. It was the alleged name of the place in India where Thomas reportedly died: Kalamene. Certain aspects of the apostolic itineraries suggest that it refers to an actual cult site for Thomas the Apostle that took shape due to Persian Christian contact with south India.

The apostolic itineraries have a fairly distinctive form, even if they exist with some variation and could easily be integrated into other texts. In general, they consist of short descriptions of where the apostles were born, had preached, and had died. Providing cursory information on the apostles, the texts reflect a widespread interest in their biographies throughout the late antique and early medieval Mediterranean. They tend to borrow heavily from one another. Nonetheless, certain patterns enable us to discern when such texts first started to treat Thomas the Apostle as having been martyred at Kalamene in India.

The earliest lists and itineraries consist of the so-called Verona fragment and a text whose author is called 'Anonymus I', both generally dated to the fourth and fifth centuries. Noticeably, the Verona fragment and the early manuscripts of Anonymus I do not refer to Thomas' burial site in India. But during the sixth century, certain manuscripts of Anonymus I began to place

Thomas' martyr site at Kalamene, and other apostolic itineraries followed suit. In some cases, the itineraries even specified that Thomas was originally buried at Kalamene but that his body was now at Edessa. Here, I provide a list of the attestations for Thomas' travels from the various itineraries.

Verona Fragment (fifth–sixth centuries): no reference[27]
Anonymus I, early manuscripts (fifth century): no reference[28]
Anonymus I, later manuscripts (sixth century): Kalamine[29]
Anonymus II (sixth century): Kalamatide[30]
Liber epiphanii (sixth century): Calamina[31]
De ortu et obitu prophetarum et apostolorum (sixth century): Calamina, transfer to

Edessa[32]

Breviarium apostolorum (sixth century): Calamina[33]
Isidore of Seville, *De ortu et obitu patrum* (c. 600): Calaminia[34]
De ortu et obitu patriarcharum et apostolorum (eighth century): Calaminice[35]
ps. Epiphanius (late eighth century): Kalamene[36]
ps. Dorotheus (late eighth century): Kalame[37]
ps. Sophronius (ninth century): Kalamine[38]
ps. Hippolytus (ninth century): Kalamene[39]
BL 17193, syr. (874 CE): Qalamiya, transfer to Edessa[40]

On the basis of their shared information and very similar phrasings, it is evident that the later lists generally culled their material from the earlier ones. But the question remains: how were references to Kalamene introduced to the earlier texts in the first place? Whereas it may be tempting to conceive of Kalamene as some sort of creative invention, there are legitimate reasons to surmise that it reflects how Latin- and Greek-speaking Christians were referring to a sacred site for Thomas in south India about which reports had arrived in the Mediterranean.[41] First, the earliest lists make no reference to any putatively specific site of India where Thomas was buried, and nor do any prior texts that deal with the travels of Thomas the Apostle in a substantive way. The interest in a putative burial site in India only emerges in the sixth century, and this seems to reflect a shift in knowledge about cult sites devoted to Thomas the Apostle. Second, we can confirm on the basis of Cosmas Indicopleustes and various sources that Persian Christians were populating south India by the end of the fifth century, having arrived recently.[42] Third, as we will see momentarily, commercial and pilgrimage networks extending from the Roman Empire were making direct contact with south India during the sixth century and were transmitting information from the subcontinent to the Mediterranean. Both Cosmas Indicopleustes and Gregory of Tours bear witness to this.

It is also interesting to note that the anonymous text called *De ortu et obitu patrum* (the sixth century) claims that Thomas was initially buried in 'Calamina' but that his relics were transferred to Edessa.[43] Like Gregory of Tours, it appears to be reconciling the relatively recent reports of Thomas' tomb in India with the older claims of the Edessenes that they had Thomas' body. At the same time, the text arguably reflects the awareness of Christians in south India that the Edessenes boasted of having Thomas' body and that these claims were almost universally recognised by Christians in the Mediterranean world and the Middle East. The basic premise is attested, perhaps independently, by a Syriac manuscript dated to 874 CE.[44] But how did Christians in the Mediterranean world know of an empty tomb for Thomas in India? How did Christians in India learn that the Edessenes were universally recognised to possess Thomas' body? The commercial, ecclesiastical, and pilgrimages that linked Christians living in Europe, the Middle East, and India provide something of an answer.

The Networks

What sort of networks connected the Mediterranean directly to India during late antiquity? Here, we aim to identify four generic types that facilitated trust and cultural exchange to varying degrees: commercial, ecclesiastical, diplomatic, and pilgrimage. Before providing surveys of them (though with no claims of comprehensiveness), some words of clarification are in order. First, the definitions being used here are somewhat arbitrary. At a certain fundamental level, networks represent the mental projections that we impose on an infinitude of complicated human activity (or on the mental projections of the ancients regarding their own complicated activity), however grounded in data or evidence they may be.[45] Here, we define commercial networks as forms of social relations driven by commercial aspirations. Likewise, ecclesiastical networks reflect the social relations of members of Christian church dioceses as determined by church institutions and organisation. Diplomatic networks consist of connected embassies and ambassadors travelling and interacting on behalf of imperial or royal governments. Finally, pilgrimage networks reflect social relations generated by the efforts of Christians to visit remote sacred sites.

A second key point is that these networks, as just described, constitute ideal types outlined here as distinct for purposes of clarification. During the sixth century, they were enmeshed and shared many of the same players, institutions, and activities. Bishops and churchmen who embodied the connectedness of ecclesiastical networks engaged in commercial ventures or participated in pilgrimages.[46] Merchants could, as Christians, associate with church institutions or travel on pilgrimage themselves.[47] Government officials could make treaties with foreign dynasts pertinent to matters of trade and merchants' movements.[48] Christian pilgrims had a litany of commercial and ecclesiastical connections.[49] The main reason for distinguishing among the different types is that, as maintained below, commercial networking provided the armature for the forms of ecclesiastical and diplomatic networking that connected the Mediterranean to India. These in turn facilitated the endeavours of pilgrims, whose activity and networking extended farther than the limit of any single commercial or ecclesiastical network even if they were not as intensive, robust, or as conducive to 'strong' ties (reflecting multifaceted, dynamic, and frequent contact). When Theodore, the informant of Gregory of Tours, travelled from Gaul to India, he exploited a pilgrimage network made possible by a web of commercial and ecclesiastical networks. Gallic merchants were not traveling to India directly, nor did Gallic bishops like Gregory of Tours have any jurisdiction there. Their domestic commercial and ecclesiastical networks did not reach so far afield. But since Gallic merchants and churchmen did have connections to merchants and churchmen in Roman Egypt that were directly connected to India, they participated in a pilgrimage network that linked them to India, even if the ties that connected them to India were relatively 'weak' (reflecting one-dimensional and infrequent social contact).

We can begin our survey with the commercial networks. During the sixth century, a commercial network of Roman Egyptians provided a direct linkage between Alexandria and Red Sea Egypt and the Indian subcontinent; another consisting of Persians extended from lower Mesopotamia to India and east Asia. While the constitutive features of these networks do not lend themselves to social network analysis, comparative analyses of pre-modern commercial networks support that they had certain generic features, even if they were characterised by considerable variation.[50] Ancient commercial networks often consisted of people of common regional origin, with certain levels of further ethnic, religious, or civic refinement. The overall regional orientation of such networks can be explained by the significance of trust in business relations for which contracts could be difficult to enforce in a pre-modern economy. People typically forged business connections to others who were known to them personally or through other personal contacts, and these were overwhelmingly (but not exclusively) compatriots who lived overseas.

In this vein, the Roman Egyptian network consisted of commercial players from Roman Egypt who operated from Red Sea ports but maintained ties with Alexandria.[51] It extended to south India and Sri Lanka during the sixth century. We can identify the contours of this network based on the testimony of later Roman authors and inscriptions. In his description of how a figure named Frumentius had evangelised the Aksumite Ethiopians, Rufinus clarifies that Roman *negotiatores* were already there to abet him.[52] When Philostorgius specifies how Theophilus 'the Indian' arrived in south Arabia to preach, he reportedly discovered that Christian communities were already present.[53] These would seem to reflect the movement of Christian commercial players to south Arabia along established social pathways. A Greek inscription made by a Christian on the island of Socotra can be most plausibly traced to a Roman Egyptian.[54] So can a Greek inscription from a synagogue at Qane in south Arabia.[55] The travels of the figure now called Cosmas Indicopleustes demonstrate that this network was in direct contact with India during the early sixth century.[56]

Another network consisting of people living under Sasanian Persian rule extended from lower Mesopotamia and Mesene to south India during the fifth and sixth centuries.[57] It in fact persisted for centuries and enabled Palmyrenes to sail from Mesene to India in the second century and for the preacher Mani to board a ship for India at Phorat in the third.[58] Again, Cosmas Indicopleustes is a key source. According to him, Persian Christians inhabited various parts of coastal India, Sri Lanka, and Socotra during the sixth century.[59] But such a testimony is not isolated. Procopius describes how Persian merchants and sailors purchased all the silk at Indian ports before the Romans' contacts could acquire it.[60] The Syriac *History of Mar Yaunan*, now dated to the seventh century, refers to a Persian ship that sailed to 'Sin'.[61] Tang dynasty sources describe Persian merchants in east Asia.[62] The commercial network of the Persians persisted into the Islamic period, and by then, included sailors originating from Fars, if not before. The ninth-century copper plates of Quilon contain privileges conferred by an Indian dynast in the Keralan coast upon a Christian community that certainly included Persians; the texts of the plates were in Tamil, but the signatures were in Arabic, Pahlavi, and Hebrew.[63] Crosses dated to the seventh-tenth centuries and inscribed with Pahlavi bear witness to this too.[64] Intriguingly, an Arabic-writing merchant from ninth-century Fars noted that Thomas' tomb could be found in a short sailing distance from Quilon in south India.[65] He too inhabited such a network.

The epigraphic texts from the cave of Hoq on the island of Socotra add another key layer. Overwhelmingly, they consist of graffiti made by Indian sailors from the region of Gujarat. Their regional origins are demonstrated by explicit statements made by their texts or by lexical aspects that link them to Gujarat between roughly 150 CE and 400 CE. In fact, the texts from the grotto are overwhelmingly written in Sanskrit or Prakrit and reflect a network extending from the vicinity of ancient Barygaza (Bharukaccha) and Astakpra (Hastakavapra).[66] Cookware and belts excavated at Berenike in Red Sea Egypt suggest the presence of visitors from the subcontinent in late antiquity.[67] Altogether, we can envision a commercial network of travellers from Gujarat who were directly connected to south Arabia and perhaps Red Sea Egypt.

The commercial networks just described laid many of the social pathways that ecclesiastical networks exploited by the sixth century. The churchmen at Alexandria had direct contact with bishops and priests in Aksumite Ethiopia and south Arabia. The bishop of Aksum was subordinate to the metropolitan of Alexandria.[68] Theophilus 'the Indian' had initiated his mission to south Arabia from Constantinople by transiting through Alexandria.[69] The church at Edessa similarly had key linkages to both the

Mediterranean coast and to the Persian Gulf. Edessene bishops were players in (or adversaries with) an 'Antiochene' doctrinal network during the fifth century.[70] They also maintained vital connections to the church dioceses of Sasanian Mesopotamia. Its members could thus identify Persian martyrs killed during the fourth century, as reflected in a famous manuscript composed at Edessa in 411.[71] It was certainly through the mediation of the Edessenes that Greek authors like Sozomen could summarise the exploits of some of the main martyrs from this list or those who were celebrated more elaborately in Persian hagiography written in Syriac.[72] The *Acts of Mar Mari* similarly reflect how an author working at Edessa maintained connections to Christian communities in lower Mesopotamia.[73]

Likewise, the Persian Church of the East at Seleucia-Ctesiphon had key linkages to Christian communities at Edessa and in India. By the fifth century, it oversaw church dioceses throughout Persian Mesopotamia, as the records from various church councils show.[74] It also was connected to a web of monasteries reaching deep into the Persian Gulf and then India.[75] By the late fifth century, Christians in Roman territory could conceive of India as governed by the church of Persians.[76] We have seen how Cosmas Indicopleustes places Persian Christians and their priests at south India, Sri Lanka, and Socotra. By the eighth (or ninth) century, the Persian Church even oversaw congregations in the Malay Peninsula.[77]

Diplomatic networks, which too were founded on the social pathways carved by connected commercial players, also provided a vital armature. One stretched from Constantinople into the Indian Ocean, with Alexandria and its merchants and churchmen maintaining some of its key pathways. Theophilus 'the Indian', originating from Socotra, had been sent to Constantinople as a hostage during his youth. But he returned to Socotra and south Arabia upon being dispatched by Constantius II himself.[78] A law from the Theodosian Code explicitly prohibits ambassadors to south Arabia and Aksumite Ethiopia from lingering in Alexandria.[79] During his brief reign, the emperor Julian received embassies from the 'Indian' peoples called *Divi* and *Serendivi*; these probably referred to ambassadors from Socotra, not Sri Lanka.[80] Two centuries later, direct diplomatic contact with Sri Lanka was occurring.[81] If we shift our attention to the Middle East, we note that diplomatic contacts between Rome and Persia are well known.[82] The Persian king of kings maintained his own connections with various royal courts in India.[83]

On the basis of the commercial, ecclesiastical, and diplomatic networks just described, a pilgrimage network that directly linked western Europe to India took shape. Jerusalem and Palestine undoubtedly were their foremost points of attraction. But Edessa also drew pilgrims who desired to interact with the remains of Thomas the Apostle and of Abgar Ukkama. Let us take the journey of Egeria for an example. During the late fourth century, she travelled from Gaul to Palestine and visited many holy sites there. She also made a trip to Edessa, where she visited the martyrion of Thomas the Apostle. She did not travel farther east, nor did Thomas' tomb in India exist at this time.[84] In a similar vein, Gregory of Tours reports a Christian festival at Edessa that attracted pilgrims from various places. Gallic contemporaries had apparently attended it. One is tempted to see the *Itinerary from Eden*, derived from the same fourth-century source as the *Exposition of the Entire World* but specifying where Christians reportedly resided in the fourth and fifth centuries, as reflecting similar types of this knowledge. The *Itinerary from Eden* in particular locates Christians at Aksum, Diaba (Socotra), lesser India (Nubian Ethiopia), and greater India (Arabia), thus identifying for potential pilgrims where they could find brethren.[85]

But Thomas' tomb in India also drew visitors, who foremost came from Indian Ocean or Persian locations (primarily as merchants) but could also hail from the Roman Mediterranean and its hinterland (as pilgrims). At certain points, a pilgrimage network did in effect stretch from the late antique Mediterranean to India. Cosmas Indicopleustes and the author of the eleventh book attributed to him (if a different author) demonstrate the effects of the network; so do the various gold and copper coins dating of the fourth-sixth centuries found in India and Sri Lanka.[86] In the ninth century, pilgrims from the court of English king Alfred the Great reportedly visited the tomb.[87] As we have seen, the references to a burial site at 'Kalamina' reflects a burgeoning interest among pilgrims in this site during the sixth century. It was this network that Gregory's contact Theodore exploited when he travelled to India and reported to Gregory, resident in northern Gaul, on the tomb. Undoubtedly, direct ties between Europe and India were relatively weak. In other words, members of this pilgrimage network rarely exploited it to make trips to India, their knowledge of India was generally not based on personal autopsy, and the social contact that its travellers to India forged with actual people in India was mostly one-dimensional and infrequent. But due to the linkages provided by intersecting commercial, ecclesiastical, and diplomatic networks, which the figure known as Cosmas Indicopleustes too exploited, such a network did exist and facilitated this form of travel. It enabled Christian churchmen to visit holy sites far beyond where the commercial networks of their territories extended. Again, it would be inaccurate, for example, to claim that a Gallic commercial network connected Gaul directly to India. But the various connected commercial networks of Afro-Eurasia enabled the activity of churchmen who were intent on visiting sacred sites populated by apostles. These created the social pathways that Theodore, the Gallic informant of Gregory of Tours, exploited to travel to Thomas' tomb in India in the late sixth century. Empowered by the networks of the Mediterranean, he was able to journey from southern Gaul to Egypt. Passing through Alexandria, he moved to a Red Sea port. A commercial network of Roman Egyptian merchants, an ecclesiastical network anchored at Alexandria, and a diplomatic network extending from Constantinople enabled his travel to south India. There, communities of Persians abetted by commercial, ecclesiastical, and diplomatic networks that directly linked lower Mesopotamia to south India had settled. They maintained connections to the Persian Gulf and lower Mesopotamia, where Edessenes could be found. They thus knew that Thomas' relics were at Edessa and maintained his original empty tomb in south India. Theodore, in turn, carried this information to Egypt and then Gaul itself.

Conclusion

The key commercial, ecclesiastical, diplomatic, and pilgrimage linkages outlined in this chapter should be kept in mind when we reflect on the knowledge that Mediterranean Christians exhibited about Thomas' burial at Edessa and in India. A remarkable feature about the sixth-century traditions regarding Thomas is the level of coordination that they reflect. As already stated, the Mediterranean sources, when taken in aggregate, communicate general agreement that Thomas' body was at Edessa and that India held his empty tomb. Did this reflect a Mediterranean assumption that since Thomas' body was at Edessa, it could not be in India? Perhaps it did. But just as likely, Christians in south India were vaunting an empty tomb there and promoting the premise that the apostle's body had been relocated to Edessa, in coordination with the Edessene claim. This was why Gregory of Tours and various other authors reported an empty tomb in India. This perception was characterised by both continuity and transition. Throughout the medieval Islamic period, many Syriac Christian sources in fact presumed that Thomas' remains were located at Edessa.[88] Yet, some medieval visitors from Europe and the Islamic world understood Christians in south India to have Thomas' relics in a local tomb; this suggests a shift in the local traditions.[89] Nonetheless, when the Portuguese started to visit Mylapore in the sixteenth century, they found only trace remains of a corpse. They also learned that Thomas' remains were reportedly at Edessa. The belief that Thomas, who rested in Edessa, had an empty tomb in India was thus revived.

Notes

1. For example, Ephrem, *CN* 42.1–3; *Teaching of the Apostles*, lb–lh; Ambrose, *in Ps.* 45.21; Gaudentius, *Tract.* 17.11; Greg. Naz., *Or.* 33.11; John Chry. *That Christ is God*, 5–6, *Act. Apost.* 4.4, *Ioan.* 65.1; Jerome, *Ep.* 59.5; *Acts of Philip* 3.2 and 8.1; *Manichaean Psalms*, 192.15, 193.2, 194.13; Paulinus of Nola, *Carmina*, 19.80–82; Ps.-Gelasius, *HE* 3.9; Jacob of Sarug, *Homilies on Thomas*; Epiphanius, *Pan.* 47.1.5 and 61.1.5; Aug., *Sermon on the Mount* 1.20.65, *Against Adimantus* 17, *Against Faust* 22.79; Turribius, *Letter to Leo*; Gregory of Tours, *Glor. Mart.* 31; Isidore of Seville, *Births and Deaths of the Fathers* 73 and 80. See Nedungatt 2008, 177–200.
2. Gregory of Tours, *Glor. Mart.* 31.
3. Gregory of Tours, *Hist. Franc.* 8.12.
4. Gregory cites *historiam passionis eius*, which appears to refer to *Passio sancti Thomae*. Gregory of Tours, *Glor. Mart.* 31.

5. See n. 13–17 for relevant references.
6. *Itinerary of Egeria*, 17.1 and 19.2–3; Ephrem, *CN* 42.1–3; *Passio sancti Thomae* 61–62.
7. For Syrians and other people of the eastern Mediterranean in later Roman and post-Roman Gaul, see Lambrechts 1937; Pieri 2005.
8. See the translation of T. K. Joseph published by H. Hosten, which is now in Menachery 1998, 518–22. See Perczel 2006, 389–93 and now 2019, 655–60 for discussion.
9. Frykenberg 2008, 92–110; Nedungatt 2008; Ramelli 2001; McGrath 2008; Van den Bosch 2001 represent differing interpretations of the relevance of the *Acts of Thomas* and the oral traditions ranging from overwhelming acceptance to near complete dismissal.
10. Andrade 2018, esp. 1–66 and 207–37.
11. Discussed by Andrade 2018, esp. 18 and 210. The use of *anno domini* dates and Gregorian calendar months are some key examples. Perczel 2006, 389–93 and now 2019, 655–60 provides important discussion of the context of the written versions of these traditions. One can remark similarly on the oral hymns ascribed to the Christian settlers of Thomas Kinayi, which probably refers to a ninth-century phenomenon. I do not treat them in this article, but see Perczel 2019, 663–71.
12. For recent compelling work on Christian communities that productively integrates many established concepts and language of network theory, see Kloppenborg 2020. My language on strong and weak ties is based on his formulation.
13. On these topics, see Klijn 2003; Attridge 1990; Myers 2010, 29–55; Roig Lanzillotta 2015.
14. My views are outlined in detail in Andrade 2018, 28–50.
15. For what follows, Andrade 2018, 28–50; Myers 2010, 29–55; Van den Bosch 2001; McGrath 2008.
16. Andrade 2019.
17. Andrade 2018, 207–32. For recent discussion of the oral traditions and an emphasis on their early modern production (in their present form), see Perczel 2006, 389–93 and now 2019, 655–60.
18. *Passio Sancti Thomae*, 61–62.
19. Ramelli 2001, 64.
20. Jullien and Jullien 2002: 89–92. Of note are the thirteenth-century Syriac works: *The Chronicle of 1234*, 2.133 and Barhebraeus, *Ecclesiastical Chronicle* 1.33–34 and 3.4–11.
21. Nedungatt 2008, 308–9; Yule 1929, 2.358; Mundadan 1970, esp. 26–34 and 68–84, 1981, 22–36 and 1984, 36–64, 415–21; Van den Bosch 2001, 147–48.
22. From 1300–1500, various Islamic and European travellers posited that Thomas' tomb at Mylapore had his body. See Jullien and Jullien 2002: 107–09.
23. See n. 21.
24. Ephrem, *CN* 42.1–3; *Itinerary of Egeria*, 17.1 and 19.2–3.
25. Gregory of Tours, *Glor. Mart.* 31.
26. Important studies are Guignard 2015 and 2016; Dolbeau 2012, which reprints Dolbeau 1986a, 1986b, 1990, 1994, 2005. Schermann 1907 is an edition for many Greek lists.
27. Turner 1913, 64.
28. Guignard 2015, 188. Also, Turner 1913, 64.
29. Guignard 2015, 188.
30. Schermann 1907, 172, with Dolbeau 2005, 468.
31. Dolbeau 1986a, 128, now in Dolbeau 2012, 48.

32. Dolbeau 1994, 106.
33. Schermann 1907, 209.
34. Isidore of Seville, *Births and Deaths of the Fathers*, 73 and 80.
35. Fraga 1996, 68.
36. Schermann 1907, 111.
37. Schermann 1907, 156.
38. Von Gebhardt 1896, 7–8.
39. Schermann 1907, 165.
40. Van Esbroeck 1994, 144 and 188–89.
41. For theories about Kalamene, and whether it could refer to Mylapore, see Nedungatt 2010.
42. Cosmas Indicopleustes, 2.54 and 3.65, with 11.17.
43. Dolbeau 1994, 106.
44. Van Esbroeck 1994, 144 and 188–89.
45. Schor 2011, 12–13.
46. For example, Seland 2012, 2013a, 2013b.
47. Jullien and Jullien 2002, 215–21.
48. Speidel 2015.
49. For instance, McCormick 2001, 123–210.
50. Terpstra 2013, esp. 65–70, 95–125, with 2019, 33–82 (esp. 33–35) informs my thinking in this paragraph. For relevance to the Indian Ocean, see Seland 2013b. Andrade 2015, 2017, and 2018 discusses region-based networks extending from Roman Egypt and the Roman Levant.
51. Andrade 2017 and 2018, 94–135; Seland 2012; Darley 2014, with important refinements by Darley 2019 that emphasise commercial connections but ideological disconnection.
52. Rufinus, *HE* 10.9.
53. Philostorgius, *HE* 3.3–6. For arguments against conceiving of Theophilus as from India or returning to preach there, see Andrade 2018, 74–77.
54. Strauch et al. 2012, 11:28; Bukharin 2012a.
55. Bukharin 2012b, 531–34.
56. Cosmas Indicopleustes, esp. 2.30, 2.54, 3.64–65. Bowersock 2013, 22–33.
57. Andrade 2018, 207–232.
58. *Cologne Mani Codex*, 144–45. Tubach 1995.
59. Cosmas Indicopleustes, 2.54 and 3.65. See generally Isho'yabh III, *Liber epistularum* in Duval 1962, 247–55 (esp. 252) and 255–60. Whitehouse and Williamson 1973, 45–49 and Colless 1969.
60. Procopius, *Bell. Pers.* 1.20.
61. Text in Bedjan 1890–97, 1.494; Payne 2011, 99–101 and 106.
62. Schafer 1951; So 1987–1988.
63. Lambourn, Veluthat, and Tomber (forthcoming); Cereti 2009; Perczel 2019, 665–71.
64. Cereti, Olivieri, and Vazhuthanapally 2002; Perczel 2019, 671–78.
65. Reinaud 1845, 2.19.
66. Strauch et al. 2012; Strauch 2012b and 2012c (esp. 344–45).
67. Sidebotham 2011, 75, 238, 261–62.
68. Sozomen, *HE* 2.24; Socrates, *HE* 1.19; Rufinus, *HE* 10.9–10; Theodoret, *HE* 1.23–24; Athanasius, *Apology to Constantine*, 29–31.
69. Philostorgius, *HE* 3.3–6, and n. 50.
70. Schor 2011, esp. 40–132.
71. Nau 1912, 24–26.
72. Sozomen, *HE* 2.9–14. See Devos 1966.
73. *Acts of Mar Mari* generally, with Jullien and Jullien 2003.
74. *Synodicon orientale*, 1–53.

75. *Synodicon orientale*, 34; *History of Mar Yaunan*, in Bedjan 1890–1897, 1.466; Brock 1999–2000; Kozah et al. 2014. Carter 2008 and 2013.
76. Ps.-Gelasius of Cyzicus, *HE* 2.28.4 and 2.38.1. Jullien and Jullien 2002, 110.
77. Isho'yabh III, *Liber epistularum* in Duval 1962, 247–55 (esp. 252) and 255–60. Whitehouse and Williamson 1973, 45–49 and Colless 1969.
78. Philostorgius, *HE* 3.3–6, and n. 50.
79. *CT* 12.12.2.
80. Ammianus, 22.7.10. Andrade 2018, 82 discusses the evidence for both the *Divi* and *Serendivi* as being from Socotra.
81. Cosmas Indicopleustes, 11 generally.
82. Blockley 1992; Dignas and Winter 2007, and Canepa 2009.
83. Payne 2016.
84. *Itinerary of Egeria*, 17.1 and 19.2–3.
85. *Itinerary from Eden* generally. For more on this source and itinerary, Bukharin 2012b, 522–26; Andrade 2018, 77–79.
86. Walburg 2008, esp. 319–44; Krishnamurthy 2007; Darley 2014 and 2015.
87. *The Anglo-Saxon Chronicle*, 883/884 CE in Dumvile and Keynes 1983-.
88. Jullien and Jullien 2002: 89–92 treat the various testimonies.
89. Jullien and Jullien 2002: 107–09.

Bibliography

Primary Sources:

Acts of Judas Thomas (Greek). Ed. Maximilian Bonnet, *Acta Apostolorum Apocrypha*, pt. 2, vol. 2. Leipzig: H. Mendelsohn, 1903 (repr. Hildesheim: Olms, 1959), 99–291.

Acts of Judas Thomas (Syriac). Ed. and trans. William Wright, *Apocryphal Acts of the Apostles*. vols. 2. London: Williams and Norgate, 1871 (repr. Hildesheim: Olms, 1990), vol. 1, q'b–šlg, vol. 2, 146–298.

Acts of Mar Mari. Ed. and trans. Christelle Jullien and Florence Jullien. *Les Actes de Mār Māri*. Leuven: Peeters, 2003; Ed. and trans. Amir Harrak, *The Acts of Mar Mari the Apostle*. Atlanta: SBL, 2005.

Acts of Philip. Ed. François Bovon, Bertrand Bouvier, and Frédéric Amsler, *Acta Philippi*. vols. 2. Turnhout: Brepols, 1999.

Ambrose, *Commentary on Twelve Psalms* (=*in Ps.*). Ed. Michael Petschenig, *Sancti Ambrosii Opera*, pars. 6: *Explanatio Psalmorum XII*. Vienna: Tempsky, 1962.

Athanasius, Apology to Constantius. Ed. Hanns Christof Brennecke, Uta Heil, and Annette von Stockhausen, *Athanasius: Werke*, vol. 2.8: *Die Apologien*. Berlin: De Gruyter, 2006.

Augustine, *Against Adimantus*. Ed. Joseph Zycha, *Sancti Aureli Augustini de utilitate credenda; de duabus animabus; contra Fortunatum; contra Adimantum. contra epistulam fundamenti; contra Faustum*. Vienna: Tempsky, 1891.

Augustine, *Against Faustus*. Ed. Joseph Zycha, *Sancti Aureli Augustini de utilitate credenda; de duabus animabus; contra Fortunatum; contra Adimantum; contra epistulam fundamenti; contra Faustum*. Vienna: Tempsky, 1891.

Augustine, *Sermon on the Mount*. Ed. Almut Mutzenbecher, *Aurelii Augustini opera*, pt. 7.2: *De sermone domini in monte libros duos*. Turnhout: Brepols, 1967.

Barhebraeus, *Ecclesiastical Chronicle*. Ed. and trans. Jean Baptiste Abbeloos and Thomas Joseph Lamy, *Gregorii Barhebraei Chronicon Ecclesiasticum: the Ecclesiastical Chronicle of Bar Hebraeus. Leuven: Peeters*, 1872–1877. (repr. Piscataway, NJ: Gorgias Press, 2012).

Chronicle of 1234. Ed. Jean Baptiste Chabot, *Anonymi auctoris Chronicon ad annum Christi 1234 pertinens*. Paris: Typographeus Reipublicae, 1916.

Cologne Mani Codex. Ed. and trans. Ludwig Koenen and Cornelia Römer, *Der Kölner Mani-Kodex: über das Werden seines Leibes*. Cologne: Westdeutscher Verlag, 1987.

Cosmas Indicopleustes, *Christian Topography*. Ed. and trans. Wanda Wolska-Conus, *Cosmas Indicopleustès: topographie chrétienne*. vols. 3. Paris: Cerf, 1968–1973.

Cyrillona. *On the Scourges*. Ed. and trans. Carl Griffin, *The Works of Cyrillona*, 136–195. Piscataway, NJ: Gorgias Press, 2016.

Epiphanius, *Panarion*. Ed. Karl Holl, *Epiphanius*. 2nd ed. vols. 5. Berlin: Akademie Verlag, 1980–2013.

Ephrem, *Nisibene Hymns* (=*CN*). Ed. and trans. Edmund Beck, *Des heiligen Ephraem des Syrers Carmina Nisibena*. Louvain: Secretariat du CorpusSCO, 1961.

Exposition of the Entire World. Ed. and trans. Jean Rougé, *Expositio totius mundi et gentium*. Paris: Cerf, 1966.

Gaudentius of Brescia, *Tractates*. Ed. and trans. Ambruse Glück, *S. Gaudentii episcopi brixiensis tractatus*. Vienna: Hoelder-Pichler-Tempsky, 1936.

Gregory of Nazianzus, *Orations*. Ed. and trans. Claudio Moreschini, *Grégoire de Nazianze: Discours 32–37*. Paris: Cerf, 1985.

Gregory of Tours, *Glory of the Martyrs*. Ed. Bruno Krusch, *Gregorii episcopi Turonensis miracula et opera minora*. Hanover: Hahn, 1969.

Gregory of Tours, *History of the Franks*. Ed. Bruno Krusch, *Gregorii Episcopi Turonensis Historiam Libri X*. Hanover: Hahn, 1937–1951.

Isidore of Seville. *On the Births and Deaths of the Fathers. César Chaparro Gómez. Isidoro de Sevilla: De ortu et obitu patrum*. Paris: Belles Lettres, 1985.

Itinerary from Eden (Hodoiporia apo Edem). Ed. and trans. Jean Rougé, *Expositio totius mundi et gentium*, 350–355. Paris: Cerf, 1966.

Itinerary of Egeria. Ed. and trans. Pierre Maraval, *Journal de voyage: itinéraire, Egérie*. Paris: Cerf, 1997.

Jacob of Sarug, *Homilies on Thomas*. Ed. and trans. Werner Strothmann, *Drei Gedichte über den Apostel Thomas in Indien*. Wiesbaden: Harrassowitz, 1976.

Jerome, *Letters* (=*Ep.*). Ed. Isidore Hilberg, *Sancti Eusebii Hieronymi epistulae*. Ed. alt. vols. 3. in 4 pts. Vienna: VOAW, 1996.

John Chrysostom, *That Christ is God. Quod Christus sit Deus*. PG 48, cols. 813–843.

John Chrysostom, *On Acts of the Apostles (=Act. Apost.). In Acta Apostolorum*. PG 60, cols. 13–582.

John Chrysostom, *On John (=Ioan.). In Ioannum*. PG 59, cols. 23–484.

Manichaean Psalms. Ed. Charles Robert Cecil Allberry, *A Manichaean Psalm-Book*, pt. 2. Stuttgart: Kohlhammer, 1938.

Passio Sancti Thomae. Ed. Klaus Zelzer, *Die alten lateinischen Thomasakten*. Berlin: Akademie-Verlag, 1977.

Paulinus of Nola, *Carmina*. Ed. G. de Hartel, *S. Pontii Meropii Paulini Nolani opera*. 2nd ed. Vienna: VOAW, 1999.

Philostorgius, *Ecclesiastical History*. Ed. Joseph Bidez and Friedhelm Winkelmann, *Philostorgius: Kirchengeschichte*. 3rd ed. Berlin: Akademie Verlag, 1981.

Pseudo-Gelasius of Cyzicus, *Ecclesiastical History*. Ed. Günther Christian Hansen, *Anonyme Kirchengeschichte (Gelasius Cyzicanus, CPG 6034)*. Berlin: De Gruyter, 2002.

Socrates. *Ecclesiastical History (=HE).* Ed. Günther Christian Hansen, *Kirchengeschichte: Sokrates.* Berlin: Akademie Verlag, 1995.
Sozomen. *Ecclesiastical History (=HE).* Ed. Joseph Bidez and Günther Christian Hansen, *Kirchengeschichte.* 2nd ed. Berlin: Akademie Verlag, 1995.
Synodicon orientale. Ed. Jean Baptiste Chabot, *Synodicon orientale.* Paris: Librarie Klincksieck, 1902. (repr. Piscataway, NJ: Gorgias Press, 2010).
Teaching of the Apostles (Doctrina apostolorum). Ed. and trans. William Cureton, *Ancient Syriac Documents.* London: William and Norgate, 1864: kd-lh; Arthur Vööbus, *The Synodicon in the West Syrian Tradition.* 4 vols.: 1.200–11.
Theodoret. *Ecclesiastical History* (=*HE*). Ed. Leon Parmentier and Günther Christian Hansen, *Theodoret: Kirchengeschichte.* 3rd ed. Berlin: Akademie Verlag, 1998.
Turribius of Asturica. *Letter to Leo.* PL 54, col. 693–694.

Secondary Scholarship:

Andrade, Nathanael. (2015). "The Voyage of Maes Titianos and the Dynamics of Social Connectivity between the Roman Levant and Central Asia/West China," *Mediterraneo antico* 18: 41–74.
Andrade, Nathanael. (2017). "Drops of Greek in a Multilingual Sea: The Egyptian Network and its Residential Presences in the Indian Ocean," *JHS* 137: 42–66.
Andrade, Nathanael. (2018). *The Journey of Christianity to India in Late Antiquity: Networks and the Movement of Culture.* Cambridge: Cambridge University Press.
Andrade, Nathanael. (2019). "A Palmyrene Contribution to Eastern Christianity: The Name of King Gondophares," in R. Raja (ed.), 189–198. *Revisiting the Religious Life of Palmyra.* Turnhout: Brepols.
Attridge, Harold. (1990). "The Original Language of the Acts of Thomas," in Harold Attridge, John Collins, and Thomas Tobin (eds.), 241–250. *Of Scribes and Scrolls: Studies on the Hebrew Bible, Intertestamental Judaism, and Christian Origins.* Lanham, MD: University Press of America.
Bedjan, Paul. (1890–1897). *Acta Martyrum et Sanctorum Syriace.* Paris: Harrassowitz.
Blockley, R. C. (1992). *East Roman Foreign Policy: Formation and Conduct from Diocletian to Anastasius.* Leeds: Cairns.
Bowersock, G. W. (2013). *The Throne of Adulis: Red Sea Wars on the Eve of Islam.* New York: Oxford University Press.
Brock, Sebastian. (1999–2000). "Syriac Writers from Beth Qatraye," *Aram* 11–12: 85–96.
Bukharin, Mikhail. (2012a). "The Inscriptions at Hoq," in Ingo Strauch (ed.), *Foreign Sailors on Socotra: The Inscriptions and Drawings from the Cave Hoq.* Bremen: Hempen Verlag, 494–500.
Bukharin, Mikhail. (2012b). "Greeks on Socotra: Commercial Contacts and Early Christian Missions," in Ingo Strauch (ed.), *Foreign Sailors on Socotra: The Inscriptions and Drawings from the Cave Hoq.* Bremen: Hempen Verlag, 501–539.
Canepa, Matthew. (2009). *The Two Eyes of the Earth: Art and Ritual of Kingship between Rome and Sasanian Iran.* Berkeley: University of California Press.
Carter, R. A. (2008). "Christianity in the Gulf during the First Centuries of Islam," *Arabian Archaeology and Epigraphy* 19: 71–108.

Carter, R. A. (2013). "Christianity in the Gulf after the Coming of Islam: Redating the Churches and Monasteries of Bet Qatraye," in Christian Julien Robin and Jérémie Schiettecatte (eds.), 311–330. *Les préludes de l'Islam: Ruptures et Continuités dans les Civilisations du Proche-Orient, de l'Afrique Orientale, de l'Arabie et de l'Inde à la Veille de l'Islam*. Paris: Boccard.

Cereti, Carlo. (2009). "Pahlavi Signatures on the Quilon Copper Plates (Tabula Quilonensis)," in Werner Sundermann, Almut Hintze, and François de Blois (eds.), 31–50. *Exegisti monumenta: Festschrift in Honour of Nicholas Sims-Williams*. Wiesbaden: Harrassowitz.

Cereti, Carlo, Luca Olivieri, and Joseph Vazhuthanapally. (2002). "The Problem of the St. Thomas Crosses and Related Questions: Epigraphical Survey and Preliminary Research," *East and West* 52.1: 285–310.

Colless, Brian. (1969). "Persian Merchants and Missionaries in Medieval Malaya," *Journal of the Malaysian Branch of the Royal Asiatic Society* 42.2: 10–47.

Darley, Rebecca. (2014). *Indo-Byzantine Exchange, 4th to 7th Centuries: A Global History*. PhD Diss: Birmingham University.

Darley, Rebecca. (2015). "Self, Other, and the Use and Appropriation of Late Roman Coins in Peninsular India (4th to 7th Centuries CE)," in H. P. Ray (ed.), 60–86. *Negotiating Cultural Identity: Landscapes in Early Medieval South Asian History*. London: Routledge.

Darley, Rebecca. (2019). "The Tale of the Theban Scholastikos, or Journeys in a Disconnected Sea." *JLA* 12.2: 488–518.

Devos, Paul. (1966). "Sozomène et les *Actes Syriaques* de S. Siméon bar Sabba'e," *AB Journals* 84: 443–456.

Dignas, Beate and Engelbert Winter. (2007). *Rome and Persia in Late Antiquity: Neighbours and Rivals*. Cambridge: Cambridge University Press.

Dolbeau, François. (1986a). "Deux opuscules latins, relatifs aux personnages de la Bible et antérieurs à Isidore de Séville," *Revue d'Histoire des textes* 16: 83–139.

Dolbeau, François. (1986b). "Une liste ancienne d'apôtres et de disciples, traduite du grec par Moïse de Bergame," *AB Journals* 104: 299–314.

Dolbeau, François. (1990). "Une liste latine de disciples et d'apôtres," *AB Journals* 108: 50–70.

Dolbeau, François. (1994). "Nouvelles recherches sur le *de ortu et obitu prophetarum et apostolorum*," *Augustinianum* 34: 91–107.

Dolbeau, François. (2005). "Listes d'apôtres et de disciples," in Pierre Geoltrain and Jean-Daniel Kaestli (eds.), vol. 2, 455–480. *Écrits apocryphes chrétiens*. Paris: Gallimard.

Dolbeau, François. (2012). *Prophètes, apôtres et disciples dans les traditions chrétiennes d'Occident: vies brèves et listes en latin*. Brussels: Bollandistes.

Dumvile, David and Simon Keynes. (1983). *The Anglo-Saxon Chronicle: A Collaborative Edition*. Cambridge: D. S. Brewer.

Duval, Rubens. (1962). *Išo'yahb Patriarchae III Liber epistularum*. Louvain, Secrétariat du CorpusSCO.

Fraga, J. Carrecedo. (1996. *Liber de ortu et obitu patriarcharum*. Turnhout: Brepols.

Frykenberg, Robert. (2008). *Christianity in India: From Beginnings to Present*. Oxford: Oxford University Press.

Gebhardt, Oscar von. (1896). *Hieronymus De viris inlustribus in griechischer Übersetzung (der sogenannte Sophronius)*. Leipzig: Hinrichs.

Guignard, Christolphe. (2015). "La tradition grecque de la liste d'apôtres 'Anonyme I' (*BHG* 153c), avec un appendice sur la liste *BHG* 152n," *Apocrypha* 26: 171–209.

Guignard, Christolphe. (2016). "Greek Lists of the Apostles: New Findings and Open Questions," *ZAC* 20.3: 469–495.

Jullien, Christelle and Florence Jullien. (2002). *Apôtres des confins: processus missionnaires chrétiens dans l'Empire Iranien*. Bures-sur-Yvette: GECMO.

Jullien, Christelle and Florence Jullien. (2003). *Aux origins de l'Église de Perse: les Actes de Mar Mari*. Leuven: Peeters.

Klijn, A. F. J. (2003). *The Acts of Thomas: Introduction, Text, and Commentary*, 2nd ed. Leiden: Brill.

Kloppenborg, John. (2020). "Recruitment to Elective Cults: Network Structure and Ecology," *NTS* 66: 323–350.

Kozah, Mario et al. (2014). *The Syriac Writers of Qatar in the Seventh Century*. Piscataway, NJ: Gorgias Press.

Krishnamurthy, R. (2007). *Late Roman Copper Coins from South India: Karur, Madurai, and Tirukkoilur*. Chennai: Garnet.

Lambourn, Elizabeth, Kesavan Veluthat, and Roberta Tomber. (Forthcoming). *The Copper Plates from Kollam: Global and Local in Ninth Century South India*. Delhi: Primus.

Lambrechts, Pierre. (1937). "Le commerce des 'Syriens' en Gaule du Haut Empire a l'Époque Mérovingienne," *L'Antiquité Classique* 6: 35–61.

McCormick, Michael. (2001). *Origins of the Medieval Economy: Communications and Commerce A.D.* 300–900. Cambridge: Cambridge University Press.

McGrath, James. (2008). "History and Fiction in the *Acts of Thomas*: The State of the Question," *Journal for the Study of Pseudepigrapha* 17: 297–311.

Menachery, George. (1998). *The Nazranies*. Thrissur: South Asia Research Assistance Services.

Mundadan, A. M. (1970). *Sixteenth Century Traditions of St. Thomas Christians*. Bangalore: Dharmaram College.

Mundadan, A. M. (1981). "Traditions of St. Thomas and the Tomb of Mylapore," in K. J. John (ed.), 22–36. *Christian Heritage of Kerala: Grand Chevalier Prof. L. M. Pylee Felicitation Volume*. Cochin: L. M. Pylee Felicitation Committee.

Mundadan, A. M. (1984). *History of Christianity in India: From the Beginning up to the Middle of the Sixteenth Century (up to 1542)*, vol. 1. Bangalore: Theological Publications.

Myers, Susan. (2010). *Spiritual Epicleses in the Acts of Thomas*. Tübingen: Mohr Siebeck.

Nau, F. (1912). "Martyrologes et ménologes orientaux: un martyrologe et douze ménologes syriaques," *PO* 10: 7–26

Nedungatt, George. (2008). *Quest for the Historical Thomas, Apostle of India: A Re-Reading of the Evidence*. Bangalore: Theological Publications.

Nedungatt, George. (2010). "Calamina, Kalamides, and Cholamandalam: Solution of a Ridde," *Organic Chemistry Plus* 76: 181–199.

Payne, Richard. (2011). "Monks, Dinars, and Date Palms: Hagiographical Production and the Expansion of Monastic Institutions in the Early Islamic Persian Gulf," *Arabian Archaeology and Epigraphy* 22: 97–111.

Payne, Richard. (2016). "Iranian Cosmopolitanism: World Religions at the Sasanian Court," in Myles Lavan, Richard Payne, and John Weiswiler (eds.), 209–230. *Cosmopolitanism and Empire: Universal Rulers, Local Elites, and Cultural Integration in the Ancient Near East and Mediterranean*. Oxford: Oxford University Press.

Perczel, István. (2006). "Language of Religion, Language of People, Languages of Documents: The Legendary History of the St. Thomas Christians of Kerala," in Ernst Bremer et al. (eds.), 387–428. *Language of Religion, Language of People: Judaism, Medieval Christianity, and Islam*. Munich: Wilhlem Fink.

Perczel, István. (2019). "Syriac Christianity in India," in Daniel King (ed.), 653–697. *The Syriac World*. London: Routledge.

Pieri, Dominique. (2005). *Le commerce du vin oriental à l'Époque byzantine, Ve-VIIe siècles: le témoignage des amphores en Gaule*. Beirut: IFPO.

Ramelli, Ilaria. (2001). "La tradizione su Tommaso apostolo dell'India," in Cristiano Dognini and Ilaria Ramelli (eds.), 59–82. *Gli apostoli in India: nella patristica e nella letteratura sanscrita*. Milan: Medusa.

Reinaud, M. (1845). *Relations des voyages faits par les Arabes et les Persans dans l'Inde et à la Chine*. Paris: Imprimerie Royale.

Roig Lanzillotta, Lautaro. (2015). "A Syriac Original for the *Acts of Thomas?* The Hypothesis of Syriac Priority Revisited," in Ilaria Ramelli and Judith Perkins (eds.), 105–134. *Early Christian and Jewish Narrative: The Role of Religion in Shaping Narrative Forms*. Tübingen: Mohr Siebeck.

Schafer, Edward. (1951). "Iranian Merchants in T'ang Dynasty Tales," *University of California Publications in Semitic Philology* 11: 403–22.

Schermann, Theodore. (1907). *Prophetarum vitae fabulosae, Indices apostolorum discipulorumque Domini, Dorotheo, Epiphanio, Hippolyto aliisque vindicate*. Leipzig: Teubner.

Schor, Adam. (2011). *Theodoret's People: Social Networks and Religious Conflict in Late Roman Syria*. Berkeley: University of California Press.

Seland, Eivind. (2012). "Trade and Christianity in the Indian Ocean during Late Antiquity," *Journal of Late Antiquity* 5.1: 72–86.

Seland, Eivind. (2013a). "The *Liber Pontificalis* and Red Sea Trade of the Early to Mid 4th Century," in Dionisius Agius et al. (eds.), 117–26. *Navigated Spaces, Connected Places: Proceedings of the Red Sea Project V.* Oxford: Archaeopress.

Seland, Eivind. (2013b). "Networks and Social Cohesion in Ancient Indian Ocean Trade: Geography, Ethnicity, Religion," *Journal of Global History* 8.3: 373–390.

Sidebotham, Steven. (2011). *Berenike and the Ancient Maritime Spice Route*. Berkeley: University of California Press.

So, Francis K. H. (1987–1988). "Middle Easterners in the T'ang Tales," *Tamkang Review* 18.2: 259–276.

Speidel, Michael. (2015). "War, trade, and treaties: new, revised, and neglected sources of the political, diplomatic, and military aspects of imperial Rome's relations with the Red Sea basin and India, from Augustus to Diocletian," in K. S. Mathew (ed.) 83–128. *Imperial Rome, Indian Ocean Regions, and Muziris: New Perspectives on Maritime Trade*. New Delhi: Oxford University Press.

Strauch, Ingo. (2012a). *Foreign Sailors on Socotra: The Inscriptions and Drawings from the Cave Hoq*. Bremen: Hempen Verlag.

Strauch, Ingo. (2012b). The brahmi scripts of Hoq: The palaeographical perspective, in Ingo Strauch (ed.), *Foreign Sailors on Socotra: The Inscriptions and Drawings from the Cave Hoq*. Bremen: Hempen Verlag, 286–342.

Strauch, Ingo. (2012c). The Indian participants of trade: The historical perspective, in Ingo Strauch (ed.), *Foreign Sailors on Socotra: The Inscriptions and Drawings from the Cave Hoq*. Bremen: Hempen Verlag, 343–359.

Strauch, Ingo et al. (2012). Materials, in Ingo Strauch (ed.), *Foreign Sailors on Socotra: The Inscriptions and Drawings from the Cave Hoq*. Bremen: Hempen Verlag, 213–218.

Terpstra, Taco. (2013). *Trading Communities in the Roman World: A Micro-Economic and Institutional Perspective*. Leiden: Brill.

Tubach, Jürgen. (1995). "Mani und der palmyrenische Kaufmann aus Forat," *ZPE* 106: 165–169.

Turner, C. H. (1913). "A Primitive Edition of the Apostolic Constitutions and Canons: An Early List of Apostles and Disciples," *JThS* 15: 53–65.

Van den Bosch, L. P. (2001). "India and the Apostolate of St. Thomas," in Jan Bremmer (ed.), 125–148, *The Apocryphal Acts of Thomas*. Leuven: Peeters.

Van Esbroeck, Michel. (1994). "Neuf listes d'apôtres orientales," *Augustinianum* 24.1: 109–199.

Walburg, Reinhold. (2008). "Ancient Ruhuna: Sri Lankan-German Archaeological Project in the Southern Province," in *Coins and Tokens from Ancient Ceylon*, vol. 2. Mainz: Zabern.

Whitehouse, David and Andrew Williamson. (1973). "Sasanian Maritime Trade," *Iran* 11: 29–49.

Yule, Henry. (1929). *The Book of Ser Marco Polo, the Venetian, Concerning the Kingdoms and Marvels of the East*. 3rd ed., vol. 2. London: J. Murray.

Part III

Systems: the structures that bind

8 Networking Christians? The spread of Christianity in the Eastern Mediterranean

Rebecca Sweetman

Introduction

The Christianisation of the Eastern Mediterranean was a steady but gradual process, the manifestations of which appear in inscriptions and burial customs from the first century. While the remains of Christian house churches have been identified from the late first century, the most secure earliest example is that at Dura Europos, dating to the mid third century. By the time Constantine and Licinius had issued the Edict of Milan in 313 allowing Christian worship, Christianity had already had some three centuries to hone methods of conversion and integration. Following the Edict of Thessalonica in 380, Christianity became the state religion but it is probable that some churches had already been constructed before this; for example, the Cathedral of Paulinus was purportedly built in Tyre in 317[1] and locations of other Constantinian churches include Rome (313), Aquileia (313–19) and Chlef, Algeria (324).

After the construction of these first monumental churches, the impetus for church building in the east, both organic and planned, spread through a number of different interpersonal channels such as the peripatetic people involved in trade, craft, and pilgrimage. Church building, changes in burial practices, and epigraphy are some of the archaeologically visible results of Christianisation, a term that refers to both the means and end-result of conversion to Christianity. To date, the focus of Christianisation has been on the outcome of conversion, rather than on the practicalities of spread of Christianity. That is to say, studies of literary data and material culture such as architecture and iconography have enabled scholars to focus more on the impact of Christianity than on the actual process. The consequence of this has been that while a great deal has been done to counter traditional interpretations of the negative elements of conversion, such as the destruction of temples,[2] and even highlight the more positive aspects such as a peaceful transition, we are still left with more nebulous views of how and why Christianisation was so successful. Many have argued that the presence of Jewish communities in the Mediterranean provided the conduit for the spread of Christianity[3] and its success was due to its strategic and

DOI: 10.4324/9780429429217-11

largely peaceful processes that drew on existing traditions.[4] Furthermore, Kloppenborg has argued convincingly that the spread of Christianity required focused efforts of persuasion.[5] With some notable exceptions, such as Kloppenborg, to date, scholars have focused on the impact of Christianity, rather than its spread, which has led to generalisations determining the timing and extent of Christianisation in various places based on factors such as geography and existing religious context. If we take Athens as an example, it is prominent in its late adoption of Christianity, with the earliest church, the Ilissos Basilica, constructed in the first half of the fifth century. It is purportedly slow for a number of reasons, including the strength of the philosophical schools and a decline in the aristocracy. The consequences of this were a dearth of patrons for church and other buildings which in turn had an impact on the numbers of network connections that Athens belonged to; for example, those created by architects, craftspeople, and elites.[6]

Christians actively sought to convert others in the early stage, and conversion was naturally based on direct contact between people,[7] that is to say, strong ties. Strong ties are created by regular contact and reciprocity and tend to be characteristic of family links and these are ties that bind communities. Weak ties are more often connected through association such as through business or trade. The presence of random weak ties enables the easy transmission of information between separate groups or clusters and have been apparently observed in many different situations across time and space.[8] Trade routes and even pilgrims provided the conduits, and well-trodden patterns of travelling the same route helped to consolidate the connections made by weak ties. Weak ties extend beyond the core network in terms of scope and information had significant strength and broader spreads than strong ties.[9]

Recent discussions of Christianisation have focused on the false dichotomies that have been previously used to highlight differences between polytheists and Christians.[10] Furthermore, Christianity was not an exclusive religion bound to particular communities (such as Mithraism was) and it actively included diverse populations from different parts of the Empire with a range of socio-economic backgrounds, within the specific formula of religious practice.[11] Moreover, the size of the area that was converted and the diversity of the regions, communities, and individuals within makes it difficult to encapsulate the macro view of Christianisation, and instead, focus has been on individual regions. Recent developments in network analysis and its application to archaeological data make such a challenge more feasible.[12] Furthermore, it enables a means of seeing how different regions relate to each other with the common denominator of the early church as a comparator.

Studies of Christianisation have identified theoretical processes of how the impetus for church building spread; as a result of strategic change that incorporates elements of planned and unplanned results. Three phases of change in non-linear and complex groups have been identified in these strategic processes: complexity, social movement, and tipping point.[13] Complexity can be attributed to church building in busy and highly networked locations (for example, the edge of popular sanctuaries). Planned elements of strategic change can be identified in the earliest church constructions, likely through Imperial behest in any given region followed by unplanned outcomes which can be defined as social movement and tipping point.[14] Social movement is the spread of ideas through adaptive processes spurred on by striving to have what others already have. Social movement describes the spread of churches from well-networked locations such as sanctuaries or port towns (hubs) to their connected places like other busy towns. Tipping point, the final stage, defines the spread of churches to rural areas, the most conservative and least connected.[15] The aim of this chapter is to explore the links that created the planned and unplanned spread of Christianity. The starting point for this discussion is an examination of the archaeological data from urban spaces in the Eastern Empire. A much more substantial study would be required to focus on the extra-urban data and indeed the historical data, which too is quite patchy. Within the cities, focus will be on the monumentalisation of Christianity, that is to say, the construction of the earliest churches in the cities. While many cities may claim to have an early Christian population, this is not always supported through external data. It will be argued here that Christianisation was the successful product of weak and strong ties. Price argued that the spread of religion, in this case Christianity, was in part because of Christians working and travelling within and between already established groups of Christians. As such, Price would argue that the act of moving within the Empire did not create new cults independently, that there needed to be communities of like-minded people already there, to be nodes and enable the spread of ideas through strong ties.[16] Collar[17] also argued that the spread of Rabbinic Judaism in the late first to second century CE was through strong ties and was therefore reliant on the presence of like-minded groups. However, I would argue here, that the extent and success of the spread of Christianity was such that diversity and a combination of strong and weak ties were necessary for different stages of conversion.

A number of different urban spaces will be included as the central focus including Salamis, Antioch, Apamea, Ephesus, Athens, Corinth, Edessa, and Pergamon. These eight cities have been chosen for a number of different reasons. They are included in Stark's list of twenty-two of the largest cities in the first century CE based on Chandler and Fox's population density analysis published in 1974.[18] They have been selected because of their sizes and diversities within and between the cities. They are in different topographic locations, belong to different provinces, are situated at a range of distances from the Holy Land, and have a wide variety of population groups.

Following a brief summary of the situation in each of the eight cities and the evidence for the earliest church building, patronage, and pilgrimage, a synopsis of trade networks will be presented. The issue of how to connect the Christianisation processes in these cities that are so different from each other in terms of size, location, and population will be addressed by looking at networks created through strong and weak ties. This approach allows a more complex view than the traditional one which favours the idea that relative geographic location was a determining factor in how quickly an area was evangelised.[19] Furthermore, this will enable not just a better understanding of the practicalities of Christianisation, but will also highlight the importance of place, tradition, and choice in the adoption of the new religion.

Salamis

Salamis, situated on the east coast of Cyprus and of the eight towns discussed here, is the closest to Jerusalem, only taking five-and-a-half days by sea and/or land to get there.[20] In the Roman period, although Salamis was not the seat of the governor, it was favoured by Emperors such as Trajan and Hadrian who had invested in the city following damage caused by an earthquake in the late first century and the Jewish revolt in 116 CE.[21] The island thrived during the late antique period particularly from the fourth century as pilgrimage traffic began to intensify. Salamis was destroyed by earthquakes in 332 and 342 CE and it was rebuilt on a significant scale by Emperor Constantius II as a Christian city.[22]

The cities of the east coast were a natural stopping point for pilgrims travelling by sea, especially between Constantinople and the Holy Land. Cyprus was part of the Diocese of Oriens administered by Antioch; but in 431, it was given its own archbishop with the creation of the metropolitan at Salamis.[23] Being on the pilgrimage route provided an income as well as a stimulus to bolster the status of its own pilgrim sites. An example of this is the construction of the church of Agios Varnavas with a *mensa martyris* (fifth c.), near the location of the relics of the Apostle Barnabas west of Salamis. The earliest church in Salamis is also the largest on the island, Agios Epiphanius.[24] This was constructed in the fourth century, close to the Agora, along a roadway and in the centre of the city. Agios Barnabas and another small church were built in the fifth century and the Hagiasma of Nicodemus was founded in the sixth century. The church by the harbour, Kampanopetra, dates to the sixth century.[25] As noted above, Salamis attracted investment by Emperors, most significantly Constantius II. It had an excellent harbour for trade and pilgrimage traffic.

Antioch

Antioch was located at a significant point along the land route from Jerusalem to Constantinople as well as benefiting from being on the lucrative silk route from the east. It took a little under seven days to get from Jerusalem to Antioch by land and sea. The city was one of the largest in the Empire: it was provincial capital of Syria, a patriarchal see and its wealth and potential attracted a varied population of different religions and cultures from across the Empire. A significant Jewish community had grown in the city from the second century BCE. Mayer and Allen suggest that a key reason for the presence of a strong early Christian community from the first century CE was because of Antioch's location on the Mediterranean coast, the route along which the original dissemination of Christianity from the Holy Land spread.[26] Antioch was the focus of some early visits by key proselytisers such as the Apostle Peter and slightly later, Paul and Barnabas. Ammianus Marcellinus (a fourth-century Antiochan historian) noted that Christian, Jewish, and polytheistic communities lived together reasonably peacefully, although scholars such as Shepardson[27] and Sandwell[28] suggest that this too was carefully orchestrated and that the city had its share of tension and contested spaces. Some argue that the Emperor Julian's reason for staying in Antioch for so long was partly in the hope of drawing on the city's polytheistic population to bolster his anti-Christian resolve while possibly hoping for some potential friction between the different communities there.[29]

The importance of the city as a Christian centre became consolidated and imbued with stories of its important Christians, such as one of its first bishops, Ignatius, who was martyred in Rome at the beginning of the second century and those who were martyred under Diocletian and then Julian.[30] Many churches and martyria at Antioch have not yet been fully excavated or published. One of the earliest churches, the Palaia, was destroyed in 293–313 and its rebuilding was completed in 324 under bishop Vitalis.[31] Shortly afterwards, the Great Church construction was completed in 340[32] and this was followed by the foundation of the Church of St Babylas and Church of the Maccabees.[33] A baptistery was added to the church of St Babylas in the early fifth century and several churches and martyria are built after that throughout the city, suburbs, and port at Seleuceia Pieria.[34] The relics of Babylas could be viewed at one of the first martyria built in the city, near the Temple of Apollo in the 350s. The construction of the early martyria and churches provided the visual reminders of the importance of Christianity in the city even with such a diverse population. The fact that the city was on a key route meant that its standing was maintained. Furthermore, numerous pilgrims and visitors such as Egeria, Paul, and Jerome from the fourth century onwards were attracted by the promise of the sight of relics of Procopius, St Ignatius, Marinus, and Phocas in the church of Michael the Archangel, as well as the body of Symeon Stylites.[35] The city even had a common martyrium (the Koimeterion). This lay outside the gate situated at the southwest end of the main street on the way to Daphne.[36] In spite of earthquakes and fires, the churches in the city were kept in a good state of repair with several churches undergoing rebuilding under Justinian. In addition to trade and pilgrimage connections, as was the case in Salamis, Antioch had significant political and imperial links. Notably, the city was visited by Julius Caesar, several Emperors, and members of the imperial family (Germanicus, Trajan, and Julian), and benefitted from Imperial investment by Augustus, Tiberius, Hadrian, and Commodus. In the fourth century, following Julian's visit and the closure of the Great Church, Valens visited the city and re-opened the cathedral. This high-level interest in the city would have encouraged others to invest in the city as well as attracting visitors such as architects and craftspeople.

Apamea

Apamea was a metropolitan see and was located just over nine days away from Jerusalem on the land route to Constantinople at a distance of 280 miles. It was known for its very early Jewish community[37] and, according to Procopius,[38] the city had a relic of a fragment of the true cross, housed in the Eastern Cathedral. In the Roman period, the city was favoured with some imperial investment (Trajan and Marcus Aurelius). It was also famous for its neo-Platonic schools of philosophy.[39] At the same time as some cities were building churches, the Apameans, in 362 CE, laid a mosaic depicting Socrates (named in an inscription) and six sages in the building which later was replaced by the Eastern Cathedral. Balty notes that the arrangement of the philosopher and the sages is reminiscent of Christ and the disciples at the Last Supper.[40] If this is the case, it suggests that there was unproblematic shared knowledge of the Christian imagery, although the earliest church seems to date to the fifth century.[41] This is paralleled with a resurgence of interest and investment in the city with public officials inscribing their investments in the city.[42] Three churches are known to have been built in the sixth century: the atrium church, which replaced a fifth-century building which had been built on a fourth-century synagogue, the Rotunda, and the Eastern Cathedral which was built on a martyrium and incorporated a fifth-century church construction.[43] These churches were built as part of a major reconstruction of the city under Justinian. In comparison to other cities close to the Holy Land, it is interesting that the churches are founded at a relatively late date at Apamea, particularly considering how close the city was to Jerusalem. This highlights the importance of looking at factors including the composition of the population (including temporary visitors), patronage, and communication links. Apamea, like Athens, seems to have lacked distinctive investment in the city in the fourth and early fifth centuries and, like Athens, the investment which starts in the mid fifth century marks the construction of the first churches, grows steadily with the majority of churches being built in the sixth century.

Ephesus

Ephesus was a city of significant size, known as a trade emporium, and is located a travelling distance of nine days by sea from Jerusalem. The Gothic raids on Ephesus in the 260s destroyed renowned features like the Ephesian

Harbour, as well as the Temple of Artemis. However, during the fourth century, Ephesus retained much of its existing urban topography and trade appears to have continued in spite of the destruction and reconstruction of the harbour area.[44] The city made a fast recovery with plenty of investment, ranging from private citizens to imperial players. Ephesus thrived in its central position as a trading hub and as a city of patrons, and it was able to capitalise on this when it became the seat of the proconsul of Asia.

A few churches are constructed in the fourth century, such as the church of St Mary and the Basilica with mosaics in the late fourth or early fifth. Some public buildings were converted at Ephesus which drew on the integration of existing spaces.[45] The city had a multicultural population including Jewish communities, as well as Christians. Ephesus, like Salamis and Antioch, had different levels of network connections (imperial, political, and economic). The extensive re-building in the city in the fourth century by Constantine would have made it an attractive location for craftspeople thus diversifying the connections already created by trade and pilgrims.

Athens

It would have taken just over ten days to travel between Athens and the Holy Land by sea in the summer. Evidence for early Christian groups is primarily known from literary sources, including St Paul's lacklustre attempts to gain more converts in the 50s CE in Athens.[46] During the first half of the third-century Athens was renowned as a centre of culture and education but this was brought to an end by a massive attack on the city by the migrating Heruli tribal group in 267 CE. According to Frantz, the city took at least a century to recover.[47] At the end of the fourth century, Athens was supposedly invaded by Alaric and the Goths. Destructions, for example in the Herodian, are often attributed to a sack but the literary sources do not necessarily all support a destructive invasion. For example, Zosimus suggests that Alaric withdrew and left Athens unharmed while St Jerome suggests the city was taken.[48] After a hiatus, investment in the city was rekindled in the early fifth century with financing of places like the Library of Hadrian in 408–12 by Herculius, Eparch of Illyricum. The earliest church, the Ilissos Basilica, was constructed just outside one of the main eastern gates of the city in the mid fifth century. Several church constructions followed, including one in the Library of Hadrian. After these constructions, the majority of churches were built in the sixth and seventh centuries. It was only during the sixth century that many of the city temples were converted.

Some have suggested the lateness of church building in Athens is partly due to the strength of the philosophical school. There may be some weight in this suggestion, given the fact that Apamea, similarly renowned for its neo-Platonic school also has somewhat later church constructions than its neighbour Antioch. However, a pattern of disturbed imperial and elite investment in the city may have a significant role to play in this delay, which would have had a knock-on effect on the circulation of craftspeople and elite visitors to the city thereby narrowing the diversity of network connections.

Corinth

Corinth was the capital of the province of Achaea. It would have taken around sixteen days to get from Jerusalem to Corinth by sea. At Corinth, the phasing of the church construction is clear. The earliest churches are constructed in the existing cemeteries, for example, the three-aisled Kodratos Basilica,[49] which in addition may have become a martyr location in the early fifth century. As outlined in an earlier paper,[50] a tradition of visiting these locations would already have been built up by families making visits to their dead and memories of the importance of the place would be triggered through the ritual actions undertaken at the site, often in community groups. Other churches were then constructed outside the city walls such as at Kraneion and Skoutelas.[51] Regular movements of people to the extramural churches from the urban centres to these locations would have helped consolidate the sense of community and ownership with the church at the heart of this. As in other Late Antique Peloponnesian towns, one of the latest church constructions occurs on the hill in the old urban centre of Corinth (in the late sixth century), close to the former central cult site, the temple of Apollo. At that point, Christianity had already taken a firm hold; several churches in the suburbs had been in existence for at least 100 years. Corinth had extensive trade connections in the west and east and had a long tradition of investment by Emperors such as Nero, Vespasian, and later Constans and Theodosius II.[52]

Edessa

Edessa was a 16.7-day journey north of Jerusalem and was located on the Silk Road. It was reputedly one of the locations of the shroud of Christ (the Holy Mandylion).[53] Christianity spread quickly through Edessa in the late 2nd century with the conversion of Abgar VIII; however, but there had been a long association between the royal family and Christianity as in the 1st century, Abgar V had been converted by Thaddeus of Edessa, who was possibly Jude the Apostle.[54] It is also likely to have had a very early church in the city, as can be ascertained from the anonymous Chronicon Edessenum (covering the period 133 BCE to 540 CE), which mentions the destruction of the church through flooding in 201.[55] However, as Ross notes, there was little contemporary physical evidence to support this.[56] Although it was a client polity of Rome by 165, and a municipality and part of the Roman frontier under Caracalla (212/3), it was not until 242 CE that the city became fully part of the Roman Empire.[57] There were purportedly a number of Christian martyrs under Diocletian from Edessa, including Gurya, Shamona, and Habib.[58] While Edessa successfully maintained a number of different religions, it thrived as a Christian centre helping to convert communities in the surrounding areas and it was a stopping point for pilgrims, including Egeria, travelling to Carrhae (the birthplace of Abraham and location of the well where Rebecca was discovered and agreed to marry Isaac). In her letters, Egeria recalls the story of the letter sent by Jesus to Abgar in which he said that no enemy would breach the gates of the city. When the Persians laid siege to the city and cut off its water, the king read out the letter and by a miracle the water was restored, and the Persians retreated (19.16–19). Egeria noted that the letter was kept as a memorial in the city.

Because of the continuous occupation of the city, it has been difficult to define the Roman and Late Antique buildings, and many are incorporated into the ninth-century Urfa castle. Egeria was a visitor to Edessa in the 380s and provides the best description of the city. In her letters, she describes being shown the palace and statue of King Abgar by the Bishop of the city (19.16–19) as well as numerous pools of fish (19.7).[59] Other than the church which was supposedly flooded in the third century, there was a cathedral, known from a description in a poem, likely to date to the sixth century and which may be located below the Makam Ibrahim Camii.[60] The Great mosque of Urfa is supposedly built on a Christian church known as the Red church. Other churches are known to have been built including the fourth-century Church of the Confessors, to mark the martyrdom of Gurya, Shamona, and Habib, and the church of St. Stephen which was converted from a synagogue in the fifth century. The bodies of Thaddeus and Abgar were exhumed in the late fifth century and placed in a shrine in the centre of the city. Ross notes that the city would have had a lively mix of religions and philosophical beliefs such as Gnosticism in the first to early third centuries.[61] Furthermore, when Egeria made her detour to Edessa, and although she commented that it was located at the end of the world, the city seems to have been well connected through its wide range of residents.[62] The purported early endorsement of Christianity by the royal family must have helped considerably with its local spread too thereby creating a small and early Christian network.

Pergamon

Of the eight cities under discussion here, Pergamon lay at the furthest point away from Jerusalem. It would have taken 17.5 days by sea to get to the city. At Pergamon, the Upper Acropolis had been abandoned after the third century as a result of earthquakes and Gothic raids. A new fortification wall was constructed enclosing the lower city which became the new focus. The city struggled to recover, in part because of its contracted size and a lack of investment, but it eventually became an episcopal see in the Byzantine period. The first churches were constructed in the fifth century: the Red Hall, which was built inside the shell of the destroyed temple of Egyptian gods, and after this the Megalesion (i.e. the temple of the Megale Meter, mother goddess of Anatolia) located near the Philetairian gate, had a church built on its foundations.[63] A basilica with atrium was constructed in the Lower Agora.

The invading Goths caused a rapid destruction of the cultural core of Pergamon in the third century, as was the case in Athens. Although both cities constructed new walls, these walls essentially side-lined their visual centres of tradition (the Athenian Agora and the Pergamon Upper Citadel) and there was a long, slow process of recovery, perhaps because of a break in tradition of investment. Other cities such as Antioch and Salamis made fast recoveries due to rapid and significant imperial investment following their destruction.

Material networks and the spread of Christian ideas

Trade

One of the fundamental connections any city would have had would have been commercial and trade networks. Salamis was renowned as a trade centre and probably one of the busiest and best harbours on Cyprus. Its location close to the copper mines and forests provided two of its key resources. It is possible to trace at least some connections through an examination of *sigillatas* (Roman fine wares) in circulation in the second to third century.[64] In the Eastern Mediterranean, Eastern Sigillata A (ESA), manufactured along the coast between Tarsus (in Turkey) and Latakia in Syria, had the widest distribution until the late first century. At that point, Eastern Sigillata B (ESB) (made in Western Asia Minor, with workshops at Tralles) and Eastern Sigillata C (ESC) (produced around Pergamon) had a more restricted distribution than the other Eastern Sigillatas.[65] Cyprus produced its own significant quantities of *sigillata* (ESD), exported throughout the East and Mediterranean but its production petered out by the early third century.

In the heartlands of Early Christianity, ESA dominated Palestinian fine ware assemblages until the second century and Cypriot *sigillata* is popular, particularly in locations such as Jerusalem, Jericho, and surrounding areas. In Athens and Corinth, ESA dominated the pottery assemblages. This was the same at Pergamon and Ephesus with some ESB at Ephesus. In Cyprus, Cypriot *sigillata* was most common but with some ESA in the first century.

Vokaer has noted that from the fourth century onwards, all fine wares were imported into Syria and that there is a clear and definable difference between distribution of vessels between the coast (Antioch) and interior (Edessa).[66] This suggests different types of trade connections. The early Imperial levels have not been published yet in detail at Antioch and Apamea; however recent work, particularly by Agnès Vokaer, has provided some helpful results. In both cases, ESA is the most common import in the third and fourth centuries, alongside Kapitän II amphorae.[67] It is notable that in the case of Apamea, ESA is abundant but is imported from nearby locations. Malfitana's comparative distribution of *sigillata* wares highlights a similar picture in the third century.[68]

The brief analysis presented here shows that in terms of the distribution of fine wares, the cities in focus here share the same trends seen across the Mediterranean, that is with the exception of Apamea and Edessa; the former having indirect imports of ESA only.

As we saw, the earliest churches were constructed at Salamis (fourth century), Antioch (third/fourth century), and Edessa (third century). The latest are at Apamea, Athens, and Pergamon (with mid-fifth century dates) with Corinth and Ephesus in the middle. Notably, Edessa is one of the furthest cities away from Jerusalem, yet has one of the first churches. Conversely, Apamea was only nine days from Jerusalem and on a major route to Constantinople, yet its earliest church seems to be from the fifth century. This suggests that neither provincial designation, nor distance from the Holy Land makes a discernible impact on the speed of Christianisation (in contrast to Stark's view). In social network theory, the nearest neighbour is not defined by distance between hubs rather by the numbers of shared contacts.

Communities and social networks

Czachesz[69] has argued that the spread of Christianity was the result of weak ties created by three specifically Christian practices: itinerant proselytisers (typified by Paul); charity particularly to non-Christians; and women in formal positions in the church. However, I would argue that this Christian perspective on the spread is too simplistic. It does not explain, for example, why it took so long for Christianity to be adopted after Paul's missions. Furthermore, it is likely that Paul and his companions, such as Barnabas, John Mark, and Timothy, moved via established economic networks, but to places where they already had friends, family, and Christians, often organised into Christian voluntary associations.[70] When we approach Christianisation processes from the perspectives of both the converters and the subjects, it is clear that a range of relationships and contexts were involved and required a combination of strong and weak ties to create both the dynamic process as well as the environment for successful conversion and the expression of that, church building.

Environment

As noted above, diversity of population is key to the success of the first stage of church building: complexity. Most of the cities discussed here have a variety of populations, some more so than others. This is not entirely surprising given the size of the eight cities in question. Many have strong Jewish communities (Antioch, Ephesus, and Corinth) and Edessa was particularly known for its multiple different groups, including Parthians,

Syrians, Arabians, Nabateans, Armenians, Persians, Greeks, and Romans. Athens and Apamea were renowned for their philosophical schools and as cities of culture. But both were also known for their conservative views and close-knit communities which were in part based on the key characteristic of the city, the philosophical schools. It can be argued that Apamea and Athens had predominately strong ties with fewer weak ties than cities such as Ephesus and Edessa.

Interpersonal Social Network Bonds I: mission and pilgrimage

Based on the evidence of missionary activity, church building, trade, and tradition, it is argued here that Christianisation processes drew on different network connections to varying degrees of success depending on a range of factors concerning context. The initial stages of Christianisation, in the first and second centuries, were dependent on development and growth due to very personal connections (the sociological 'strong ties' mentioned above), in particular, in the form of the social networks accessed by the early missionaries such as Paul and Barnabas. Stark[71] noted that the early missionaries who were sent abroad already had family and friends in the Jewish diaspora. These were strong ties. Titus was left to work on Crete. Barnabas stayed to work at Antioch, while Paul set off on his mission to Corinth (there setting up the Christian community) and returning to Ephesus. He left his associates, Aquila and Priscilla, at Ephesus for them to convert and found the first church at Ephesus.

Conversion of the population continued throughout the centuries but with limited success.[72] Even Paul himself is well-known for the self-promotion of his own successes, rather than more obvious evidence of attainment. These missions, based on strong ties, had limited success in the spread of Christianity in the first instance. There were conversions, but they remained restricted in numbers and only grew more steadily over the next two centuries. This is reflected in mortuary and epigraphic data. Stark suggests that the rate of Christian conversion between 40 and 350 CE was at 3.42% of the population, with 0.0023% of the population being converted by 50 CE and 56.5% by 350 CE.[73] Even after the troubled third century, Kloppenborg suggests that only between 7% and 10% of the population had become Christian by 311 CE.[74] There are many possible reasons for the slow growth, including the limited interest that the powerful and elite had in Christianity, and a lack of an organised process. At Corinth, the first major church was not constructed until three centuries (at least), after Paul. When the earliest churches were constructed, they were located in existing cemeteries drawing on the community tradition of visiting these spaces[75] with no reference to the earliest attempts of Christianisation by Paul.

Once Christianity becomes the state religion, the impetus for church building spread widely through a range of strong and weak ties. The first church foundations are in cities like Salamis, Edessa, Antioch, and Ephesus which are connected to the Holy Land through a range of channels including pilgrimage, economy (from trade to craftspeople to military), and imperial attention. As noted above, each of these cities benefits from imperial investment at key stages of the Christianisation process, which would have likely encouraged further investment by local elites through a desire to emulate the Emperor or King. Within these cities, the churches are situated in strategic locations, such as by gates into the city (in Edessa, the Church of the Confessors was by the Gate of the Hours; the Palia church in Antioch was by the Beroea Gate); by popular locations such as Baths and Gymnasia (Church of St Mary in Ephesus); and in highly visible locations such as the Antiochan island on which the imperial palace also stood (the Great Church). The earliest church locations were planned and not haphazard and were often the result of imperial or elite investment. These churches were meant to be seen and act as vehicles to encourage others to convert. Once the church was built, a number of methods were incorporated to encourage the community to use it including maintaining familiar elements of procession or imagery.[76]

Pilgrimage and patronage became increasingly important as a means of gaining and maintaining social cachet in Christian society, particularly as the old elite families had to compete with new influential groups which arose as a result of the changing power structures of the church. In addition to investment and patronage, stories about one's pilgrimage were considered valuable in securing a place in the power structure of the church. Pilgrims created physical interpersonal networks through their travels and communications, and imagined abstract networks through the stories they told about their pilgrimages, thereby connecting places and people. As well as patronage, the imperial family also set much of the trends for pilgrimage and both became networks associated with the elite.

Interpersonal Social Network Bonds II: tradition and disruption

The delay in church construction in places such as Apamea, Athens, and Pergamon can be attributed to a number of reasons. In the fourth century, the main focus of these cities were as philosophical centres and the elites and visitors to the schools tended to be networked through strong ties. In the cases of Athens and Pergamon, their traditional centres had been divided off from the city centres and this, along with a lack of investment in the city, would have had an impact on social networks.

In Athens, the post-Herulian wall (late third century) contracted the city, and what had been the central focus of the city for centuries, the Agora, was for the first time, outside the city walls. The same applies to Pergamon. This was a fundamental change in the city's identity: centuries of dedications, political machinations, and personal interactions were excluded from the civic core and so represents a complete break in the traditional topographies of the cities. Furthermore, in the case of Athens, there was a break in the tradition of investment in the city, as when their rural landholdings became depleted in the third-fourth centuries, the wealthy landowners had to turn to different sources of income. Education offered a more reliable (if less lucrative) business prospect, and Athens became increasingly known as a city of education and philosophy. Unlike other cities discussed here, there is limited investment in Athens or Pergamon in the fourth century.

Chronologically, Pergamon is closer to Athens in terms of church foundation, regardless of its provincial designation or proximity to Ephesus. Athens and Pergamon were not significant trade emporia or social network hubs; they were cultural capitals reliant on their philosophical schools. As such, with the destruction and annexation of the Agora in Athens and the upper citadel in Pergamon, there was a rupture to the spaces of memory and tradition within the city which in turn meant a break in patronage of and investment in monumental building. The practical consequence of the contraction of the cities was the reduction in craftspeople travelling to work in these cities thereby curtailing the numbers of weak ties. Because of the break in tradition of place and of investment, impetus for church building must have been a conscious decision and perhaps a deliberate move to attract people to the city, particularly in the case of Athens, as the strength of the philosophical schools waned. The earliest church in Athens is that of the Ilissos, on the eastern edge of the city. This church is reputed to have been constructed at the behest of Athenian-born Aelia Eudocia (401–460), who was married to the Emperor Theodosius II. At Pergamon, the earliest church seems to be that known as the Red Hall, possibly dedicated to St John which was constructed within a second-century temple of Serapis. This means that when the first churches were built, they were the result of direct social involvement in the city. They were the result of strong ties and personal imperial investment.

In cities such as Salamis, Antioch, and Ephesus, epigraphic data shows the tradition of investment continued unabated throughout the third and fourth centuries and illustrates how important the elite were in helping to rebuild the city and in doing so, also attracted craftspeople to the city. Their locations on important trade and pilgrimage routes, as well as attracting pilgrims themselves, meant also that ideas and importantly, external funding, continued to flow. This helped to keep the momentum of public investment in buildings up without the kind of interruptions endured by cities such as Athens and Pergamon.

Conclusions

In the early years of Christian spread, missionaries like Paul and Barnabas had some success in converting like-minded groups, friends, relatives, and acquaintances based on strong ties but these were limited gains as exhibited in the resistance to their mission in places such as Athens and Corinth

(Paul) and on Cyprus (Barnabas). St Luke's version of their travel through Cyprus (Acts 13:4–12) highlights the process of preaching to the Jewish communities first, and then Gentiles, however, as Paget notes, his reluctance to discuss the reactions to their preaching suggests a muted reception to the pair.[77] Barnabas had family on Cyprus but even that does not appear to have eased their mission.[78] To be successful, a sustained combination of strong and weak ties were necessary and required reinforcement through path dependence. Weak ties were created through traders and craftspeople who travelled along the well-established sea and land routes in the East. These were further reinforced by pilgrims who would have travelled the same conventional paths. Whiting's work on 'braided networks' illustrates how these routes used for multiple reasons (economic, political, administrative, military, and religious) are interdependent on each other for expedient travel.[79]

These networks can draw on a range of processes simultaneously and their flexibility and mix of planned and organic change led to the successful Christianisation of the Mediterranean lands. In all, Christianisation processes are the result of evolving networks of strong and weak ties that change over time, which can be highlighted when seen in the context the construction of the earliest monumental churches. The impetus for the spread of Christianity comes from personal and strong ties. For example, some of the earliest churches appear to have been constructed at Imperial or elite behest. The Katapoliani church on Paros is a good example, which is said to have been founded on the orders of Empress Eleni after she had visited the island. As such, the first phase of church building, complexity, is planned and undertaken in busy locations (ports and sanctuaries) which are already known to the investors. But for the construction to be successful and then to spread to towns through social movement, which is more organic, a combination of strong and weak ties is required. Those weak ties were created through economic networks, especially through trade and craftspeople. If cities were lacking in new building, investment, and patronage, like Athens, Apamea, and Pergamon, they would have been deficient in at least some of the weak ties to stimulate church building until a targeted attempt was made to construct the church, as seen in the case of Aelia Eudocia's purported instruction for the foundation of the Ilissos Basilica in Athens.

Once the first churches were constructed, they became more important over time and they increased in number, and communities began to orientate their towns around the churches. Christianity maintained its importance in these urban spaces which were further enhanced by offering a wide range of visitors and pilgrims something worth seeing and in the case of cities such as Salamis, Ephesus, and Antioch and even Edessa, these were martyria. A multiplicity of visitors including imperial family and politicians, bankers, traders, craftspeople, and pilgrims helped to sustain and perpetuate the weak ties in these cities making the monumental expression of Christianity significantly more obvious than the less-connected cities that were reliant on strong ties.

Notes

1. Patrich 2016: 273.
2. See, for example, Sweetman 2015a & b and Ward 2016.
3. Stark 1991: 82; Ekeke 2013.
4. Sweetman 2015a.
5. Kloppenborg 2020.
6. Sweetman 2017.
7. Stark 2001: 106.
8. Granovetter 1983.
9. Price, 'Religious Mobility': 10.
10. Frankfurter 2018: 7–15.
11. Bennema 2015: 43.
12. Brughmans 2010; Collar 2013; Collar et al. 2015; Kloppenborg 2020.
13. Sweetman 2015b.
14. Sweetman 2015b.
15. Sweetman 2015b.
16. Price, 'Religious Mobility': 10.
17. 2013.
18. Stark 1991: 78–9.
19. Stark 1991: 80.
20. All travel times are based on the Orbis route calculations using the fastest time in Summer. https://orbis.stanford.edu/.
21. Karageorghis 1969: 185, 195.
22. Karageorghis 1969: 190.
23. Rautman 2002.
24. Karageorghis 1969 :197.
25. Roux 1998.
26. Mayer and Allen 2012: 3.
27. Shepardson 2014.
28. Sandwell 2007.
29. Teitler 2013: 263–88; Downey 1939: 306–8.
30. Mayer and Allen, 2012.
31. Mayer and Allen 2012: 100.
32. According to Eusebius' Panegyric of Constantine on the anniversary of his 30th birthday, was constructed under Constantine (as part of a development in Antioch and Jerusalem). It is likely to have been completed under Constantius II (Mayer and Allen 2012: 70).
33. Mayer and Allen 2012: 124.
34. Finally numbering thirty-one throughout the city, suburbs and port (Mayer and Allen 2012).
35. Mayer and Allen 2012: 125.
36. Mayer and Allen 2012: 86.
37. Trebilco 1991: 86.
38. *HE* 4.26.
39. Ward 2016.
40. Balty 1977.
41. Ward 2016: 85.
42. For example, Ward (2020: 85) notes an inscription in mosaic which notes the construction of a portico in 469 by a Julianos.
43. Ward 2020: 85.
44. Foss 2010.
45. Foss 2010.
46. Acts 17: 34; Frantz 1965: 188.

47. Frantz 1965: 189.
48. Camp 2001: 231
49. Stikas 1962: 51–6.
50. Sweetman 2017a.
51. Pallas 1976 and Pallas 1955, respectively.
52. Brown 2018.
53. Eusebius' History of the Church (1.13.5–1.13.22).
54. Drijvers 1980: 195–6. Ross (2001: 118) suggests that the connection of the royal family was a later addition to the story of the early conversion of the city. However, he goes on to note that the early Christianity is indicative of the diversity of the city at the time.
55. Ross 2001: 117 & Krautheimer 1986: 302. Krautheimer goes on to suggest that the church may have been a house shrine, perhaps similar to that at Dura Europos.
56. Ross 2001: 117.
57. Ross 2001: 83.
58. Ross 2001.
59. Wilkinson 2006.
60. Krautheimer 1986: 219, 240.
61. Ross 2001: 118.
62. Hunt 2000: 35.
63. Nohlen 1998.
64. Brughmans and Poblome 2016 undertook computational modelling of the main Roman fine wares (ESA, ESB, ESC, and ESD) to shed light on relationships between Roman commercial centres and practices of the traders who worked between them.
65. Brughmans and Poblome 2016, fig. 1.
66. Vokaer 2015.
67. The origin of the Kapitän II is debated. Most commonly it is ascribed to the Aegean but some suggest the Black Sea. I am very grateful to Agnès Vokaer for sharing her insights on the material with me.
68. Malfitana 2002.
69. Czachesz 2011: 142–9.
70. Kloppenborg 2019.
71. Stark 1991: 82.
72. Acts 19:8.
73. Stark 1997.
74. Kloppenborg 2020: 324.
75. Sweetman 2015a.
76. Sweetman 2015b.
77. Paget 2020: 40–41.
78. Paget 2020: 40.
79. Whiting 2020.

Bibliography

Assmann, J. (2011). 'Communicative and Cultural Memory,' in P. Meusburger, M. Heffernan and E. Wunder (eds.). *Cultural Memories. The Geographical Point of View*, 15–28. Knowledge and Space 4. Heidelberg: Springer.

Balty, J. (1977), *Mosaïques de Syrie*. Brussels: Musées royaux d'art et d'histoire.

Balty, J. (1988). 'Apamea in Syria in the Second and Third Centuries A.D.,' *The Journal of Roman Studies*, 78: 91–104. Doi:10.2307/301452

Barnard, L. W. (1968). 'The Origins and Emergence of the Church in Edessa during the first two centuries AD,' *Vigiliae Christianae* 22(3): 161–75.

Bennema, C. (2015). 'Early Christian Identity Formation amidst Conflict,' *Journal of Early Christian History* 5(1): 26–48.

Brown, A. (2018). *Corinth in Late Antiquity*. London: I. B. Tauris.

Brughmans, T. (2010). 'Connecting the Dots: Towards Archaeological Network Analysis,' *Oxford Journal of Archaeology* 29(3): 277–303.

Brughmans, T. and J. Poblome. (2016). 'Roman Bazaar or Market Economy? Explaining Tableware Distributions through Computational Modelling,' *Antiquity* 350: 393–408.

Camp, J. (2001). *The Archaeology of Athens*. Yale: Yale University Press.

Chandler, T. and G. Fox. (1974). *3000 Years of Urban Growth*. New York: Academic press.

Czachesz, I. (2011). 'Women, Charity, and Mobility in Early Christianity: Weak Links and the Historical Transformation of Religions.' in I. Czachesz and T. Biró, *Changing Minds: Religion and Cognition through the Ages*, 129–54. Peeters: Leuven.

Collar, A. (2013). *Religious Networks in the Roman Empire. The spread of new ideas*. Cambridge: Cambridge University Press.

Collar, A., F. Coward, T. Brughmans, B.J. Mills. (2015). 'Networks in Archaeology: Phenomena, Abstraction, Representation,' *Journal of Archaeological Method and Theory* 22(1): 1–32.

Di Genua, A., E. Marinelli, I. Polverari and D. Repice. (2014). 'Judas, Thaddeus, Addai: Possible Connections with the Vicissitudes of the Edessan and Constantinopolitan Mandylion and any Research Perspectives,' *Bari* 4–5, Settembre 2014: 1-6.

Downey, G. (1939). 'Julian the Apostate at Antioch,' *Church History* 8(4): 303–15.

Drijvers, H.J.W. (1980). *Cults and Beliefs at Edessa*. Leiden: Brill.

Eidinow, E. (2011). 'Networks and Narratives: A Model for Ancient Greek Religion,' *Kernos* 24: 9–38.

Ekeke, E. (2013). 'The Greco-Roman and Jewish Contributions to the Growth of Earliest Christianity,' *American Journal of Social Issues and Humanities* 3: 24–33.

Erin, Ş. (2016). 'Diffusion of Islam in the United States: Comparative Social Conversion Social Networks,' *International Review of Social Research* 6(4): 221–32.

Foss, C. (1997). 'Syria in Transition, A. D. 550–750: An Archaeological Approach,' *Dumbarton Oaks Papers* 51: 189–269. doi:10.2307/1291765

Foss, C. (2010). *Ephesus after Antiquity. A Late Antique Byzantine and Turkish City*. Cambridge: Cambridge University Press.

Frankfurter, D. (2018). *Christianizing Egypt: Syncretism and Local World in Late Antiquity*. Princeton: Princeton University Press.

Frantz, A. (1965). 'From Paganism to Christianity in the temples of Athens,' *DOP* 19: 187–205.

Granovetter, M. (1983), 'The Strength of Weak Ties: A Network Theory Revisited,' *Sociological Theory* 1: 201–33.

Hunt, E. D. (2000). 'The Itinerary of Egeria: Reliving the Bible in Fourth Century Palestine,' in R. Swanson (ed.) *Holy Land, Holy Lands, and Christian History. Papers read at the 1998 Summer Meeting and the 1999 Winter Meeting of the Ecclesiastical History Society*, 34–54. Suffolk: Boydell Press.

Karageorghis, V. (1969). *Salamis in Cyprus. Homeric, Hellenistic and Roman*. London: Thames and Hudson.

Karageroghis, V. (1999). *Excavating at Salamis in Cyprus*. 1952–74. Athens: The A.G. Leventis Foundation.

Kennedy, H. (2000). 'The Early Development of Church Architecture in Syria and Jordan c. 300 – c. 750),' in R. Swanson (ed.) *Holy Land, Holy Lands, and Christian History. Papers read at the 1998 Summer Meeting and the 1999 Winter Meeting of the Ecclesiastical History Society*, 1–33. Suffolk: Boydell Press.

Kloppenborg, J. (2019). *Christ's Associations: Connecting and Belonging in the Ancient City*. New Haven; London: Yale University Press.

Kloppenborg, J. (2020). 'Recruitment to Elective Cults: Network Structure and Ecology,' *New Testament Studies* 66(3): 323–50. doi:10.1017/S0028688519000511

Koester, H. ed. (1998). *Pergamon, Citadel of the Gods: Archaeological Record, Literary Description, and Religious Development*. Harvard Theological Studies 46. Harrisburg: Trinity Press International.

Krautheimer, R. (1986). *Early Christian and Byzantine Architecture*, rev. ed. Yale: Yale University Press.

Malfitana, D. (2002). 'Eastern Terra Sigillata Wares in the Eastern Mediterranean. Notes on an Initial Quantitative Analysis,' in *Céramiques hellénistiques et romaines, productions et diffusion en Méditerranée orientale (Chypre, Égypte et côte syro-palestinienne). [Actes du colloque tenu à la Maison de l'Orient méditerranéen Jean Pouilloux du 2 au 4 mars 2000]*, 133–157. Lyon: Maison de l'Orient et de la Méditerranée Jean Pouilloux. (Travaux de la Maison de l'Orient méditerranéen 35).

Mayer, W. and P. Allen. (2012). *The Churches of Syrian Antioch (300–638). Late Antique History and Religion*, 5. Leuven: Peeters Publisher.

Millar, F. (1993). *The Roman Near East: 31 BC – AD 337*. Harvard: Harvard University Press.

Meusburger, P., M. Heffernan and E. Wunder (eds.). (2011). *Cultural Memories. The Geographical Point of View*. Knowledge and Space 4. Heidelberg: Springer.

Nohlen, K. (1998). 'The 'Red Hall' (Kizil Avlu) in Pergamon,' in Koester (ed.) *Pergamon, Citadel of the Gods: Archaeological Record, Literary Description, and Religious Development*, 77–110. Harvard Theological Studies 46. Harrisburg: Trinity Press International.

Paget, J. C. (2020). 'Cyprus in the New Testament and Beyond,' in L. Nasrallah, A. Luijendijk and C. Bakirtzis (eds.) *From Roman to Early Christian Cyprus*, 33–54.. Wiseenschaftliche Untersuchungen zum Neuen Testament 437. Tübingen: Mohr Siebeck.

Pallas, D. (1955). ''Ανασκαφή τῆς βασιλικῆς τῆς Σκουτέλας,' *Prakt* 1960:193–200.

Pallas, D. (1976). ''Ανασκαφή τῆς βασιλικῆς τοῦ Κρανείου ἐν Κορίνθῳ,' *Prakt* 1978:163–95.

Patrich, J. (2016). 'The Early Christianization of the Holy Land – The archaeological Evidence,' in O. Brandt and G. Castiglia (eds.) *Acta XVI Congress Internationalis Archaeologiae Christianae Romae 22–28.9.2013. Costantino e I Costantinidi l'innovazione Constantiniana le sue radici e I suoi sviluppi. Pars I*: 265–93. Oxford: Pontificio Istituto Di Archeologia Cristiana.

Potts, D. ed. (2012). *A Companion to the Archaeology of the Ancient Near East, Volume 1*. Chichester: Wiley-Blackwell.

Price, S. (2012). 'Religious Mobility in the Roman Empire,' *JRS* 102: 1–19.

Rautman, M. (2002). 'From Polytheism to Christianity in the Temples of Cyprus,' in C. Callaway (ed.) *Ancient Journeys. A Festschrift in Honour of Eugene Numa Lane* (The STOA, available online http://www.stoa.org/hopper/text.jsp?doc=Stoa:-text:2001.01.0014 accessed September 2018.).

Roux, G. (1998). *Salamine de Chypre - Tome 15, La Basilique de la Campanopétra*. Paris: De Boccard.

Ross, S. (2001). *Roman Edessa: Politics and Culture on the Eastern Fringes of the Roman Empire, 114–242 C.E.* London: Routledge.

Sandwell, I. (2007). *Religious Identity in Late Antiquity: Greeks, Jews and Christians in Antioch.* Cambridge: Cambridge University Press.

Salzman, M. (2004). *The Making of a Christian Aristocracy Social and Religious Change in the Western Roman Empire.* Cambridge MA: Harvard University Press.

Shepardson, C. (2014). *Controlling Contested Places: Late Antique Antioch and the Spatial Politics of Religious Controversy.* Berkley: University of California Press.

Stark, R. (1991). 'Christianizing the Urban Empire: An Analysis Based on 22 Greco-Roman cities,' *Sociological Analysis* 52(1): 77–88.

Stark, R. (1997). *The Rise of Christianity: How the Obscure, Marginal Jesus Movement Became the Dominant Religious Force in the Western World in a Few Centuries.* San Francisco: Harper Collins.

Stark, R. (2001). 'Efforts to Christianize Europe,' *Journal of Contemporary Religion* 16(1): 105–23.

Stikas, E. 1962. 'Κοιμητηριακή Βασιλική Παλαιάς Κορίνθου,' *Prakt* 1962:51–6.

Stroumsa, G. (2009). *The End of Sacrifice. Religious Transformation in Late Antiquity.* Chicago: University of Chicago Press.

Sweetman, R.J. (2015a). 'Memory, Tradition and Christianization of the Peloponnese,' *AJA* 119(4): 501–31.

Sweetman, R.J. (2015b). 'The Christianization of the Peloponnese: The Case for Emergent Change,' *ABSA* 110: 285–319.

Sweetman, R. J. (2017). 'Networks and Church Building in the Aegean: Crete, Cyprus, Lycia and the Peloponnese,' *ABSA*112: 207–66.

Sweetman, R. J., A. Devlin and N. Piree Iliou. (2018). 'The Cyclades in the Late Antique Period. Churches, Networks and Christianization,' in A. Angliker and J. Tully (eds.) *Cycladic Archaeology: New Approaches and Discoveries*, 215–38. Oxford: Oxbow.

Sweetman, R. J. (2019). 'Macedonia, Greece and the Aegean,' in W. Caraher, T. Davies, and D. Pettegrew, (eds.) *The Oxford Handbook of Early Christian Archaeology.* Oxford: Oxford University Press.

Teitler, H. C. (2013). 'Ammianus, Libanius, Chrysostomus, and the Martyrs of Antioch,' *Vigiliae Christianae* 67: 263–88.

Trebilco, P. (1991). *Jewish Communities in Asia Minor.* Cambridge: Cambridge University Press.

Vaage, L. ed. (2006). *Religious Rivalries in the Early Roman Empire and the Rise of Christianity. Studies in Christianity and Judaism* 18. Waterloo: Wilfrid Laurier University Press.

Vokaer, A. (2015). 'Pottery Production and Exchange in Late Antique Syria (Fourth to Eight Century A.D.). A Study of Some Imported and Local Wares,' *Late Antique Archaeology* 10(1): 567–606.

Wang, T. M. Zhou, Z. Fu and X. He. (2017). 'Link Prediction in Evolving Networks Based on Popularity of Nodes,' *Nature. Scientific Reports* 7: 7147.

Wallace-Hadrill, D.S. (1982). *Christian Antioch: A Study of Early Christian Thought in the East.* Cambridge: Cambridge University Press.

Ward, W. (2016). 'The 363 Earthquake and the End of Public Paganism in the Southern Transjordan,' *Journal of Late Antiquity* 9(1): 132–70.

Whiting, M. (2020). 'Braided Networks. Pilgrimage and the Economics of Travel Infrastructure in the Late Antique Holy Land,' in A. Collar & T. M. Kristensen (eds.) *Pilgrimage and Economy in the Ancient Mediterranean*, 62–90. Leiden/Boston: Brill.

Wilkinson, J. (2006). *Egeria's Travels*. Oxford: Aris & Phillips.

Woolf, G. (2016). 'Only Connect? Network Analysis and Religious Change in the Roman World,' *Hélade* 2(2): 43–58.

9 Strong ties and ecclesiastical law in the Later Roman Empire

Kilian Mallon

Among the strongest ties in antiquity were the relations of religion and community.[1] In the proceedings of episcopal councils and in imperial laws of the fourth to sixth centuries CE, bishops and other officials developed a complex set of regulations to shape the nature of strong and weak ties between bishops, clergy, and laity. In this chapter, I examine these legal and regulatory sources to establish an outline of the ideal system that these regulatory efforts tried to create. It was not just the weak ties of evangelism, coerced conversion, and demography that consolidated and created the late Roman Christian world and served as the foundation for transmission of religious ideas, but also a series of carefully regulated strong and weak ties in ecclesiastical law (canon law).[2] While previous models of transmission have focused on weak ties or non-ecclesiastical relationships, it is important to consider how ecclesiastical law maintained and regulated the kinds of strong ties needed for a new religion to succeed through granting or denying access to work and worship within churches. Strong ties were maintained through the organisation of responsibilities within each church, regulations that established close interpersonal relationships within the clergy, placed restrictions on the mobility of the bishops and clergy, and established the threat of excommunication—the revocation of both strong and weak ties to the community—as the core punishment for violation of all ecclesiastical law.

The question of networks and ties in the early Church has been a highly productive area of research in recent years.[3] Scholars have examined numerous examples of ecclesiastical networks, often with epistolary evidence. This chapter, however, takes a macro-level (and deliberately abstracting) view of the regulatory system, by viewing the kinds of networks that bishops tried to create for themselves, as opposed to small-scale studies in particular contexts. It thus sits, hopefully not too awkwardly, between the worlds of the legal historian and the sociologist. It is not an empirical analysis of any specific Christian social network, but rather an exploration of ideal normative multi-authored structure and its likely impact on social interaction if it was adhered to in Late Antiquity. This study is thus intended to raise questions and prompt further discussion about the nature, intent, and social impact of ecclesiastical law-making, about the divergence between ideal and real-life

DOI: 10.4324/9780429429217-12

episcopal networks, and about the capacity of lawmakers to engineer social reality in Late Antiquity. Because of the normativity of the regulations (see below), this study will rely on concepts from Social Network Analysis to hypothesise a model of social networks and transmission that can be compared to existing and future datasets, and accordingly there is no direct network mapping or quantitative analysis.

The standard models of Christian conversion—the transmission of Christianity across social networks and through generations—are evangelical, socio-political, and demographic. In the first model, Christianity spread according to missionary activities. People convinced other people to believe in the Christian message that they heard from missionaries.[4] In the second, elites decided to convert and forced the new religion on their subjects by decree or the need of the lower classes to emulate the elites to advance socially. The imperial and Church authorities used legal privileges to make conversion to Christianity an attractive (even necessary) choice for anyone who wished to be socially mobile. These mostly secular laws gave Christians political, and status privileges while pagans were discriminated against and eventually punished for holding to the old religions.[5] In the third model, demography was the decisive factor. Christian families were larger and Christian mothers taught their children to be Christian.[6] The first two models describe mostly weak-tie relationships; hearing a speaker in a provincial forum, having a distant friend introduce you to a convincing Christian believer, or simply converting to get a better chance of promotion within the imperial bureaucracy. The demographic model focuses on strong ties, where conversion relies on the raising of children within the family but does not account for the importance of relationships within the clergy itself.

These three models form key elements in the story of Christianisation in the Roman Empire. The model proposed here will suggest that, if they were followed, ecclesiastical regulations managed and promoted the kinds of strong ties needed for a new religion to succeed through granting or denying access to the work and worship within churches. Strong ties are the primary focus of canon law. While strong ties are not a good vector of transmission, they do create the solidarity and trust at the core of an institution on which other forms of transmission rely. Weak ties, rather than serving as vectors for evangelism, are treated in ecclesiastical regulation as dangerous. As they were seen as vectors for the transmission of heresy or disobedience, weak ties had to be curtailed and controlled. If rapid information flow helped Christianity spread, it was not always seen that way by the bishops who were, in theory, supposed to lead the transmission.

Strong and weak ties

The term strong tie refers to close relationships and frequent interactions between people. The term weak tie, on the other hand, refers to distant relationships and infrequent interactions between people.[7] Dividing strong from weak ties in a given social network is a matter of context. As Granovetter states 'the strength of a tie is a (probably linear) combination of the amount of time, the emotional intensity, the intimacy (mutually confiding), and the reciprocal services which characterise the tie'.[8] Strong ties generally involve more time, more frequency of interactions, and more sharing of intimate information. Importantly, they tend to produce tight-knit clusters. In Granovetter's formulation, weak ties are key to transferring information between strong-tie groups, because otherwise they tend to pass information around amongst themselves rather than to people outside the group. Counter-intuitively, new information, whether news from far away or heretical religious ideas, spreads most rapidly through networks which have many weak ties than through those with many strong ties but few weak ties. In strongly tied groups, information gets stuck again and again in the same groups.

How can we distinguish strong from weak ties in Late Antique Christian communities? To establish a baseline, let us take the definition of a strong tie as a relationship (hierarchical or non-hierarchical) that involves face-to-face discussion, based on family connections or an external goal (rather than casual chit-chat), occurring on average on more than one day per week or, if on one day only, for a long time. This places the bishop and clergy in the strong-tie category but the weekly lay church-goer in the weak-tie category. It also places clergy and bishops from other dioceses in the weak-tie category also. One thing this does leave out is differences in emotional impact. A lay person could see a charismatic bishop once a week and it would have a greater emotional impact (and be more memorable) for this person than daily contact might for a deacon, but that is not important to the overall weight of the relationships over the long term, which is undoubtedly more intimate for the strong-tie holder than the weak-tie holder.

Churches as multi-purpose spaces and strong-tie centres

In Late Antiquity, churches were multi-purpose spaces. The fact that they were so multi-purpose, housing social, economic, cultural, religious, administrative, financial, and judicial roles, made them centres of strong-tie relationships. How did this happen? Pre-Christian temples of the Roman world were dedicated to worship of the gods through ritual and sacrifice. They had religious workers (priests) chosen from local elites, who also had to manage the temporal economic affairs of the institution. This daily management was a key part of the life of any temple. Many temples were also centres of economic networks of pilgrimage or hosted traders who set up stalls to sell animals and religious paraphernalia.[9] Any physical structure required building and maintenance; temples were no different to churches in this respect, and so they would requisition slaves or hire wage-labourers as needed. Most temples were not, generally, the centre of political life in their cities; this was the agora or forum, and the basilica that was the seed of the Christian church in later years. The leaders of a city would have held both religious and political roles, perhaps even concurrently, even in the course of carrying out a unified single office. Any clear separation between religious, economic, and political roles is next-to-impossible. For bishops to gather religious, economic, and political powers into a single portfolio is not odd in the Roman context.

In this context, bishops created a new religious space, the church, that used the previously political space of the basilica[10] as its physical and spiritual architecture.[11] Bishops also had strict control over an exclusive and perennial labour force, the clergy, who carried out both economic and religious duties.[12] Political, social, religious, and economic roles of the bishop thus merged in churches and were set within a larger hierarchical universal religious organisation. This was a consequential innovation to the ancient world after the early fourth century and one that sets the stage for ecclesiastical law to have effect.

Regulatory sources in late Roman Christianity

It is important to be aware of the origins, benefits, and limitations of the regulatory texts of Late Antiquity. By the time Christianity was legalised in 311 and 313 CE, the Christian population had grown to a formidable size. If demographic models of the Christian population are broadly accurate, Christians made up approximately 10% of the Roman population by 300–312 CE and were reaching towards 50% by 350 CE.[13] A once fringe group had become a significant political force. Legalisation gave bishops a new confidence to organise more openly and to design and regulate Christian life.

Ecclesiastical law, or canon law, emerged as a set of regulations issued at general provincial synods or pan-Mediterranean ecumenical councils for the organisation and running of the Church.[14] Ecclesiastical councils were convened with the goal of consolidating centralised hierarchical power in the hands of mainstream bishops by setting out strict regulations and punishing any deviance in belief or practice. Scholarly interest has focused on their theological disputes, but those disputes were always rooted in problems of discord and disunity in the real world, with factions often forming according to geographic proximity, and were far from purely intellectual or philosophical enterprises.

There were a series of important foundational legal and regulatory councils between the early fourth century and the mid-sixth century. This includes the first four ecumenical councils: Nicaea (325 CE), Constantinople I (381 CE), Ephesus (431 CE), and Chalcedon (451 CE).[15] Since they were organised and presided over by the reigning emperor, these councils had authority across the Roman world. Other important councils or synods (more provincial versions of councils) were held to solve local disputes, to have local efficacy, or to lay groundwork for larger pan-Mediterranean councils; for instance, the Synod of Elvira in Spain in the early 300s CE, the Council of Arles in Gaul in 314 CE, a precursor to the Council of Nicaea, and the Synod of Neo-Caesarea in Asia Minor in 314 or 315 CE.[16] Bishops were more independent in this period than in the Medieval period and these councils were ostensibly agreed between peers. The bishops of more important cities such as Rome and Constantinople had become more respected and powerful than the average bishop. Nevertheless, power struggles and bitter disputes were a common feature of the ecumenical councils.[17] Part of the problem for the councils was that they were not, in theory, supposed to innovate, but merely to clarify what was meant in the scriptures and patristic texts.[18] This was more obviously a problem for theological disputes, but it meant that it was never possible for radical change to ecclesiastical organisation to happen quickly. This period of canon law, therefore, saw rapid evolution of the role of bishops and the social world of Christianity, and councils were usually working hard to catch up with changing regional and generational needs.

By the start of the sixth century, there was a large body of canon law but no standard compilations of it. In the West in particular, the codification of the monk and scholar, Dionysius Exiguus, around the start of the sixth century, quickly became a key reference work, though there were numerous other early compilers as late as the reign of Justinian.[19] Dionysius Exiguus' collection was one of the most important sources for early medieval canon jurists. He and other less important compilers gathered the pronouncements of synods and councils that had occurred since the early fourth century, as well as the likely apocryphal apostolic regulations. The supposedly earliest material consists of the Didache, the Apostolic Church-ordinance (*Constitutio Ecclesiastica Apostolorum*), the Didascalia, the Apostolic Canons, and the Apostolic Constitutions.[20] Although claimed in antiquity to be from the first century, all probably date to the turn of the fourth and fifth centuries.[21] The Apostolic Constitutions only ever applied to the Eastern Empire. In addition to ecclesiastical councils, imperial authorities collected relevant ecclesiastical regulations in the *Codex Theodosianus* and the *Corpus Juris Civilis*.[22]

There are some important caveats to using this body of material. Canon law of this period is nascent; it is far less complex, consistent, and clear in scope and procedure than later canon law would become. Regulatory texts (primarily canon laws and secular laws pertaining to the Church) are normative in the sense that they attempted to establish a set of rules that the community would have to obey. Lawmakers aspired to shape the world how they would like it would be. While theological or jurisprudential debates were important in this period, lawmakers lived in a real social world; their laws both reflected social norms in that society and shaped those social norms too. Certain regulations were only ever meant to be limited to certain areas; others were intended to be universal. In both cases, however, it is a struggle to know the actual geographic reach of any set of regulations.

There are complex histories of zealous infighting, personal rivalries, and non-sensical violence at the heart of each of the councils and synods, which gave each a distinctive flavour and purpose.[23] Moreover, we can only guess to what extent the councils and local synods were obeyed. Establishment of Christianity as state religion for all in 380 CE would have made canon law more powerful, at least in theory. Unfortunately, we do not have records of synods or disciplinary canons of Arians or other versions of Christianity, if there were any, to compare with these sources. Likewise, the canons were transmitted in later collections of laws issued at various times, which may have introduced inconsistencies and contradictions. Nevertheless, these early canon laws provide invaluable insight to the overall agenda and practices of the Church in these centuries.

Creating the bonds of financial and administrative responsibility

The modelling of relationships on the family—father, mother, brother, and sister—is well known in Christian communities from the time of the era of the New Testament.[24] The metaphors of kinship are infused in Christian social relations. Canon law did not regulate the use of terms, but it did recognise the value that the close ties of family, even metaphorically, could have for structuring an organisation. As such, canon law relied on a familial framework for regulating the actual roles of financial and administrative responsibility. These administrative roles were structured within the clergy and used regular contact that can be defined as strong ties. The bishop as *paterfamilias* aided by his household was a useful analogy, but the legal rules to establish the organisation of the Church required considerable debate and codification. Bishops were, ostensibly at least, all equal to one another under the primacy of their provincial metropolitan. There were a number of sees with special respect and privileges, which led to a number of high-profile disputes until Justinian decreed them once and for all to be Rome, Constantinople, Jerusalem, Antioch, and Alexandria.[25] The primary non-ritual role of the ordinary bishop was financial management.

When ecumenical councils began to regulate ecclesiastical responsibilities, they started at the top, with the bishops, who could already have been described as the central node in highly centralised networks. As Christianity became legal, churches could begin to legally hold property. Bishops were able to gather vast quantities of wealth through donations which they could accumulate, reinvest, or redistribute.[26] Wealth could be stored either in monetary form or translated into material wealth in the form of ecclesiastical real estate, purchase of relics, or the commissioning of works of art. A church was also a landowner and it could reinvest profits by buying new land or enslaved persons. A bishop could also find ways to use wealth to promote the further accumulation of wealth. For example, he could take legal action over wills or pay for widows' upkeep in return for their inheritances. In turn, a bishop could redistribute wealth through charity and through projects such as hiring artisans and workers. In this respect, his role was modelled on the father. The people closest to him would be involved in so many different religious and economic tasks that strong ties would be formed. But many people, such as those who were members of the laity or worked on ecclesiastical estates, would have much less frequent and intimate contact with the bishop.

The Council of Antioch in 341 CE stated most explicitly that it was the responsibility of the bishop to manage the economic ties of his church. God was, naturally, the final overseer and judge and the bishop was his second-in-line for church property management.[27] Very early on, the Council of Ancyra of c. 314 CE, for instance, gave bishops full authority over distribution of the fruits to the poor. But in effect he has total control to decide his own overheads and the cost of his staff. Expenditures included food, clothing, construction, travel, and the purchase of objects for the liturgy, but the canons left it to the bishops' discretion. A major concern of the Council of Antioch (341 CE) was that, given how varied and complicated a bishop's holdings in a region could be, in the event of the sudden death of the bishop, not all that knowledge would be lost. As such, the Council required bishops to make their lower-ranked clergy (presbyters, deacons, subdeacons, and deaconesses) fully aware about what was in his possession. The ecumenical councils were concerned about making sure that ecclesiastical gains were not lost after a bishop's death. Generally, already in the fourth century, councils had permitted bishops more rights to empower representatives to do this work on their behalf.[28] A so-called *oeconomus* could be chosen from among the bishop's own clergy to help manage finances and provide accountability. This system gave bishops power similar to that of a *paterfamilias* but gave their representatives knowledge and experience of the estate so that they could be responsible for it much as an heir could handle the estate of a deceased father. [29]

A small but tight-knit cluster of strong ties surrounded the bishop, and these interactions would have dominated day-to-day life for the clergy of a community, but a wide network of weak ties was connected through the bishop also. The bishop was judge over many social and economic affairs, centred on the *audientia episcopalis*, a court presided over by the bishop, which appeared in the early fourth century and survived the decline of Roman civil authorities in the West.[30] They could offer their community the valuable role of arbitration of disputes, which placed them in a prominent position of power and influence.[31] Bishops also had the financial power to offer loans and guarantees to young people in the community who needed a start in life. Bishops of course carried out the important spiritual roles of instructing Christians, leading worship, performing the sacraments, and overseeing the moral character of the community. Bishops were responsible for the administration of their own territories, which were usually not bigger than one city, but this did not mean they had no external economic or social contacts. Bishops, since they were frequently members of the propertied elite, maintained social and economic bonds with their own friends and families across the Mediterranean.[32] They also had some potential for very infrequent inter-regional influence by sending charity to other areas during outbreaks of famine and disease.[33] Some bishops maintained active lines of communication with the military; a regulation in the *Codex Justinianus* prevents military grain rations from being sent to chapels, clerics, or monasteries in need.[34] This suggests that, ordinarily and commonly, military officers would give some of their rations to the Church in times of shortage. These kinds of ties perfectly fit the weak-tie model.

Within the strong-tie inner circle, however, the clergy—presbyters, deacons, deaconesses, and subdeacons—were empowered by regulation to fill in for the bishop when needed. Presbyters, originally hardly distinct from bishops, were his direct subordinate and bore the same powers. Deacons had the right to manage finances, incoming and outgoing, as well as to give communion and preach. They were prevented from leading the holy sacrifice and not allowed to perform baptism, a right reserved only for bishops and deaconesses.[35] The bishops and clergy maintained close and frequent contacts among themselves, and also, by standing in for each other in a range of social, economic, and religious contacts with the lay community, produced and reproduced the social networks that made up the Christian world.

Deaconesses are an especially interesting category of persons who may have acted as brokerage nodes in the ecclesiastical network, meaning that they consistently acted as go-betweens for information within the larger community network. They existed primarily in the East (though most likely not in Egypt) until the sixth century, when they start to appear in the Latin West to the great consternation of the conservative bishops of Gaul.[36] The most well-known is the noblewoman, Olympias the Younger, renowned in her time for her charitable works. There is enormous variation in the sources about what the word means and what the expected roles of deaconesses were. In general, it appears that they were allowed to perform baptism, but only on women, due to the fact that it was morally inappropriate for a male bishop to perform immersion baptism on a woman.[37] Deaconesses were often widowed women or the wives of bishops and were allowed to minister widely in the community, as long as they ministered to women. They welcomed women to the church, and instructed catechumens wishing to enter the community. Significantly, they were responsible for looking after sick women, and women in childbirth, which would have granted them a strong level of connection to people beyond the clergy. Various laws regulated their minimum age, at first sixty, then forty, then fifty.[38] Clark has argued that this age limit was a benchmark of when women had left their sexually active years and were thus too old to have further children.[39] New children could have serious negative effects on any carefully agreed inheritance deals made between deaconesses and bishops, whereby they were provided with an income in return for a donation of property.

Admission to the clergy developed gradually and also was local in character. However, the local administration and layout of tasks was never explicitly laid out. Bishops were elected by the people and confirmed by the bishops of their immediate area, a process that became increasingly dominated by established elites.[40] The community could refuse to accept a bishop, perhaps because they were against him on account of his beliefs or morals. There were many moral restrictions on who could work for a church, enforced by removing from office those who broke the rules. This was one of the major benefits of having closely interconnected social networks; everyone knew everyone else's business. The Synod of Elvira, for instance, set out a series of moral requirements for the clergy. The regulations limited admission to the clergy in a province to those baptised there, so that their past would likely be known to everyone. Subdeacons could not have sinned sexually in youth. Marriage and penance were seen as an antidote for men who had committed sexual sins earlier, preferable to life in the church. Serious sins required penance from a bishop and not a lower-ranking member of the clergy such a deacon or subdeacon. The clergy in charge of the altar were ordered to refrain from sexual activity with their wives and from having children.[41] Freed slaves were only allowed to be ordained if their masters had died.[42] Women who were not a clergyman's virgin sister or daughter were banned from living in a bishop's residence, which was usually next to or on the church grounds. These moral requirements were picked up later by many councils, in particular the Council of Nicaea, with the clear purpose of creating a close-knit clergy made of morally upstanding and, as a result, financially responsible people. It was this kind of close-knit network in each city of the Christian world that would provide the foundation for consolidation and transmission of the Church's message and would prevent any risk of immorality or heresy (see below).

There is no evidence that certain types of labour were preferred or regulated by councils. This is perhaps surprising given the fact that narrative historical texts and archaeological research have demonstrated the strong economic ties between ecclesiastical communities and local agricultural production, as well as the immense donations in property, including enslaved people, and land that wealthy individuals donated to churches.[43] Churches and monasteries were in charge of, for example, oil and wine production on estates across the Mediterranean.[44] As ecclesiastical regulations focused primarily on the strong ties of the clergy itself, little attention was paid to the ties that had to exist between agricultural or construction labourers and local churches. Late Antique society used a diverse set of labour forms, including state-conscripted labour (corvée), slave labour, community-organised labour, labour within the family, as well as market-based wage-labour.[45] In that context, important ties would have formed between the clergy and people who worked the ecclesiastical estates (e.g. estate supervisors) or artisans and construction workers who built and maintained the physical structures of the church.[46] If they did form, they were not specified as thoroughly by canon law in the same way as clerical relationships, perhaps because, as less-prestigious ties, they were not as important to the successful functioning of a complex organisation as the more high-stakes relationships between bishop and clergy. This is not to say that that these rules did not exist, written or unwritten, but they are not the central concern of the regulatory sources. These ties may have been strong or weak depending on the situation at hand but are not within the scope of regulation.

Due to the constraints of the surviving data, the account above is necessarily simplified. Nevertheless, the administrative and spiritual roles that were regulated, the lives of bishops, deacons, and deaconesses, show how valuable it was for bishops to have strong ties with their clergy, as well as methods to minimise poor conduct and conflict.

Restrictions on clerical mobility

The successful transmission of Christianity relied on far and wide dissemination, but this came with risks. Any successful route of transmission for conversion could also transmit the alternative ideas and new interpretations of theology or religious practice known as heresy to the orthodox believers. Any successful concentration of Christians in one city could, given enough time, produce an illicit idea. This is much more likely to be a feature of weak-tie networks, which have a much higher rate of information exchange and likelihood for mutation. The highly centralised strong-tie networks that the bishop and clergy created would have hindered new information from spreading or emerging. Whether the laity's networks were made up mostly of a mass of weak ties or small groups of strong ties with little information-sharing was not important; the laity were not the source of the councils' fears. The councils were more terrified of contagion spread by the clergy. Heresy against official dogma greatly upset the Church, and the councils began to identify this problem as a transmission of ideas enabled by high levels of personal mobility among the clergy. Canon law demonstrates that they were particularly afraid of the lower levels of clerics, deacons, and monks, who might more easily develop dangerous new ideas and have the energy to spread them widely. While explicit punishments were set out for heresy (see below), canon law also devised a variety of checks on mobility that were designed to reduce the prevalence and risk of interactions. While it was impossible (nor desirable) to eliminate all contact between communities, but it was possible to ensure that they had the oversight of the bishops when they did occur. Restrictions on mobility also had the secondary effect that they placed the clergy even more under the panoptic control of the bishop, who was able to consolidate his immediate network as he saw fit.

The Council of Antioch in 341 CE stated that if a priest or deacon had spent a lengthy time in another jurisdiction, he had to refrain from ministering, at the risk of losing his position.[47] Setting up a long-term base in another area presented the risk that heretical ideas could be formed and spread. In 363 CE, at Laodicea, a council legislated that a priest or clergyman could only travel outside of his own bishopric or diocese if he possessed letters of permission from his bishop.[48] The bishop, as head of each church, was the local arbiter of mobility.

The Council of Chalcedon in 450 CE wrote a long list of further regulations. They declared that a cleric had to be fixed to a certain church, martyrium, or monastery; in other words, it placed a ban on a common practice of ordaining priests without fixed location, and a strict ban, on punishment of excommunication, on clerics leaving their home bishop to set up permanent residency elsewhere.[49] It did not matter if the cleric had good grounds for moving or not, ordination was not easily transferable. Ordination was tied to a specific church and although a clergyman was a clergyman wherever he went, he required the consent of his bishop to be ordained in a new church. The laws imposed a strong localism to the emerging social networks, much as was happening in the secular world of the Late Roman Empire, where children were increasingly legally bound to keep the professions of their fathers and stay in their home districts. In this sense, the clergy were treated as if they were workers bound to an estate.

By the end of the fifth century, the growing number of highly mobile monks was becoming a concern. A canon of 471 CE stated that monks would not be allowed to travel freely. Only the monastic stewards (*responsales*) could travel to cities to conduct strictly necessary business, lest they spread (or learn) any false dogma in conversation with heretics.[50]

There were also restrictions on bishops. In that case, however, it was much less about the fear of heresy spreading as it was about territoriality and competition. Bishops spent a lot of time building the social and political networks they needed to maintain their positions. It was not acceptable for them to have to compete with outsiders looking to take their positions. To prevent competition, as early as the fourth century bishops were prevented from exercising their power outside of their own diocese.[51] This restriction was tempered, however, by canons such Canon 19 of the Synod of Arles, which allowed for bishops on pilgrimage to say mass in another diocese. In fact, the canon required that they be given a place to do so.[52] This was something of a compromise and exception. Bishops evidently wanted themselves to have the right to say mass when they were on pilgrimage more than they wanted to stop any bishop from ever saying mass in their territory. Perhaps it was worth the risk. Perhaps they imagined pilgrimage as a rare event that involved a visiting bishop meeting the local bishop in cooperation rather than unilaterally causing unrest in another's territory. The regulations issued by the Council of Ancyra in 314 CE stated that bishops who were working in one parish could return to churches in which they had previously worked without any penalty, as long as they did not create unrest or revolt.[53] The Council of Antioch in 341 CE banned bishops from going to another parish to ordain clergy.[54] The Council of Constantinople in 381 CE required an invitation before travel. There was still a lot of confusion over power in this period, and a hierarchy that, as canon law shows, was still finding its feet. The metropolitan bishop, at the top of the hierarchy, was used as the arbiter of power. If needed, the metropolitan could override these restrictions to invite a bishop to travel and ordain.

For the bishops themselves, an important goal of restrictions on mobility was preventing them from stepping on anyone's toes. Canon law regulated clerical mobility, however, to prevent the more plentiful (and less mature and disciplined) lower ranks of clergy from having too many strong and weak ties. The less ties they had, the less opportunity for them to create, hear, or spread dangerous ideas that could then infect their own or other communities. Bishops and their clergy were not completely barred from travelling, but continual and elaborate checks on movement were put in place. In many respects, sedition against the dogma of the Church was now the same as sedition against the state, and thus, those who were likely to spread it, whether they in fact were orthodox in their views or not, were watched and constrained suspiciously by their bishops.

Punishment and excommunication

The motivating factor for controlling mobility to dampen the effects of weak ties was bishops' fear of losing control over Christian belief. In turn, ecclesiastical law then used fear to maintain control over the community. The lay community was kept under the eye of the bishop through fear of excommunication, violence, and fines.

Violence and fines were powerful instruments of control, especially as the line between ecclesiastical and secular punishment began to blur more and more in the fourth century. Gaddis has shown how this created two images of episcopal violence. On the one hand, bishops could use violence righteously to protect the honour of the Church and God. In one episode at the Council of Ephesus (341 CE), Nestorius supposedly placed the council's copies of the Gospels on the ground, a great sacrilege, and received a swift blow and admonition from a lower-ranking monk in return.[55] On the other hand, the so-called 'tyrant bishops' were felt to be abusing their power for their own personal gain, something which councils were eager to restrain.[56] Although the clergy could not strike members of the laity, some must have used that tactic to ensure compliance.[57] Anyone who felt that they had been treated unjustly was allowed to raise the issue with other bishops or with the secular courts.

The power of clergy to investigate morality and belief and to carry out trials and violence against both clergy and laity was not just implied or hidden in the shadows. It was part of a package of duties given by God and legal avenues given the secular authorities. One especially memorable story was reported to Augustine by an ascetic living in the Balearic Islands named Consentius, who was on a self-appointed mission to expose an epidemic of heretics across north-eastern Spain.[58] According to Consentius, another ascetic named Fronto, from Tarragona, had joined him in his anti-heretical quest. Fronto used some fake heretical writings to trick a woman named Severa into revealing that her relative Severus, a wealthy presbyter in a nearby city, possessed illicit books. Fronto brought them both to trial. Another local bishop named Agapius wished to protect Severus and, at the start of the *audientia episcopalis*, he tried to strangle Fronto, then he beat him, and then he turned the public crowd against Fronto, who called for him to be stoned right there and then. Each time, it was up to the presiding bishop to put a rapid halt to the violence and allow the court to continue. Whether it could be committed by a bishop, a court, a zealot, or a mob, suspicion, accusations, and violence were rife in this new Christian social world. Yet violence was not a prescribed feature of the ecclesiastical legal system. Was it merely assumed that the bishop had the same customary rights to violence as a *paterfamilias* (which were not unrestricted)? Or, was the system trying to maintain healthy hierarchical strong-tie relations within the clergy by preventing the build-up of resentment and conflict? It is hard to say, but it is noteworthy that, when violence broke out in ecclesiastical court as in the above case, it was between people connected by weak ties.

Beyond physical violence, fines were another tool for keeping the community in line. An edict of Valentinian and Valens collected from 372 CE punished the leaders of meetings of Manichaeans with heavy fines. A more social punishment, however, was devised for the attendees of such a meeting. The edict threatened them with a different punishment, exclusion from the community through excommunication or anathematisation.[59] Canon law widely applied this novel punishment for breaking regulations and for moral transgressions, especially for holding heretical beliefs. This punishment was applied primarily to the clergy, including the bishops themselves, and occasionally the laity. By the fifth or sixth century, excommunication was distinguished more clearly from anathematisation, the former a separation from worship, the latter a more substantial separation between the offender and the community at large. However, due to the social nature of Christian worship, both threatened Christians with exclusion from the strong and weak ties of their community. A lay person could lose not just their connections to the bishop and clergy, which for most people would have been weak ties at best, but more intimate relationships too. As Christianity became the majority religion in many parts of the Mediterranean, the Church could frighten people with a wide-reaching threat.

Many activities could produce excommunication or merely prevent a believer from joining the community in the first place. Certain professions were deemed illicit and had to be abandoned before entering into the community. Charioteers and pantomimes, for instance, had to give up their professions.[60] Women were required to stop spending time around hair-dressers and men with long hair.[61] At Arles in 314 CE, a synod banned participation of Christians in gladiatorial games, chariot racing, and employment in the theatre.[62] The Council of Neocaesarea (315 CE) would excommunicate a woman for marrying two brothers and a catechumen for falling into sin (not specified).[63] Likewise, the Synod of Elvira, one of the more specific of the sets of canons, stated that any Christian who played dice would be excommunicated for one year.[64] This synod also prohibited landlords from making offerings using rents that had previously been offered for pagan sacrifice.[65] In other words, social and economic ties with pagans were to be broken or at least kept away from the Church. The Synod of Gangra, one of the more mysterious fourth-century synods, among a series of specific requirements, appears to have ordered that anyone who performed religious services outside the church, anyone who criticised the Christian community, anyone who wears the trappings of an ascetic instead of regular clothes, or anyone who helped a slave to run away would be anathematised.[66]

Naturally, many councils threatened the clergy with excommunication for preaching heretic beliefs. Fear of the transmission of illicit spiritual and moral acts is at the very core of these regulations. The Apostolic Constitutions recommended that even those who prayed with some who had been excommunicated should also be excommunicated. Composing an alternative statement of faith in opposition to the Nicene Creed or expounding Nestorianism would also result in immediate removal from office and separation from the community.[67] But the clergy, including bishops, were required to follow behavioural codes also. The rulings of Neocaesarea proposed that, if a presbyter married, he would be removed from office. If he committed fornication, however, he would be excommunicated outright.[68] Clergy who took the fruits of the church (donations) without permission would also be subject to punishment.[69] The aim was to ensure that correct forms of hierarchy and behaviour were maintained. Strong ties tend to be associated with positive relationships, such as friendships. But not all close relationships are positive. In a community with such a strong moral ideology, if a deacon was caught stealing from the church, for example, it could seriously damage the strong-tie network itself, as well as any other networks for which that deacon served as a bridge or broker to the community at large. It is thus possible to see that the need to maintain the larger community network structured the focus of ecclesiastical law on correct behaviour.

Each of these sets of canons was perhaps only applied to a limited area at a particular time. But they were effective for at least one Christian community for a period of time. More importantly, they indicate the range of people and activities that could fall under the eye of the Church and be warranted for exclusion from the religious community. The possible psychological and social impacts of such a culture of threat of exclusion are profound. If someone was excommunicated, only the bishop who excommunicated them could reintegrate them.[70] The canons issued at Antioch in 341 CE prohibited anyone who had been excommunicated from even moving to a new area. In this way, bishops retained considerable control over laypersons.[71] In pagan practice, the 'control of space' was achieved through 'communication and boundaries' and the 'systematisation' of religious areas.[72] Ritual pollution could bar someone from entry to a temple but generally not from the community itself. There were always other temples and other religious communities to join. There was, in effect, no easy way or one religious figure to completely exclude one person from worship and community outright. For Christians, there was no alternative to communion, the commensal ritual of sharing the bread and wine in memory of the Last Supper. If they were excommunicated, they could not, legally, simply attend a different church. Bishops had thus developed a distinctly new system of fear and control and placed it directly in their own hands.

Rejection from the Church was a terrifying spiritual prospect, but the threat was not merely religious. It has a social impact. Churches were places which defined who was an insider and who was an outsider in the broader community. Absence was noticed. At their best, they could offer sanctuary, at worst, social exclusion.[73] It is easy to imagine how powerful was the fear of loss of business contacts, political alliances, marriage prospects for one's children, even merely the ending of friendships. How frequently excommunication was used as a punishment or to what degree people really did avoid their excommunicated or anathematised neighbours is impossible to say, but it is clear that the threat of it was pervasive and profound.

It is likely that long-term exclusion from the social ties of Christian life was the greatest threat. Spiritual life could continue without necessarily entering the physical church. The Synod of Elvira decreed that magistrates serving as *duumvir* were barred from physically entering the church in the year they were in office.[74] Constantine's initiative changed the relationship between Christians working as magistrates for the state and their access to church spaces. Note that it did not excommunicate them, it just separated them form the physical space of the church. Just ten years later, at the Synod of Arles, this rule was altered to allow Christians to serve as governors and as local magistrates, on the condition that they present letters of recommendation to the bishop (equal to their rank) in their province or city and adhere to expected Christian morality.[75] Who was present in a church was changed fundamentally, as full representatives of the state were a newly integral part of the church.

Excommunication was designed to hold together sets of strong-tie networks linked by weak ties by threatening exclusion from them. As more and more of the population became Christian, the threat became more serious. Excommunication was, after all, a very different proposition in a city where half of the population were non-Christian or of even a different sect; in a universally Christian population, excommunication was more serious. Since we have no data on rates or on the real impact, we can only guess whether it became rarer once the Church became universal and no longer had to defend itself from outside influences, or whether it became a true oppression on the peoples of Late Antiquity.

Conclusion

The evidence for Christian regulation of social life is complex, but certain patterns emerge from the ecclesiastical regulations. The creators of ecclesiastical regulation were aware of the importance of close, frequent relationships. They put day-to-day management of churches at the centre of their legal writings and this in turn helped place churches at the centre of communities across the Mediterranean. There was no central Church authority strong enough to control every minute detail of a bishop's rule over his territory. The laws therefore make little attempt to dictate clear standards for the use of finances. A bishop could decide at his own discretion and according to his own means and needs, for example, whether to build a modest chapel or a large basilica. Regulations secured the importance and power of the office of bishop and ensured that inter-personal disputes or abuses would disrupt the Christian network as little as possible. Similarly, lawmakers believed that excommunication functioned well as a preventative. Ordinary Christians would be highly fearful of such a punishment precisely because it would force them out of their own networks. In the ancient world, that was no small threat. Nonetheless, at the core of the ecclesiastical law was the desire to preserve a centralised network with the bishop at the centre surrounded by a small group of clerical strong ties working closely with him on a wide range of tasks. Dekker's study of networks in early Byzantine Egypt, for instance, show networks highly centralised around a bishop and broken into many strong-tie subgroups.[76] In such networks, the clergy would themselves have ties to the community, some strong and some weak, but were frequently, as is the case of deaconesses, meant to act as bridges between the Church. How this played out differently in reality to the normative regulations is another matter, but the regulations created by bishops imagine the network working in this way.

These laws also show that their makers were aware that Christianity was a set of beliefs that was susceptible to dramatic transformation if it spread too widely and too quickly. A consistent concern of these regulations is reducing and limiting the risk of contagion. They show a strong awareness of the nature of the transmission of ideas through different kinds of ties. They knew that ideas changed the more they were transmitted, and that ideas had the potential to spread from person to person very rapidly out side of the small strong-tie networks at the heart of ecclesiastical life. Even though personal relationships are less strong, weak-tie networks spread information more quickly.[77] On the surface, it appears to go against the evangelic mission of bishops. Episcopal law, if followed, reigned in the rate of transmission. Christianity still could spread fast, but not at the fastest possible rate. But bishops were concerned with far more than just transmission. Regulation was concerned with moral and political questions more than efficiency of evangelism, and when weak-tie networks were regulated, it was due to fear of too much transmission (heresy) rather than too little. Heavy restrictions on the mobility of the clergy were the surest way to limit the creation of weak ties and the probability of heresies starting and to prevent them spreading widely if they did erupt. Nevertheless, many alternative visions of Christianity did emerge and spread widely in the fourth and fifth centuries, such as Apollinarism, Sabellianism, and the Gnostic-Manichean heresy of Priscillianism, all of which were denounced by synods and councils. We are left only to speculate how many more might have appeared otherwise.

This study has focused on three aspects of bishop's management of churches and bishops' relationships with community: the network of clerical responsibilities, clerical mobility, and excommunication. Ecumenical councils and synods created something new in the fourth to sixth centuries; a nascent body of written regulations that established interconnected strong ties in a specific centralised political, social, economic, and religious space, the church. While the laws regulate at a high and generic level, every individual church was a unique combination of social and economic decisions and hierarchical relationships. The growing Christian society was thus not just a set of beliefs about the divine, but a set of social networks that bishops continually worked to control using ecclesiastical law.

Notes

1. Portions of this chapter are adapted from the author's doctoral dissertation, *Religion at the End of Empire: Church Construction, Labor Mobilization, and Social Power in the Late Antique Western Mediterranean* (Stanford University 2019). I am very grateful to Anna Collar and Christina Williamson for their insightful comments on this chapter. All errors are, of course, my own.
2. The term canon law is standard for internal Church regulation. The term ecclesiastical law, however, can be helpful to avoid confusion with later medieval and modern canon law, to recognise the existence of ecclesiastical regulation in the late Roman secular codes, and to emphasise the role of this law in establishing a new ecclesiastical institution, not just setting out moral guidelines.
3. See Clark 1992; Mratschek 2002; Ruffini 2008; Schor 2011; Hillner 2016; Dekker 2018; chapters in Cvetković and Gemeinhardt 2019.
4. Lizzi 1990; see also discussion in Sweetman, this volume.
5. Salzman 1993; Barnard 1995.
6. Hopkins 1998; Stark 1997.
7. Granovetter 1973; Krackhardt 2003.
8. Granovetter 1973: 1361.
9. Collar and Kristensen 2020.
10. Basilicas were quasi-religious spaces by connection to the divinity of the emperor, but they were not places of direct worship of God or the gods.
11. It is important to be specific about differentiating the Church from a church (institutional versus physical) and the Church versus Christianity (a relatively unified single institution versus the various other sects of the Christian faith). 'The Church' refers to the Orthodox or Catholic Church as a pan-Mediterranean institution. The word 'church' in lowercase refers primarily to a physical building used for communal worship. In this context, the word also referred (and continues to refer) to the community of worshippers who attended the

physical building. Church buildings played a role in forming social and economic bonds and mediating between the divine, the imperial system, and the local group or individual.

12. There are a wide range of roles bishops served that are not covered here. For example, pilgrimage was a considerably important economic engagement. For the economics of pilgrimage, see Collar and Kristensen 2020. Likewise, although the rise and development of monasteries in this period is important, this is reserved for future analysis.
13. Hopkins 1998 and Stark 1997. Hopkins broadly agrees with Stark regarding numbers.
14. For the history of canon law, see Hefele 1872; Naz 1924–65; Gaudemet 1985; Van de Wiel 1991; Somerville and Brasington 1998; Kéry 1999; Ferme 2007; Corcoran 2016.
15. I have not included the 'Robber Council' of Ephesus (449 CE), which made little impact after its revocation by the Council of Chalcedon in 451.
16. Hefele 1872. The years of the synods are unclear. There are a number of synods devoted primarily to the growing trinitarian (Arian) controversy that are not discussed here, e.g. the Synod of Alexandria, 320 CE and the Synod of Bithynia, c. 320 CE. For more on early canon law, see Hess 2002; Ohme 2012.
17. Gaddis 2005; Davis 1990.
18. Gaddis 2005: 312–14.
19. *Patrologia Latina* 67. Other sources of law existed that were less influential in antiquity, including Isidore of Seville's fifth to ninth books of the *Etymologiae*, which are themselves based on the *Lex Romana Visigothorum* of 506 CE. Macrides 1990; Clark 1993: 12.
20. The *Apostolic Constitutions* (with the *Apostolic Canons*) were not confirmed by a council. They were more hortatory than regulatory, see Peterson 1907.
21. Hefele 1872: 453.
22. The *Codex Theodosianus* was a collection of earlier decrees published in 439 CE. The *Corpus Juris Civilis* was compiled from 529–534 CE on the initiative of Justinian I. Blume et al. 2016. The latter contains the Code (*Codex Justinianus*), the Digest, the Institutes, and the Novels (*Novellae Constitutiones*).
23. See, among many others, Hefele 1872; Naz 1924–65; Gaudemet 1985; Van de Wiel 1991; Somerville and Brasington 1998; Kéry 1999; Gaddis 2005; Ferme 2007.
24. Sandnes 1994.
25. *Novellae Constitutiones* 131.
26. Brown 2012.
27. Antioch Can. 24.
28. Jones 1964.
29. I have excluded the rules of Benedict, Augustine, and Columba here since they refer exclusively to monastic communities, though they bear many of the same features.
30. Azzara 2002: 87. Episcopal courts were legalised by Constantine (*Sirmondian Constitutions* 1).
31. Azzara 2002: 87–88.
32. Brown 2012.
33. Stathakopoulos 2017.
34. *Codex Justinianus* 1.2.20.
35. This simplified account betrays considerable variation and debate among scholars over the precise nature of powers and relationships but is a general pattern that serves as a heuristic.
36. Ysebaert 1991: 428–30. See also, Karras 2004 and Martimort 1986.

37. See the *Didascalia*, Councils of Nicaea and Chalcedon, and the *Corpus Juris Civilis*. There have been studies of the day-to-day work of deacons and deaconesses in, for example, Elm 1994 and Clark 1993, as well as commentary on social issues in the moral writings of figures like Augustine, Hunter 2007.
38. Clark 1993. *Codex Theodosianus* 16.2.27; Chalcedon Can. 15; *Novellae Constitutiones* 6.6.
39. Clark 1993: 89.
40. Norton 2007.
41. Elvira Can. 24, 27, 30, 31, 33. Hefele, 1872: 145–150, who also notes that the wording of the Latin actually confused and orders the opposite.
42. Elvira Can. 80.
43. Brown 2012.
44. For a clear example, see Askanazi and Aviam 2017 on the 'tiller monk' and the interconnectedness of pastoral and agricultural labour at Nessana in the sixth and seventh centuries.
45. Banaji 2001; Harper 2011.
46. See for example, vague and distant descriptions of workers in the *Liber Pontificalis Ecclesiae Ravennatis*, particularly, §41.
47. Antioch Can. 3.
48. Grey 2004: 25–40; Chalcedon Can. 20; Laodicea Can. 41.
49. Chalcedon Can. 6 and Can. 23.
50. *Codex Justinianus* 1.3.29.
51. Antioch (341 CE) Can. 13 forbade bishops from ordaining clergy outside their home diocese and Sardica, Can. 11 (14 in Latin) forbid bishops from lingering and preaching in others' territories. Hefele 1872: 194.
52. Arles (314 CE) Can 19. *Ut peregrinis episcopis locus sacrificandi detur: De episcopis peregrinis qui in urbem solent venire, placuit iis locum dare ut offerant.* The use of *solent* here, however, is curious, because it may imply that it was not once-off. In other words, it was not bishops on pilgrimage passing through but bishops who came to a city often.
53. Ancyra Can. 18.
54. Antioch (341 CE) Can. 13.
55. Gaddis 2005: 252; *Life of Shenoute* 128–130.
56. Gaddis 2005: 283–322.
57. *Apostolic Canons* Can. 28.
58. Augustine *Ep.* 11* (Divjak 1981 numbering); Kulikowski 2002: 295–320.
59. *Codex Theodosianus* 16.5.3.
60. Elvira Can. 62.
61. Elvira Can. 67.
62. Arles (314 CE) Can. 3. Hefele 1872: 185–86. Many readings of the text have been made with quite different interpretations. Some argue for *in pace*, *in proelio*, *in bello*, meaning everything from banning the games to using weapons except in war, or banning conscientious objection in wartime. Hefele argues, based on the fact that Constantine banned the games in 325 using similar language and the bans of the following two canons, and I follow his interpretation, even though the text says the opposite. Some corruption is likely.
63. Neocaesarea Can. 2; 5.
64. Elvira Can. 79.
65. Elvira Can. 40.
66. Gangra Can. 3; 5; 6.
67. Ephesus Can. 7.
68. Neocaesarea Can. 1.
69. Gangra Can. 7.

70. Elvira Can. 53.
71. Antioch Can. 6.
72. Rüpke 2001: 174–85.
73. Gaddis 2005: 216.
74. Elvira Can. 56. See, Hefele 1872: 187–88.
75. Elvira Can. 56; Arles (314 CE) Can. 7.
76. Dekker 2018.
77. Granovetter 1973.

Bibliography

Ashkenazi, J., and Aviam, M. (2017). "Monasteries and Villages: Rural Economy and Religious Interdependency in Late Antique Palestine." *Vigiliae Christianae* 71, 2: 117–133.

Azzara, C. (2002). "Ecclesiastical Institutions." In *Italy in the Early Middle Ages: 476–1000*, ed. C. La Rocca, 85–101. Oxford: Oxford University Press.

Banaji, J. (2001). *Agrarian Change in Late Antiquity: Gold, Labour, and Aristocratic Dominance: Gold, Labour, and Aristocratic Dominance.* Oxford: Oxford University Press.

Barnard, L. (1995). "The Criminalisation of Heresy in the Later Roman Empire: A Sociopolitical Device?" *The Journal of Legal History* 16, 2: 121–146.

Blume, F. H., Frier, B. W., Connolly, S., Corcoran, S., Crawford, M., Dillon, J. N., Kehoe, D. P., Lenski, N., McGinn, T. A. J., Pazdernik, C. F., Salway, B., and Kearley, T. (2016). *The Codex of Justinian: A New Annotated Translation, with Parallel Latin and Greek Text Based on a Translation by Justice Fred H. Blume.* Cambridge: Cambridge University Press.

Brown, P. (2012). *Through the Eye of a Needle: Wealth, the Fall of Rome, and the Making of Christianity in the West, 350–550 AD.* Princeton: Princeton University Press.

Clark, E. A. (1992). "Elite Networks and Heresy Accusations: Towards a Social Description of the Origenist Controversy." *Semeia* 56: 79–97.

Clark, G. (1993). *Women in Late Antiquity: Pagan and Christian Life-Styles.* Oxford: Clarendon Press.

Collar, A., and Kristensen, T. M. (2020). *Pilgrimage and Economy in the Ancient Mediterranean.* Leiden: Brill.

Corcoran, S. (2016). "Roman Law in Ravenna." In *Ravenna: Its Role in Early Medieval Change and Exchange*, eds. J. Herrin and J. Nelson, 163–199. London: University of London Inst.

Cvetković, C. A., and Gemeinhardt, P. (2019). *Episcopal Networks in Late Antiquity: Connection and Communication Across Boundaries.* Berlin: De Gruyter.

Davis, L. (1990). *The First Seven Ecumenical Councils (325–787): Their History and Theology.* Collegeville: Liturgical Press.

Dekker, R. (2018). *Episcopal Networks and Authority in Late Antique Egypt: Bishops of the Theban Region at Work.* Orientalia lovaniensia Analecta, 264. Leuven: Peeters.

Divjak, J. (1981). *Sancti Aurelii Augustini Opera: Epistolae Ex Duobus Codicibus Nuper in Lucem Prolatae.* Vienna: Holder-Pichler-Tempsky.

Elm, S. (1994). *Virgins of God: The Making of Asceticism in Late Antiquity.* Oxford: Clarendon Press.

Ferme, B. E. (2007). *Introduction to the History of the Sources of Canon Law: The Ancient Law up to the Decretum of Gratian.* Translated by William J. King. Montreal: Wilson and Lafleur.

Gaddis, M. (2005). *There is No Crime for Those Who Have Christ: Religious Violence in the Christian Roman Empire.* Berkeley: University of California Press.

Gaudemet, J. (1985). *Les sources du droit de l'église en Occident: Du IIe au VIIe siècle.* Paris: Éditions du CERF.

Granovetter, M. S. (1973). "The Strength of Weak Ties." *American Journal of Sociology* 78, 6: 1360–1380.

Grey, C. (2004). "Letters of Recommendation and the Circulation of Rural Labourers in the Late Roman West." In *Travel, Communication and Geography in Late Antiquity: Sacred and Profane*, eds. L. Ellis and F. L. Kidner, 25–40. Aldershot/Burlington: Ashgate Publishing.

Harper, K. (2011). *Slavery in the Late Roman world, AD 275–425.* Cambridge: Cambridge University Press.

Hefele, C. J. (1872–1894). *A History of the Christian Councils.* Translated by William Clark. Edinburgh: T & T Clark.

Hess, H. (2002). *The Early Development of Canon Law and the Council of Serdica.* Oxford: Oxford University Press.

Hillner, J. (2016). "Approaches to Clerical Exile in Late Antiquity: Strategies, Experiences, Memories and Social Networks." In *Clerical Exile in Late Antiquity*, eds. J. Hillner, J. Ulrich, and J. Engberg, 11–43. New York: Peter Lang.

Hopkins, K. (1998). "Christian Number and its Implications." *Journal of Early Christian Studies* 6, 2: 185–226.

Hunter, D. G. (2007). *Marriage, Celibacy, and Heresy in Ancient Christianity: The Jovinianist Controversy.* Oxford: Oxford University Press.

Jones, A. H. M. (1986). *The Later Roman Empire, 284–602: A Social, Economic and Administrative Survey.* London: Taylor & Francis.

Karras, V. A. (2004). "Female Deacons in the Byzantine Church." *Church History* 73, 2: 272–316.

Kéry, L. (1999). *Canonical Collections of the Early Middle Ages (ca. 400–1140).* Washington, D.C.: Catholic University of America Press.

Krackhardt, D. (2003). "The Strength of Strong Ties: The Importance of *Philos* in Organisations." In *Networks in the Knowledge Economy*, eds. R. Cross, A. Parker, and L. Sasson, 82–108. Oxford: Oxford University Press.

Kulikowski, M. (2002). "Fronto, the Bishops, and the Crowd: Episcopal Justice and Communal Violence in Fifth-century Tarraconensis." *Early Medieval Europe* 11, 4: 295–320.

Lizzi, R. (1990). "Ambrose's Contemporaries and the Christianization of Northern Italy." *Journal of Roman Studies* 80: 156–173.

Macrides, R. J. (1990). "*Nomos* and *Kanon* on Paper and Court." In *Church and People in Byzantium*, ed. R. Morris, 61–85. Birmingham: Centre for Byzantine, Ottoman and Modern Greek Studies.

Martimort, A. G. (1986). *Deaconesses: An Historical Study.* San Francisco: Ignatius Press.

Mratschek, S. (2002). *Der Briefwechsel des Paulinus von Nola: Kommunikation und soziale Kon- takte zwischen christlichen Intellektuellen.* Göttingen: Vandenhoeck & Ruprecht.

Naz, R., et al. (1924–1965). *Dictionnaire de droit canonique*, vol. 7. Paris: Letouzey et Ané.

Norton, P. (2007). *Episcopal Elections 250–600: Hierarchy and Popular Will in Late Antiquity*. Oxford: Oxford University Press.

Ohme, H. (2012). "Sources of the Greek Canon Law to the Quinisext Council (691/2). Councils and Church Fathers." In *The History of Byzantine and Eastern Canon Law to 1500*, eds. W. Hartmann and K. Pennington, 24–114. Washington, DC: Catholic University of America Press.

Percival, H. R., Schaff, P., and Wace, H. (1900). "The Nicene and Post-Nicene Fathers." *In The Seven Ecumenical Councils of the Undivided Church*, vol. XIV. Revised and edited for New Advent by Kevin Knight <http://www.newad-vent.org/fathers/3805.htm>. 6 February 2017.

Peterson, J. B. (1907). "Apostolic Constitutions." In *The Catholic Encyclopedia*, vol. 1. New York: Robert Appleton Company. http://www.newadvent.org/cathen/01636a.htm, 10 December 2018.

Ruffini, G. (2008). *Social Networks in Byzantine Egypt*. Cambridge: Cambridge University Press.

Rüpke, J. (2007). *Religion of the Romans*. Cambridge: Polity.

Salzman, M. R. (1993). "The Evidence for the Conversion of the Roman Empire to Christianity in Book 16 of the 'Theodosian Code'." *Historia: Zeitschrift für Alte Geschichte H.* 3: 362–378.

Sandnes, K. O. (1994). *A New Family: Conversion and Ecclesiology in the Early Church with Cross-Cultural Comparisons*. Berlin: Lang.

Schor, A. (2011). *Theodoret's People: Social Networks and Religious Conflict in Late Roman Syria*. Berkeley: University of California Press.

Somerville, R., and Brasington, B. (1998). *Prefaces to Canon Law Books in Latin Christianity: Selected Translations, 500–1245*. New Haven, CT: Yale University Press.

Stark, R. (1997). *The Rise of Christianity: How the Obscure, Marginal Jesus Movement Became the Dominant Religious Force*. New York: Harper Collins.

Stathakopoulos, D. C. (2017). *Famine and Pestilence in the Late Roman and Early Byzantine Empire: A Systematic Survey of Subsistence Crises and Epidemics*. London: Routledge.

Van de Wiel, C. (1991). *History of Canon Law*. Leuven: Peeters Publishers.

Ysebaert, J. (1991). "The Deaconesses in the Western Church of Late Antiquity and their Origin." In *Eulogia: Mélanges offerts à Antoon R. Bastiaensen à l'occasion de son soixante-cinquième anniversaire*, eds. G. J. M. Bartelink, A. Hilhorst, and C. H. Kneepkens, 421–436. Steenbrugis: Abbatia S. Petri-Brepols; The Hague, Nijhoff.

10 Orthodox and heterodox networks: lollardy, neighbourhood, and topography in early fifteenth-century Bristol[1]

Esther Lewis

Introduction

The port town of Bristol has often been regarded as a centre of heretical belief and activity by historians of the later Middle Ages.[2] From the late-fourteenth century until the English Reformation in the 1530s, the town experienced many instances of heresy trials concerned with a form of heresy called lollardy. This was a popular adoption of John Wyclif's ideas. He was a late fourteenth-century Oxford theologian who questioned the transubstantiation of the bread and wine at the Eucharist, the power of saints, the authority of the clergy and the Catholic Church, and advocated for the Bible to be translated into English. These points of doctrine opposed the teachings of the Catholic Church. There is evidence of Wyclif's followers in Bristol at the end of the fourteenth century, individuals who were tried for heresy elsewhere in the country reporting that they learnt about lollardy in Bristol, many examples of individuals and groups being tried for heresy within the town's jurisdiction, and a large contingent from Bristol going to the Oldcastle Revolt of 1414.[3]

However, in more recent years Clive Burgess has countered the idea of Bristol as a heretical centre. He argues instead that there was never any organised lollardy in late medieval Bristol and that those who were tried were simply 'networks of disaffected individuals'.[4] His counterargument has set the historiographical debate at an impasse. This chapter gives a different perspective on this debate through an engagement with social network analysis (SNA). It will ask three main questions: how does SNA add to the historiographical debate between historians of heresy in late medieval Bristol; how do orthodox networks of piety function in comparison to heterodox networks of piety; and what social relationships did heretics have with the laity in their immediate neighbourhoods? These questions aim to provide a new perspective on the debate outlined above, but also to the way that heresiologists situate heresy within any given society.[5] This chapter will pursue these questions by first investigating how orthodox networks, for which we have more evidence, functioned in Bristol. This will then be compared with heterodox networks to shed further light on

DOI: 10.4324/9780429429217-13

which factors influenced pious association and networks (both orthodox and heterodox) within the town. SNA will then be used to consider how those tried for heresy were integrated into the community around them. It will consider that they were more than simply 'disaffected individuals' but were instead a product of the wider pious culture.

It is well established in heresy studies that heretics formed networks. [6] The label 'networks of heresy' describes the relationships and associations that heretics made among themselves within which they discussed their beliefs, practiced their faith and, in the case of lollards, read books. These networks were informal and were created through existing networks of friendship, association, and vocation. Lollardy has been investigated as a national network from Wyclif's early followers who left Oxford to preach his ideas around the country, to networks of sedition, book exchanges, and links between small regional communities in the decades before the Reformation.[7] Lollard networks have also been investigated in local contexts, and they have been found to be integrated into the communities around them that might be considered 'orthodox'.[8] The Coventry heresy trials of the early sixteenth century demonstrate that lollards continued to attend church and were part of their parish communities. They participated in the group ritual of the mass through their attendance (even if they turned their eyes away from the host at the point of elevation).[9] This recognition of the relationship between medieval lay people who might be labelled as 'orthodox' and those who might be considered 'heterodox' has led to studies of heresy and mainstream piety borrowing from each other.[10] The increasing appreciation of the relationship between orthodoxy and heterodoxy provokes the following question for this present study: what would a study of seemingly 'orthodox' documents tell us about heretical activity?

The surviving last will and testaments of some of Bristol's inhabitants provide an opportunity to reconstruct social and pious networks across the town, creating a framework through which ideas could potentially have travelled. For the purpose of this chapter, I will focus on the broader social networks in the first quarter of the fifteenth century to create a context in which to consider the surviving evidence of heresy trials. The patchy survival of wills in late medieval Bristol, and the fact that not everybody made a will, means that we cannot reconstruct a complete network.[11] Nonetheless, the networks that can be reconstructed are useful, even if they are not the entire picture.

Orthodox and heterodox networks

Traditionally, historians have viewed orthodox pieties as governed and shaped by the boundaries of the parish.[12] More recently, the parish has come to be viewed as a permeable boundary to an individual's devotional life, with wider factors such as neighbourhood or topography being seen as equally influential.[13] Figure 10.1 demonstrates that some testators were looking beyond their parishes in their wills and were leaving legacies and bequests to religious institutions elsewhere in the town. It also shows why and how last wills and testaments lend themselves nicely to a SNA: they are sources rich with data on an individual's social network and, when multiple wills are viewed together, show the complex social networks of a group of people or a locality. When preparing to die, late medieval testators would leave payments in the form of money or objects to religious institutions (such as parishes, monasteries, or friaries) that they felt they had a connection with or obligation to in life. The motivation behind these bequests were twofold: to pay off any potential debts and to encourage the clergy within the institution to pray for the testator's soul after they died. Both would ensure that the testators' soul would speed through purgatory to the bountiful rewards of Heaven.[14] When we know the locations of these medieval institutions, it gives a glimpse of the testator's movements and associations. Similarly, the testator would also bequeath valuables to their family, friends, and associates, providing data on their social networks.

Wills were generally made by the wealthiest in society who had something of material value to leave.[15] Therefore, they cannot be taken as an indicator of everybody's piety in the town but can be taken as a suggestion for how some of the residents of Bristol practised their religion. The wills used in this present study come from two probate courts: the Great Orphans Court, which sat in the Guildhall of Bristol, and the Prerogative Court of Canterbury, which had jurisdiction across the provenance of Canterbury.[16] The Great Orphans Court looked after the underage heirs of deceased burgesses of the town (men who had the full rights and freedoms of the town) while the Prerogative Court of Canterbury dealt with testators who left more than £5 in their wills or had land in more than one diocese. Both courts are therefore selective and reflect the dying wishes of wealthier testators. Other Bristol wills would also have been proved in the Guildhall, and we know that they were proved from several examples that survive in other municipal manuscripts, such as the Great Red Book of Bristol and parish archives.[17] However, most of these have been lost. Wills only survive because they were important to the activities of the burgesses, and because of this, only the wills of the wealthiest in the town survive. Requests for burial, legacies for tithes forgotten and bequests to the upkeep of the church's fabric have all been taken as bequests that demonstrate a meaningful relationship between a testator and institution and have been put into this visualisation.[18]

Figure 10.1 is a visualisation of social networks between testators and religious institutions found in last wills and testaments from Bristol between 1400 and 1426. This graph represents the social networks of 181 testators, allowing us to see patterns in devotional associations and social connections within the locality.[19] Some of the 181 wills left legacies to more than one parish church, others left legacies to other religious institutions such as hospitals, chapels, or friaries.[20] Some nodes represent testators (people

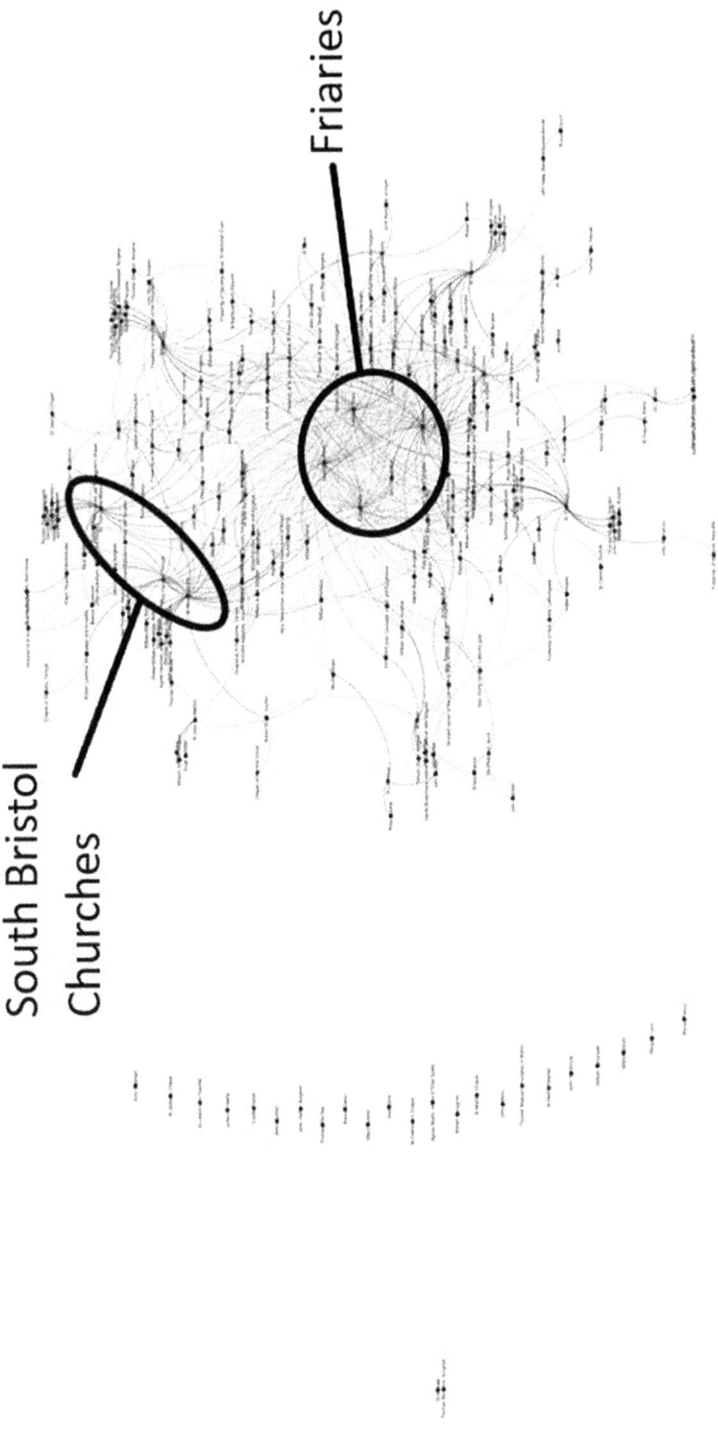

Figure 10.1 Bequests from testators to pious institutions in late medieval Bristol.

who made wills) while others represent parish churches and other institutions such as chapels, hospitals, and friaries. The links between the nodes are directed from the testator to the institution and represent a request for burial in or legacy left to the target institution.[21]

Topography and neighbourhood were important factors in a testator's association with more than one parish church or religious institution. Figure 10.1 shows a strong association between the three churches in the southern suburb of Bristol: St Mary Redcliffe, St Thomas the Martyr, and the Church of the Holy Cross of Temple (see Figure 10.2). The strongest connection is between St Thomas the Martyr and St Mary Redcliffe. Eight testators leave bequests to both of these churches. But there is also a network between these two churches and the Church of the Holy Cross of Temple. It demonstrates that testators who left legacies to one of these churches often left legacies to another church within the suburb. It also shows that they were engaging with other pious institutions in the suburb such as St Katherine's Hospital in Bedminster and St John the Baptist's Hospital in Redcliffe. These nodes are also nearby in the network. There is also a strong association with the Chapel of the Assumption that sat on the Bristol Bridge over the River Avon. This suggests that the topography of the town, with the physical boundary of the river, influenced social and religious movement around the town. In the southern suburb at least, the River Avon meant that an individual was less likely to have a relationship with religious institutions in the north of the town as the only route was across the Bristol Bridge, compared to the many bridges across the Frome to the northern suburb. There was nothing preventing the laity from crossing this bridge, but it is possible that the singularity of the route between north and south meant that there was a divide in the town's social networks.[22]

Figure 10.1 also suggests that neighbourhood was an important factor in an individual's movement around the town. Figure 10.3 filters the ego network of the parish church of St Peter, in the centre of Bristol that was close to the castle.[23] Testators who left legacies to St Peter's also left bequests to the two closest churches to St Peter's: St Mary le Port and SS Philip and Jacob. Figure 10.3 also shows that testators had relationships with St Mary Redcliffe and St James' Priory. These larger churches appear to have attracted attention from testators across Bristol; the size of these churches appears to have prompted testators to leave bequests to them despite being outside of their immediate neighbourhoods.[24] Nonetheless, the relationship between St Peter's, St Mary le Port, SS Philip, and Jacob shows that neighbourhood also played a role in the pious networks formed by the laity. This is also seen to be the case in other parts of the town too. Testators from the parish church of Holy Trinity had connections to the other parish churches of St Ewen and All Saints that were sited around the crossroad with Holy Trinity, as well as some of the other parish churches that were nearby. In this case, it was not the physical topography that appeared to have governed the pious networks formed by the laity, but the proximity of some parish

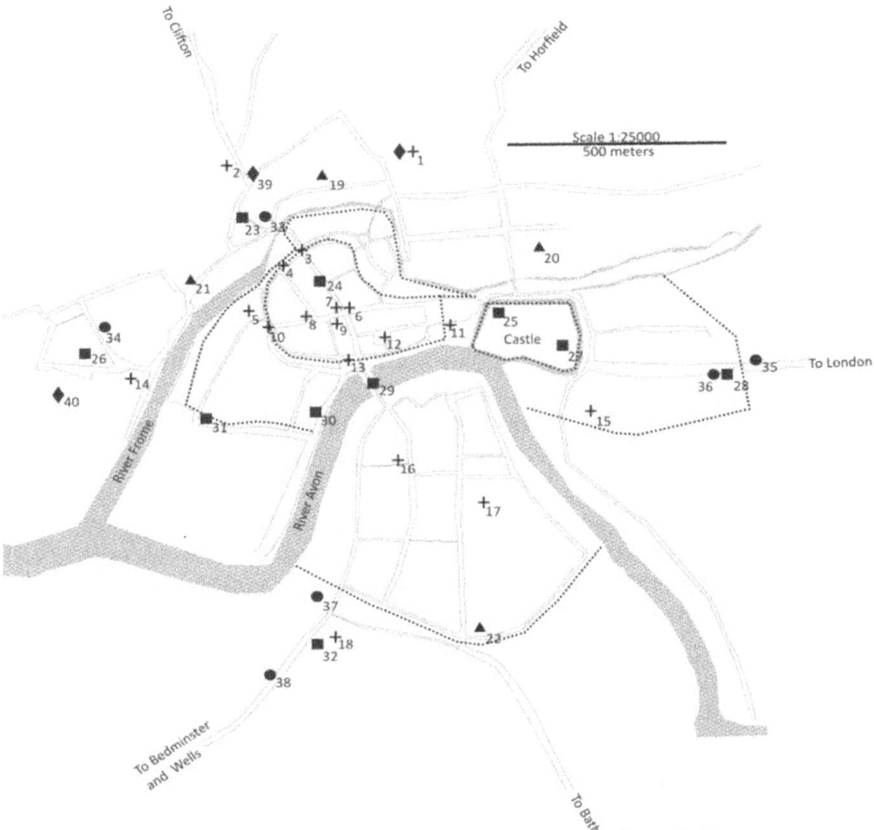

Parish Churches
1. St James' Priory
2. St Michael
3. St John the Baptist
4. St Lawrence
5. St Stephen
6. Holy Trinity
7. St Ewen
8. St Werburgh
9. All Saints
10. St Leonard
11. St Peter
12. St Mary le Port
13. St Nicholas
14. St Augustine the Less
15. SS Philip and Jacob
16. St Thomas the Martyr
17. Holy Cross of Temple
18. St Mary Redcliffe

Friaries
19. Franciscans
20. Dominicans
21. Carmelites
22. Austin Friars

Chapels
23. Three Kings of Cologne
24. St George
25. St Martin
26. St Jordan
27. Castle's Chapel
28. Holy Trinity at Lafford's Gate
29. Assumption of the Blessed Virgin Mary
30. St John the Evangelist
31. St Clement
32. Holy Ghost

Hospitals
33. St Bartholomew's Hospital
34. St Mark's/Gaunt's Hospital
35. St Lawrence's Hospital
36. Holy Trinity's Hospital, Lafford' Gate
37. St John's Hospital
38. St Mary Magdalene's Hospital

Monastic Houses
39. St Mary Magdalen's Nunnery
40. St Augustine's Abbey
41. St James' Priory

Key
.... Town walls
+ Parish churches
▲ Friaries
● Hospitals
◆ Monastic Houses
■ Chapels

Figure 10.2 A map of late medieval Bristol.

churches to each other that allowed the laity to have relationships with other parish churches within their neighbourhoods.

In comparison with Figure 10.3, Figure 10.4 demonstrates that topography and neighbourhood affected each parish in a different way depending on that particular parish's location. Figure 10.4 is the ego network of St Nicholas, which was located in the centre of the town next to the Bristol Bridge which led to the southern suburb. The testamentary network shows that testators from St Nicholas formed pious networks within their immediate neighbourhood, but their proximity to the bridge meant that there were also connections to religious institutions in the south of the town. Alice Wodeford of St Nicholas requested to be buried in St Thomas' next to her first husband, demonstrating that both churches were part of her pious network. William Yong of Holy Cross of Temple left legacies to St Mary Redcliffe and St Nicholas. Within the immediate neighbourhood, testators left legacies to the chapel of St John the Evangelist, which lay in the same parish, and there was also a relationship with the Chapel and Fraternity of the Assumption of the Blessed Virgin Mary on the Bristol Bridge. This demonstrates that there are some subtleties within the network graph, and that although generally the topography of the town and the river created a north/south divide in the social networks, the parish of St Nicholas appears to have been an exception to this rule with more connections to the south. This was probably a result of its proximity to the bridge.

However, there are some limitations to these network graphs. Some issues with the survival of evidence means that the networks represented here are unlikely to be the whole picture. A particular set of wills have survived for late medieval Bristol that means that we only have the wills of the wealthiest in the town. As mentioned above, this is because only material from the Great Orphans Court and the Prerogative Court of Canterbury has survived: both had requirements of wealth and status. Although we do not have access to all of the networks that might have been revealed by a complete set of wills, the surviving evidence allows us to see at least some of the networks from which we at least begin to make some deductions. The only parish that has no connections to the rest of the graph is St Michael's. This is because there is a very poor survival rate for wills from this parish; the reason is unclear. This demonstrates that the survival of evidence impacts the way that we can analyse social networks, but absence of evidence should not be taken as evidence of absence. We cannot conclusively say that a node which is seemingly unconnected, like St Michael's church, was socially isolated, because the dataset is incomplete.[25] Likewise, the nodes representing the Castle Chapel, St Clement's Chapel, and St Martin's Chapel are unconnected to the network graph. This does not mean that they were unused, but simply suggests that they were not engaged with on the same level as some of the other chapels in the town by the burgesses whose wills survive.

The friaries are also very prominent in Figure 10.1: the four nodes at the centre of the network are the Franciscan, Carmelite, Austin, and Dominican

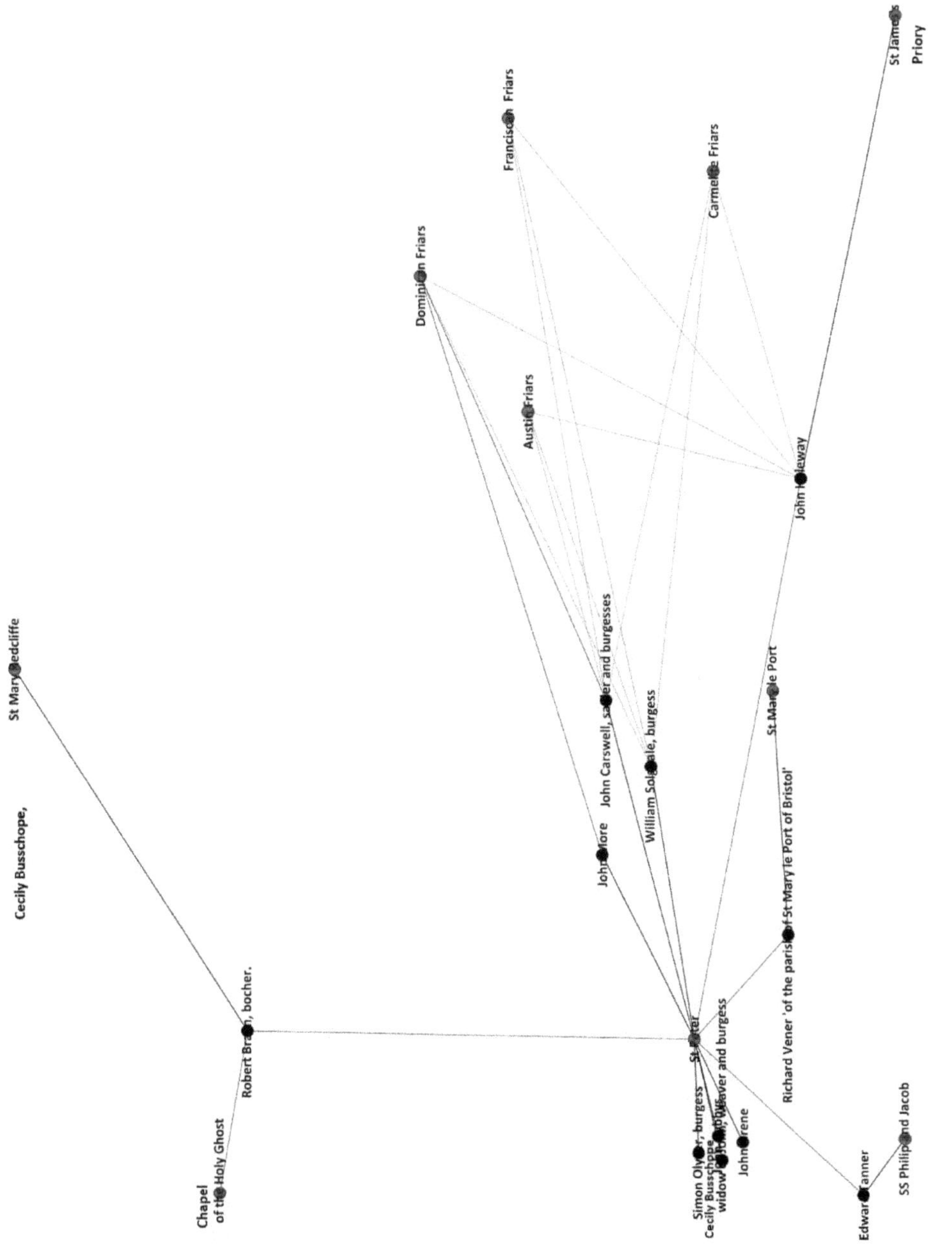

Figure 10.3 Ego network of St Peter's parish church.

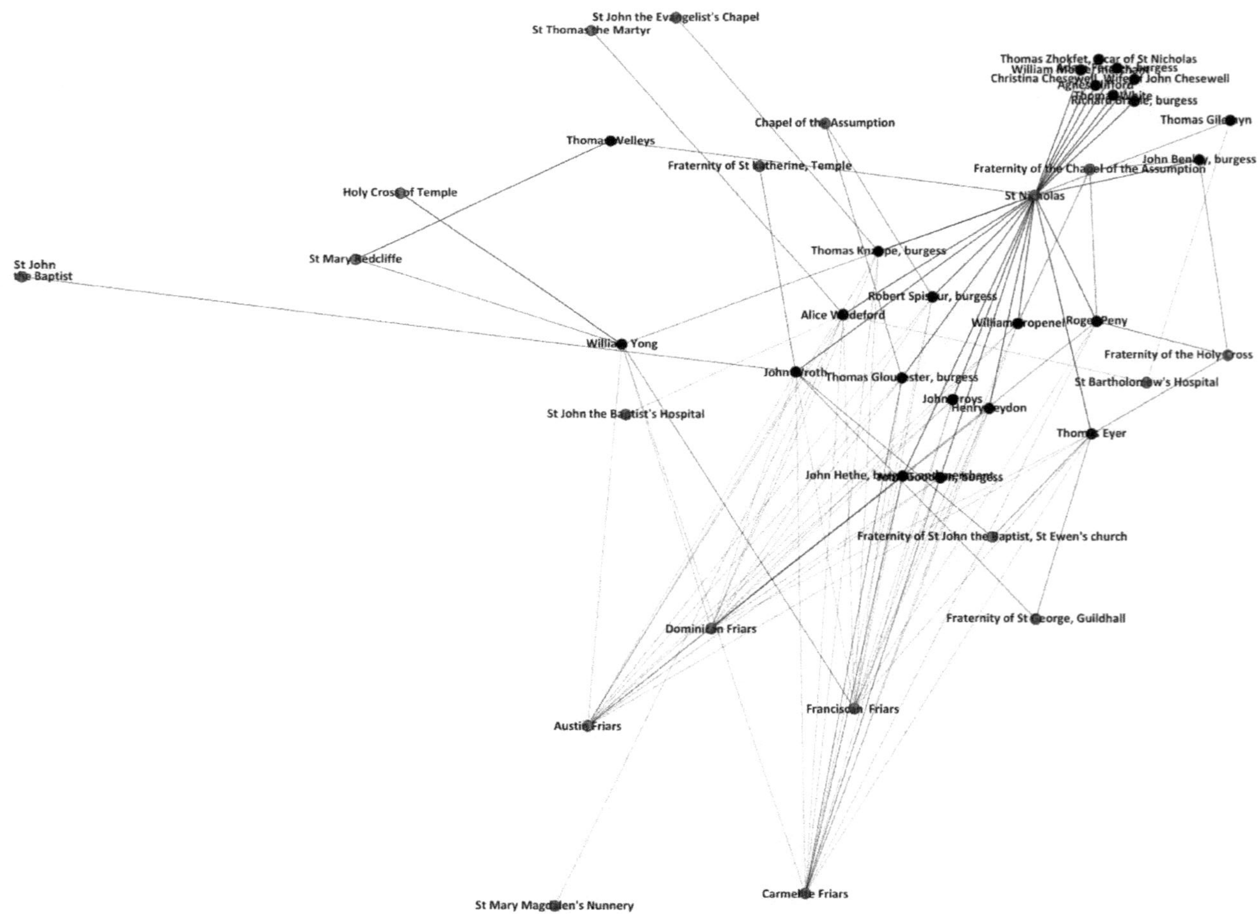

Figure 10.4 Ego Network of St Nicholas' parish church.

friaries that lay in Bristol's suburbs. At an early stage in this research, I experimented with weighting the edges between testators and friaries so that these connections would not be valued in the same way as other connections in the graph. This was because it was common for a testator to call on the friars to attend their funerals and pray for their soul as it departed to begin its passage through purgatory. In many cases, this bequest was generic and did not reflect a relationship between the testator and the friars in life.[26] I therefore planned to give the friars a weighting of 0.5 while the rest of the edges were weighted 1.0. After running the data through Gephi with and without these weightings, I discovered that it made very little difference to the outputs and, more importantly, ran the risk of distorting the network. By downgrading or upgrading connections through weighting, I was running the risk of putting my own assumptions into the analysis and changing the outcome.

Generic bequests to each of the four orders of mendicant friars in Bristol means that the friaries have a high in-degree centrality within the network. In-degree centrality is the amount of connections directed towards a node and, in this case, a high in-degree centrality means that the friars attracted a lot of bequests. The Carmelite friars have an in-degree of 86, the Franciscan friars have an in-degree of 78, and the Austin and Dominican friars have an in-degree of 76 each. This is much higher than every other type of pious institution in the town. Parish churches in the southern suburb, for example, have in-degrees ranging between 18 and 29. This is perhaps a limitation in using testamentary evidence for network analysis as there is a strong emphasis of the spiritual economy, but not all of the connections to the friars are solely for their mortuary function. Some testators named specific friars, singled out a particular friary or requested burial there. All of these types of bequests demonstrate a connection in life. Although prominent in Figure 10.1, the friaries should not be taken as dominant within the pious landscape of late medieval Bristol or integral to lay pious networks. The cluster is indicative of testator's collective use of the friaries as part of the rituals of dying, and should continue to be studied qualitatively as well as quantitatively.[27]

The nodes that represent institutions have not been arranged geographically in Figure 10.1, but it is interesting that most of the religious institutions that were close by in real life are also close to each other in this network graph. The clusters that have been identified through network analysis show how the topography of late medieval Bristol, and the neighbourhoods that this topography formed, influenced the way that these testators moved around the town and made pious relationships with these institutions. In the close urban context of late medieval Bristol, it is reasonable to assume that a testator would have visited the institution to which they left a legacy, whether it was their home parish church, another parish church, or an extra-parochial institution such as a friary or hospital.[28]

So far, the network only sheds light on orthodox or mainstream pious associations, but the strong connections between the three parishes in southern Bristol give an interesting perspective on the debate on heresy in Bristol outlined in the introduction. At the same time that the orthodox networks present in the wills were being fostered, there were three instances of heresy in Bristol that were prosecuted by the Bishops of Bath and Wells and of Worcester. In 1413, Richard Devenyssh was arrested and found guilty of lollardy after he nailed a lollard pamphlet to the choir pews in the Church of the Holy Cross of Temple. He did his penance in the parish of Holy Cross and was to repeat it the following Sunday in the neighbouring parish of Redcliffe.[29] This might suggest that the ecclesiastical authorities understood that heresy may well have spread through the social networks in the southern neighbourhood and wanted people to see Devenyssh doing his penance in neighbouring parishes. The next year, thirty-three Bristolians followed the lollard chaplain, Walter Blake, to the Oldcastle Revolt. These people came from all three of the parishes in the southern suburb. After this, a group of eight people who were connected to the Oldcastle rebels were tried from the parish of St Thomas the Martyr by the Bishop of Bath and Wells.[30] This was probably the end of one heretical group's activity in south Bristol. The next documented instance of a heresy trial was in 1420. It involved the rector of Holy Trinity in the north of the town. The Sheriff, David Ruddok, arrested Thomas Drayton and William Taillour and sent them to the Bishop of Worcester.[31] Drayton and Taillour had moved to Bristol a few months before, and so cannot be regarded as part of the networks that are outlined in the graph in Figure 10.1 because of their newness to the town. The 1413 and 1414 examples suggest that dissenting networks were also functioning within the topographical confines of the southern suburb in a similar manner to mainstream pious networks in the early fifteenth century.

The comparison between the wills and heresy trials is particularly important when thinking about networks of heterodoxy and orthodoxy. One might expect the different types of evidence to give different perspectives on the question of heresy in late medieval Bristol. The evidence from Bishop Bubwith's register, and subsequent registers from the diocese of Bath and Wells, is limited by the diocesan boundary that ran along the River Avon. The registers present heresy as a problem within the southern suburb, and not something that successfully infiltrated the rest of the town, corroborating much of what previous historians have suggested about the nature of dissent in Bristol.[32] In the early fifteenth century, at least, the wills confirm that topography, suburb, and neighbourhood affected the social networks within which people operated. They demonstrate the extent to which the networks present in medieval documents are governed by jurisdiction and the formal structures at play in the production of the document. The bishops' registers of both dioceses can only provide information on the nature of heresy within the confines of the diocese, but both stop providing information at the diocesan boundary on the banks of the River Avon. The wills, however, bridge the gap between north and south Bristol, as they were all within the jurisdiction of the town and its commonalty, and the jurisdiction of the province of Canterbury. This means that there are social networks present in these documents across the town and even outside of it. It is the comparison between these two that suggests that topography shaped the way that people socialised (although there are always exceptions to this), and that religious ideas and devotional practices functioned in relation to the space in which the laity lived.

It also suggests that there were social networks to support the conversations that dissenters were probably having. The way that these individuals were associating was consistent with how the rest of the laity in the suburb were associating with each other. If we take the trials that took place after the Oldcastle revolt in 1414 as an example, it would suggest that the people tried by the Bishop of Bath and Wells who were all from St Thomas the Martyr could reasonably have been in close contact with the people from the other two neighbouring parishes in the suburb who had been part of the Oldcastle Revolt. This is because the SNA of the wills suggests that there was a great amount of contact between the laity of the three parishes. It is also possible, then, that Richard Devenyssh, who was tried for heresy a year previously, might have been part of the same group or may, at least, have known of the other lollards in the suburb. The testamentary evidence provides a network framework through which to interpret the evidence of heresy trials in late medieval Bristol.

Dissenters and their mainstream neighbours

So far, the visualisation of pious networks in Bristol has provided a general overview and indicated a correlation between topography and pious association in the town. To explore the relationship between heterodox and orthodox networks in greater depth, I am going to focus on a case study: the heresy trials that occurred in Bristol after the Oldcastle Revolt of 1414. Sir John Oldcastle, a recently ennobled knight, and others plotted to forcefully overthrow the King and the Archbishop of Canterbury, unite both seats of power, and rule England. They were motivated by the ideas of John Wyclif who had already been declared a heretic by the Catholic Church. On the night of 9 January 1414, rebels from across the country gathered in St Giles' field just outside London. However, their plot had been foiled. They were met by forces loyal to the King and found the city gates had been closed, stopping them from getting in and, more importantly, preventing the support they had anticipated from the city-dwellers getting out. As previously mentioned, a large contingent of Bristolians went to the Oldcastle Revolt. It is estimated that thirty-three men followed lollard chaplain Walter Blake, another prominent leader of the revolt, to St Giles' field.[33] Those men were arrested and tried by the King's Bench. However, back in Bristol, eight people were arrested by the Mayor, Thomas Norton, and put in the King's Prison. They were suspected of heresy simply because of their association with the Oldcastle rebels, and because of rumours that were circulating about them in the southern suburb of Bristol. The Bishop of Bath and Wells, Nicholas Bubwith, and his commissioners tried them. Their trials lasted through 1414 and into 1415, but by 1415 all eight had managed to purge themselves. This meant that they were not found guilty of the crime of heresy, but this does not mean that they were innocent of dissenting beliefs or practices, only that they could not be proved to be guilty.[34] It is likely that these eight people had been involved with heresy in the suburb, and so an analysis of their social networks can shed some light on the integration of dissenters into the suburb around them.[35]

Previous histories of lollardy in early fifteenth-century Bristol have presented this group as isolated from the rest of the burgess society around it. Most historians note that the main connection between the eight people tried by the Bishop and the rest of burgess society was through Christina More. She was the patroness of Walter Blake, the lollard chaplain who led the group from Bristol to St Giles' field, and employed another Oldcastle rebel, James Mersshe, as her servant. Through her marriage to William More, a burgess who died in 1411, she is often presented as the critical link between burgess society and the group of dissenters.[36] If we apply some of the theory behind SNA here, her position as the heretics' link to burgess society could be viewed in several ways. If we apply Ronald Burt's theory of 'structural holes', then it is possible that her strong links to both groups might have put her in a good position. Burt argues that it can be advantageous for an individual to be the link between two clusters within the network. If the individual has strong ties with each group, where they are not just acquainted with each group but are trusted and have a good reputation, they could be the means through which information is transferred between the two groups. This could give them a competitive advantage over each group because they gained knowledge from both groups and could filter what knowledge was transferred.[37] In this scenario, ideas are able to spread because the social connections are built on trust. Christina More may have been able to reap the economic and social benefits of being a member of the burgess society while receiving the spiritual benefits, as she may have seen them, from her relationship with the group of dissenters led by Blake. If we regard her relationship with both the burgesses of the town and the dissenting group as strong ties then she may have been the bridge between the two. It is only through strong connections, trust, and a good reputation that her connection to the burgesses could be advantageous to the Bristol heretics and to herself.

However, Mark Granovetter asserts that in large networks it is rare 'that a specific tie provides the *only* path between the two points'. [38] Indeed, we do not have a complete network for the late medieval town and the lack of survival of evidence from certain groups has meant that the networks that we can see are subject to survival bias. However, we should consider the ramification of the discovery of her ties to dissenters for her husband's reputation, had the group been uncovered before his death in 1411. It may have affected the relationships he had with other reputable men of the town. Beyond her position as mistress of two of the Oldcastle rebels, and her husband's status as burgess, we have no concrete evidence about the strength of her ties to each group. It is also entirely possible that she had drawn closer to the heretics after her husband's death, receding from burgess society. However, the discussion does not need to hinge solely on her as she was not the only person who had ties to the surrounding neighbourhood and to the group that went to the Oldcastle Revolt. Others in the group of eight had a relationship with the wider neighbourhood and the rebels who went to St Giles' field. John Colchester, for example, was one of the masters of the barbers' guild. Robert Harryes was a weaver whose servant went to St Giles' field along with Mersshe and Blake.[39] The following network analysis also demonstrates that these individuals were integrated into burgess society to a greater extent than has been understood by previous historians. This makes it possible that heresy was more prevalent in the southern suburb than suggested previously, as the transmission of religious ideas are more likely to stick where they have multiple entry points or connections as Centola and Macy have argued in their 'wide bridges' argument.[40]

Figure 10.5 is an ego network of the parish of St Thomas the Martyr. It is constructed from the last wills and testaments from the parish of St Thomas the Martyr between 1400 and 1426, shipping records from the late fourteenth century, and some of the connections present in the trial record in Bubwith's register.[41] The graph tells us about social life and dynamics in the parish in the early fifteenth century.[42] The shipping records look forward in time while the last wills and testaments look back, creating a general context within which to place the deponents who were tried in 1414. Like the network presented in Figure 10.1, the strong relationship between the three parish churches in the southern suburb is clearly present in Figure 10.5. Despite being an ego network of St Thomas the Martyr, the churches of St Mary Redcliffe, and the Holy Cross of Temple also appear close to the centre of the graph. Clusters appear in the graph. The largest cluster represent a group of merchants from the parish who had joint interests in ships leaving Bristol harbour in the late fourteenth century. Although they are grouped together in the compurgation certificates that give an overview of their trial, Figure 10.5 does not include edges between the eight people tried for heresy. Edges have only been maintained between the smaller groups in which they were tried. The eight were put on trial in three groups: the first comprised of Edmund Broun, Robert Harryes, Robert Wykeham, and James Mostardmaker, the second group was John Jordan (also known as Jordan Corverser), John Ruell, and John Colchester, and the final person to be tried was Christina More. This prevents a false sense of uniformity being imposed on the eight people within the context of the network graph. It is likely that the eight were divided into smaller groups for trial because there was an association between them. We can only say with certainty that each individual knew the others that they were tried with. There is no proof that they met the rest of the eight. Therefore, edges are maintained here because there was some association between them, even if we cannot know what these associations were.

Figure 10.5 demonstrates that the people who were tried in 1414 were more integrated into burgess society than previously thought. Their association with the burgess society around them did not just depend on the More household, but was also facilitated through Edmund Broun. He is mentioned in three wills. He is mentioned in the will of John Palmer who was from the parish of SS Philip and Jacob, along with William Hurdeman of St Thomas' parish.[43] Palmer appears to have been well connected to parishioners of St Thomas, appearing as executor to several of their wills in the late fourteenth century.[44] Broun is then mentioned in William Hurdeman's will of 1417.[45] Hurdeman was a burgess and merchant in the parish of St Thomas, and was part of the cluster of merchants in Figure 10.5. Broun's third connection was to David Ruddok, a merchant, burgess, and holder of common office in the town. He was named overseer of Ruddok's 1426 will. These connections demonstrate that Broun had relationships with others within burgess society. It is also possible that he may have been connected with John Broun who also features in the network. John Broun appears to have been from the neighbouring parish of Temple. He is connected to David Ruddok and William Hurdeman through the shipping records. There is no link in the surviving evidence between Edmund and John Broun, and so it cannot be said with any certainty that they were connected. However, their shared surname, mutual associates, and habitation in the same neighbourhood may indicate a familial connection between the two. If there was a relationship between the two, Broun would be pulled further into the burgess society around him. Nonetheless, the example of Edmund Broun demonstrates that it is possible that those tried by Bubwith were integrated into the communities around them in a similar way as has been found in other places such as East Anglia, Kent, and Salisbury.[46]

Other circumstantial evidence suggests that there was dissenting activity in the southern suburb in the society of burgesses prior to 1414, supporting the notion that those tried in 1414 were representative of wider dissenting activity. The accusations made against the deponents may be reflective of a wider religious culture within the southern suburb of Bristol and some sympathy towards the activities in which they were accused of participating. Christina More's late husband, William More, was a burgess and merchant. He was probably a man with a good reputation despite his possible engagement with lollard chaplain Walter Blake who was in his employ at the time of writing his will in 1411.[47] The example of More might suggest that there was lollard activity in the suburb at the time of his death.

There are other examples that demonstrate engagement with or sympathies towards some lollard practices. In 1404, John Bount, a merchant and burgess of St Mary Redcliffe, left vernacular gospels in his will to John Caunterbury. Caunterbury was from the neighbouring parish of St Thomas the Martyr. Bount also left him 40 shillings for his 'good and true affection'.[48] It is possible that Bount was using the gospels for orthodox devotion (which was fashionable for many high-status individuals at the time).[49] He left the gospels before the Constitutions of Archbishop Arundel which restricted the laity's access to vernacular literature in 1409. Leaving them in his will in 1404, then, was less controversial than they would have been a few years later. Here, they are being viewed with the benefit of hindsight provided by later events such as Oldcastle's revolt. However, Bount's will stated that the gospels were in the possession of the Wiltshire gentleman William Stourton. Kightly raised the possibility that Stourton may have held dissenting beliefs. In his will, Stourton called his own body 'putrid' and requested to be buried naked. This shares similarities with some of the requests made in the wills of other known lollards.[50] However, rather than demonstrating dissent, Brown has suggested that Stourton's piety was merely more penitential than most.[51] He also suggests that Stourton may have known Sir Thomas Broke the elder who had connections to Oldcastle.[52] The link between Broke and Stourton has led Brown to suggest that Stourton could have been involved with lollards at court in the late fourteenth and early fifteenth century at the same time as Bount's gospels were in his possession.[53] Although this evidence is circumstantial, speculative and is viewed with the benefit of hindsight, it indicates a local context to what Bubwith might have meant when he said Broun, Wykeham, Harryes, and Mostardmaker were known to read 'books of lollards'.[54]

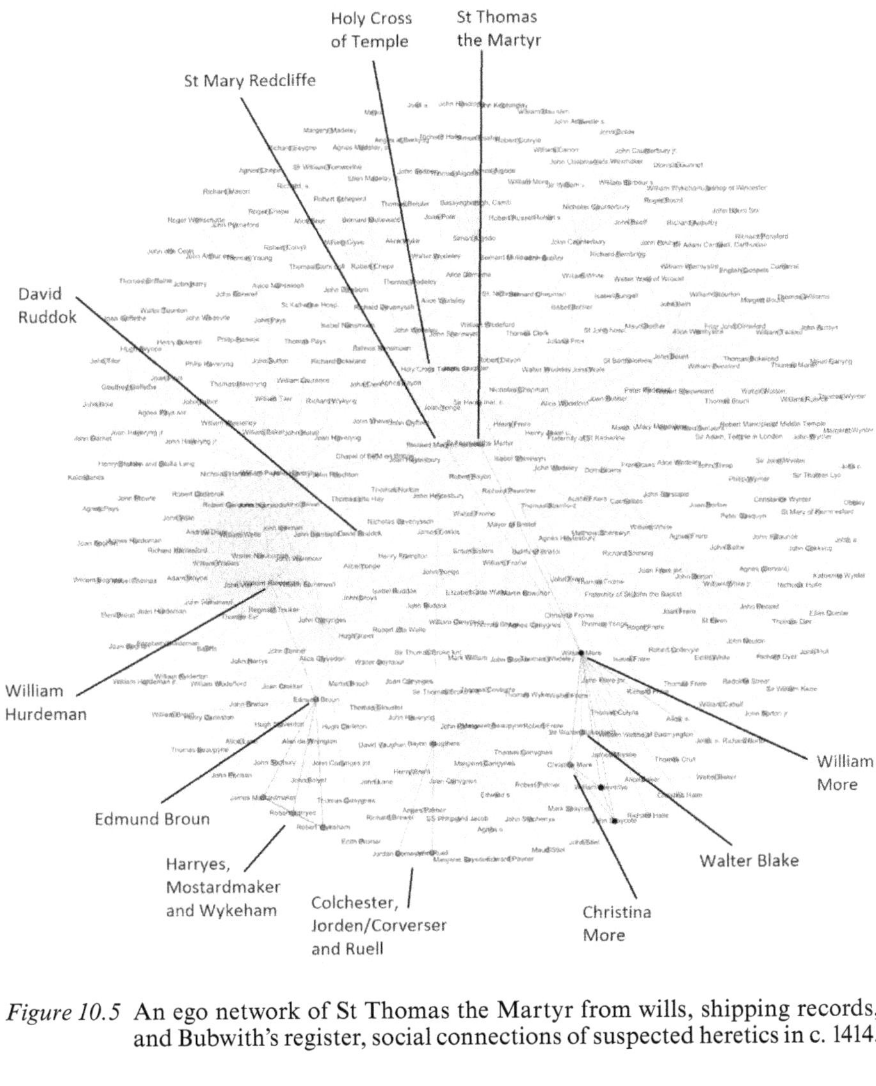

Figure 10.5 An ego network of St Thomas the Martyr from wills, shipping records, and Bubwith's register, social connections of suspected heretics in c. 1414.

Conclusion

Through a SNA of burgess society in late medieval Bristol, this chapter has investigated how identifying and visualising the social networks found in orthodox sources can give a new perspective on heresy in the town. It has identified existing social networks in late medieval Bristol through which potentially dissenting ideas may have travelled, and how topography and neighbourhood appear to have had a large role in shaping these networks. The correlation between the networks of heresy identified through trials and networks found in wills suggest that topography and neighbourhood shaped mainstream and dissenting social networks in the same way, and that both types of piety were being shaped by the same factors. It suggests that, in contrast to previous work on the parish, orthodox networks were able to extend throughout neighbourhoods. It also suggests that heterodox networks were able to function within a locality, at least in Bristol before 1414, and that a dissenting community could be maintained. The similarity between the shape of heterodox and orthodox networks suggests that we should perhaps focus on pieties, rather than heterodoxy and orthodoxy, as a focus of analysis of late medieval lay religion. By focusing on the parish of St Thomas the Martyr, it has also demonstrated that suspected heretics in 1414 were more connected to burgess society than has been recognised by previous historians.

These findings add a different perspective to the historiographical debate through the use of network analysis to think about channels of communication in the town. Once heretics cease to be viewed as an isolated group, they can be considered in relation to the communities around them. By not maintaining the edges between the eight people tried in 1414 by Bishop Bubwith, we can stop viewing those eight people as a cluster within a network graph. The bishop's register does not prove any previous association between the eight, only that they were tried in the same process (and so edges were maintained between these subgroups).[55] If we view them as individuals who were connected to the society around them at different points, as Figure 10.5 shows, then we get a different view of lollardy in south Bristol before 1414. This new perspective suggests that dissent was present at lots of different points within the southern suburb. In the social networks of the burgesses, which have been the focus of this chapter's investigation, dissent was present through the More household and through Edmund Broun. The rest of the individuals tried in 1414 are on the very edge of the graph. This is because the likes of Colchester or Corverser were not part of the burgess society, but a different set of sources might demonstrate that they were well integrated into the craft guilds (of which they were both masters).[56] Unfortunately, the lack of surviving evidence means that this reconstruction is not possible, but by assessing each of the eight individually rather than as a closed group of dissenters it is apparent that lollardy was spread throughout the suburb at different points. This raises the possibility that lollardy was present in Bristol in the early fifteenth century, and that it was more than a small network of a few disgruntled individuals as has been previously suggested.

There are, of course, limitations to the research that can be done with SNA in this instance. One clear example is the impact that the lack of surviving evidence for the parish of St Michael has on the network graphs produced. The friaries appear as dominant within Bristol's pious networks but this is a product of their pious function (although this may have fostered actual links and interactions on some occasions). It demonstrates the importance of analysing the results that SNA produces and understanding how the genre of the source under study may influence the type of networks that will be identified.

Nonetheless, SNA has, in this case, enabled the analysis of a large amount of information about interactions between different actors. The outputs of the SNA highlight the importance of neighbourhood and topography when thinking about pious networks in the local context of Bristol. The River Avon, which formed the boundary between the southern suburb and the rest of the town, and between the two dioceses, is highlighted as an important dividing line in the laity's pious engagement with religious institutions across the town, at least during the first quarter of the fifteenth century.[57] The findings add weight to the argument that there was lollardy in late medieval Bristol before 1414, although the researcher should always continue to consider these findings in relations to a wider contextual background. It produces fresh food for thought in the debate about dissent, unorthodoxy, and heresy in late medieval Bristol.

Notes

1. This research was undertaken with the generous funding of Midlands 3 Cities Doctoral Training Partnership and the Arts and Humanities Research Council. It draws on research from my PhD and most recent article in Nottingham Medieval Studies, Lewis 2019: 169–190; Lewis 2020.
2. McFarlane 1952; Thomson 1965; Kightly 1975; Rex 2002; Fleming 2016: 15–44; French 2001: 8.
3. Examples of dissent in fifteenth-century Bristol can be found in the following: Kightly 1975: chapter 4; Scott Holmes 1914; Scott Holmes vol 1, 1915; Scott Holmes vol 2, 1915; Maxwell-Lyte and Dawes 1934; Maxwell-Lyte 1937; Maxwell-Lyte 1939. Examples of connections to Bristol elsewhere in the country can be found in Tanner and McSheffrey 2003: 102–104, 109–113, 124–127, 146–150, 191–192, 223–227; For further details of Oldcastle's revolt see Strohm 1998: chapter 3; Catto 1987: 46–52; Thomson 1965: 5–19.
4. Burgess 2003: 55, 61. Instead of a 'hotbed of heresy', as Katherine French has argued, Burgess argues for Bristol as a town of Catholic orthodoxy. There is, admittedly, less surviving evidence of heresy trials in Bristol than one might expect from a town with such a dissenting reputation (although this might be a result of the split in ecclesiastical jurisdiction between the dioceses of Bath and Wells and the dioceses of Worcester which was marked by the River Avon: the larger river in Figure 2). The jurisdiction is used as an explanation by Hudson 1988: 122; McFarlane 1952: 127; Rex 2002: 98; Sacks 1986: 148; Kightly 1975: 217.
5. Often they see it as underground or subversive and situate it as such.
6. For example, Roach and Ormerod 2004: 645–652.

7. Jurkowski 2005: 201–206; Jurkowski 2011: 261–278.
8. Lutton 1997; Brown 1995; Jurkowski 2007, 120–152.
9. Roger Brown is accused of doing this in his trial in 1486. Tanner and McSheffrey 2003: 68–69.
10. Forrest 2011: 121–134; Somerset 2014: 15.
11. Burgess 1990: 14–33.
12. Duffy 2001; Mason 1976: 17–29; Owen 1971: 104; Thomson 1960.
13. Colson 2011; Wilson 2014; Atkinson 2016: 149–168; Lewis 2020: chapter 2.
14. Sheenan 1963; Grannum and Taylor 2004: chapter 1.
15. Grannum and Taylor 2004: 2.
16. The probate book of the Great Orphan Court is called the Great Orphan's book. Bristol Archives: JOr/1/1. The probate books of the Prerogative Court of Canterbury are kept at The National Archives: PROB 11.
17. For example, Veale 1931–1953.
18. Middleton-Stewart has argued that tithes forgotten reflect a testator's public duty to their parish church. Middleton-Stewart 2001: 43. It was the responsibility of the laity to look after the fabric of the nave of their home parish church. Drew 1954: 11–12. Burial requests can be regarded as demonstrating a connection with the church in question.
19. There were several more wills from Bristol, but these were not included in the analysis because they did not hold any information about pious networks. They were either nuncupative or briefly named executors.
20. These graphs were created by inputting the data from the last wills and testaments into Excel spreadsheets and then importing these into Gephi. Each testator and each religious institution was given a node number, and the relationships between them were plotted to create the edges. I was interested in the connections between testator and institution, and so recorded each bequest of money or material objects to religious institutions. I then used the atlas layout in Gephi so that the nodes were more spaced out and patterns could be better identified in a 2D printed format. By doing this, devotional giving was visible in the wider context of the locality, and commonalities and anomalies could be observed. In my wider work about lay piety, I have considered the quantitative overview that network analysis provides with qualitative analysis of case studies. See Lewis 2020.
21. The nodes were directed from the testator to the institution because the will can only show the relationship from the perspective of the testator. This also allows in-degree centrality to be considered, discussed below.
22. This has been seen to be the case for the divide in ecclesiastical jurisdictions and the survival of heresy trials in the town. See footnote 4.
23. An ego network shows the network based around a specific node. It shows the nodes directly connected to the focal node. Ego networks can be set to varying degrees of connectivity. A degree of 1 will show the focal node and the nodes that there are connected. A degree of 2 will show the focal node, connected nodes and the nodes that the connected nodes are connected to. The following ego networks are set to a degree of 2 degrees of connectivity.
24. Lewis 2020: 85–89.
25. St Michael's parish church is explored further in Lewis 2020: 65–67.
26. This has been argued by Clive Burgess in Burgess 2010: 73–96.
27. Lewis 2020: 120–128.
28. This is reasonable to assume in most cases except for bequests or legacies that serve a mortuary function which will be discussed below. I have explored this further in my PhD thesis, Lewis 2020: chapters 2–3.
29. Scott Holmes 1914: 142–143.
30. Scott Holmes 1914: 283–290, 298.
31. Forrest 2005: 179; Worcestershire Archive and Archaeology Service: 716.093/5/iii, ff. 13 r, 16v–18v; Kightly 1975: 250–264.

32. Hudson 1988: 122; McFarlane 1952: 127; Rex 2002: 98; Sacks 1986: 148; Kightly 1975: 217.
33. Kightly 1975: 242.
34. Helmholz 1983: 20, 23; Bellamy 1973: 144; Emmison 1973: 291–300; Houlbrooke 1979: 45.
35. Anne Hudson and Henry Kelly both argue that the group were probably guilty, but inadequate questioning and failings in procedure meant that the group were able to purge themselves. Hudson 1973: 147–148; Kelly 1998: 290.
36. Kightly 1975: 245–248; Fleming 2016: 40.
37. Burt 1992: 2. Burt's argument relates to the world of business based on social interactions but the concept can be applied to other social situations too. See also for example the chapter by Williamson, this volume.
38. Granovetter 1973: 1364.
39. Kightly 1975: 249.
40. Centola and Macy 2007: 702–734.
41. Scott Holmes 1914: 283–290, 298; Bristol Archives: JOr/1/1 ff. 75r–152r; The National Archives: PROB 11/3; The National Archives: PROB 11/2a–b; Carus-Wilson 1933.
42. For these graphs, I have used the layout 'Fruchterman-Reingold' in Gephi. This layout puts nodes that have a greater centrality near the middle of the graph and those with a lesser centrality on the periphery. This is a helpful layout when thinking about a community such as a parish or neighbourhood.
43. Bristol Archives: JOr/1/1 f. 85.
44. Wadley 1886: 8, 32, 50.
45. Bristol Archives: JOr/1/1 ff. 131v–132r.
46. On Kent, Lutton 2006; on East Anglia, Jurkowski 2007; on the dioceses of Salisbury, Brown 1995: 218.
47. Bristol Archives: JOr/1/1 f.155.
48. The National Archives: PROB 11/2A/139
49. Galloway 2013: 226–227.
50. Kightly 1975: 237; McFarlane 1972: 210–215.
51. Brown 1995: 203–204.
52. Sir Thomas Broke had organised the marriage of his son to Sir John Oldcastle's step-daughter.
53. Brown 1995: 209.
54. Scott Holmes 1914: 283–290.
55. This is perhaps a weakness of this methodology: that there has been an editorial process behind inputting the raw data into the program. In the work I have undertaken using network analysis, I have been careful to think through the ramifications of any editing and to show this clearly in my work.
56. Kightly 1975: 249.
57. The north/south divide appears to have changed in the later Middle Ages, as evidence from deponents in heresy trials in Coventry revealed that by the later fifteenth century, there were networks of heretics that stretched across the river. Lewis 2020: chapter 5.

Bibliography

Primary Sources

Carus-Wilson, E. M. 1933. *The Overseas Trade of Bristol in the Later Middle Ages*. Bristol: Bristol Record Society.

Maxwell-Lyte, H. C. 1937. *The Registers of Robert Stillington, Bishop of Bath & Wells, 1466–1491 and Richard Fox, Bishop of Bath and Wells, 1492–1494*. Frome: Somerset Record Society.

Maxwell-Lyte, H. C. 1939. *The Registers of Oliver King, Bishop of Bath and Wells, 1496–1503 and Hadrain de Castello, Bishop of Bath and Wells, 1503–1518*. Frome: Somerset Record Society.

Maxwell-Lyte, H. C., and Dawes, M. C. B. 1934. *The Register of Thomas Bekynton, Bishop of Bath & Wells, 1443–1465, Vol 1*. Frome: Somerset Record Society.

Scott Holmes, T. 1914. *The Register of Nicholas Bubwith, Bishop of Bath & Wells, 1407–1424, Vol 1*. London: Somerset Record Society.

Scott Holmes, T. 1915. *The Register of John Stafford, Bishop of Bath & Wells, 1425–1443, Vol 1*. London: Somerset Record Society.

Scott Holmes, T. 1915. *The Register of John Stafford, Bishop of Bath & Wells, 1425–1443, Vol 2*. London: Somerset Record Society.

Tanner, N., and McSheffrey, S. 2003. *Lollards of Coventry, 1486–1522*. London: Cambridge University Press.

The Great Orphans Book of Wills, JOr/1/1. Bristol: Bristol Archives.

The Probate Registers of the Prerogative Court of Canterbury, PROB 11. Kew: The National Archives.

The Register of Bishop Philip Morgan (1419–1426), 716.093/5/iii. Worcester: Worcestershire Archive and Archaeology Service.

Veale, E. W. W. 1931–1953. *The Great Red Book of Bristol, 4 vols*. Bristol: Corporation.

Wadley, T. P. 1886. *Notes and Abstracts of Wills Contained in the Volume Entitles The Great Orphan Book and Book of Wills in the Council House at Bristol*. Bristol: Bristol & Gloucestershire Society.

Secondary Sources

Atkinson, N. 2016. 'Seeing Sound: Mapping the Florentine Soundscape.' In *Mapping Space, Sense, and the Movement in Florence. Historical GIS and the Early Modern City*, eds. N. Terpstra, and C. Rose. 149–168. London: Routledge.

Bellamy, J. 1973. *Crime and public order in England in the Later Middle Ages*. London: Routledge and Kegan Paul.

Brown, A. 1995. *Popular Piety in Late Medieval England: The Diocese of Salisbury, 1250–1550*. Oxford: University of Oxford Press.

Burgess, C. 1990. 'Late Medieval Wills and Pious Convention: Testamentary Evidence Reconsidered.' In *Profit, Piety and the Professions in Later Medieval England*, ed. M. Hicks, 14–33. Gloucester: Sutton.

Burgess, C. 2003. 'A Hotbed of Heresy? Fifteenth-Century Bristol and Lollards in Perspective.' In *The Fifteenth Century III: Authority and Subversion*, ed. L. Clark, 43–62. Woodbridge: Boydell and Brewer.

Burgess, C. 2010. 'Friars and the Parish in Late Medieval Bristol: Observations and Possibilities.' In *The Friars in Medieval Britain: Proceedings of the 2007 Harlaxton Symposium*, ed. N. Rogers, 73–96. Donnington: Shaun Tyas.

Burt, R. S. 1992. *Structural Holes: The Social Structure of Competition*. Cambridge, MA: Harvard University Press.

Catto, J. 1987. 'Dissidents in an Age of Faith? Wycliffe and the Lollards.' *History Today* 37: 46–52.

Centola, D., and Macy, M. 2007. 'Complex Contagions and the Weakness of Long Ties.' *American Journal of Sociology* 113/3: 702–734.

Colson, J. 2011. *Local Communities in Fifteenth Century London: Craft, Parish and Neighbourhood*. Unpublished PhD thesis: Royal Holloway University of London.

Drew, C. 1954. *Early Parochial Organization in England: The Origins of the Office of Churchwarden*. London: St Anthony's Press.

Duffy, E. 2001. *Voices of Morebath*. London: Yale University Press.

Emmison, F. G. 1973. *Elizabethan Life: Morals & the Church Courts: Mainly from Essex Archidiaconal Records*. Chelmsford: Essex County Council.

Fleming, P. 2016. 'Women in Bristol, 1373–1660.' In *Women and the City: Bristol 1373–2000*, ed. M. Dresser, 15–44. Bristol: Redcliffe Press.

Forrest, I. 2005. *The Detection of Heresy in Late Medieval England*. Oxford: Oxford University Press.

Forrest, I. 2011. 'Lollardy and Late Medieval History.' In *Wycliffite Controversies*, eds. M. Bose, and J. P. Hornbeck, 121–134. Turnhout: Brepols.

French, K. 2001. *The People of the Parish: Community Life in a Late Medieval English Diocese*. Philadelphia: University of Philadelphia Press.

Galloway, A. 2013. "Writing Heresy', Apostasy and Anticlericalism in Medieval England.' In *A Companion to British Literature, Medieval Literature*, eds. R. DeMaria, H. Change, and S. Zacher, 215–231. Chichester: Wiley Blackwell.

Grannum, K., and Taylor, N. 2004. *Wills and Other Probate Records*. Chippenham: The National Archives.

Granovetter, M. S. 1973. 'The Strength of Weak Ties.' *American Journal of Sociology* 78/6: 1360–1380.

Helmholz, R. H. 1983. 'Crime, Compurgation and the Courts of the Medieval Church.' *Law and History Review* 1: 1–26.

Houlbrooke, R. A. 1979. *Church Courts and the People during the English Reformation, 1520–1570*. Oxford: Oxford University Press.

Hudson, A. 1973. 'The Examination of Lollards.' *Historical Research* 46: 146–159.

Hudson, A. 1988. *The Premature Reformation: Wycliffite Texts and Lollard History*. Oxford: Clarendon Press.

Jurkowski, M. 2005. 'Lollard Book Producers in London in 1414.' In *Text and Controversy from Wyclif to Bale: Essays in Honour of Anne Hudson*, eds. A. M. Hutchingson, and H. Barr, 201–226. Turnhout: Brepols.

Jurkowski, M. 2007. 'Lollardy and Social Status in East Anglia.' *Speculum* 82: 120–52.

Jurkowski, M. 2011. *'Lollard Networks.' In Wycliffite Controversies*, eds. M. Bose, and J. P. Hornbeck, 261–278. Turnhout: Brepols.

Kelly, H. 1998. 'Lollard Inquisitions: Due and Undue Process.' In *The Devil, Heresy and Witchcraft in the Middle Ages: Essays in Honor of Jeffrey B. Russell*, ed. A. Ferreiro, 279–303. Leiden: Brill.

Kightly, C. 1975. *The Early Lollards: A Survey a Popular Lollard Activity in England, 1382–1428*. Unpublished PhD thesis: University of York.

Lewis, E. 2019. 'Resistance, Self-Defence or Sticking up for your Friends?: A Discussion of Purgation in the Prosecution of Fifteenth-Century Lollardy.' *Nottingham Medieval Studies* 63: 169–190.

Lewis, E. 2020. *Networks and Neighbourhoods: Devotional Practices and Attitudes towards the Church in Late Medieval Bristol, 1400–1500*. Unpublished PhD thesis: University of Nottingham.

Lutton, R. 1997. ‘Connections between Lollards, Townsfolk and Gentry in Tenterden in the Late Fifteenth and Early Sixteenth Centuries.’ In *Lollardy and the Gentry in the Late Middle Ages*, eds. M. Aston, and C. Richmond, 199–228. Stroud: Sutton Publishing.

Lutton, R. 2006. *Lollardy and Orthodox Religion in Pre-Reformation England.* Woodbridge: Boydell and Brewer.

Mason, E. 1976. ‘The Role of the English Parishioner, 1100–1500.’ *Journal of Ecclesiastical History* 27: 17–29.

McFarlane, K. B. 1952. *John Wycliffe and the Beginnings of English Nonconformity.* London: English Universities Press.

McFarlane, K. B. 1972. *Lancastrian Kings and Lollard Knights.* Oxford: Clarendon Press.

Middleton-Stewart, J. 2001. *Inward Purity and Outward Splendour: Death and Remembrance in the Deanery of Dunwich, Suffolk, 1370–1547.* Woodbridge: Boydell Press.

Owen, D. M. 1971. *Church and Society in Medieval Lincolnshire.* Lincoln: Lincolnshire Local History Society.

Rex, R. 2002. *The Lollards.* Basingstoke: Palgrave Macmillan.

Roach, A. P., and P. Ormerod. 2004. ‘The Medieval Inquisition: Scale-free Networks and the Suppression of Heresy.’ *Physica A: Statistical Mechanics and its Applications* 339: 645–652.

Sacks, D. H. 1986. ‘The Demise of the Martyrs: The Feasts of St. Clement and St. Katherine in Bristol.’ *Social History* 11: 141–169.

Sheenan, M. 1963. *The Will in Medieval England.* Toronto: Pontifical Institute of Medieval Studies.

Somerset, F. 2014. *Feeling Like Saints: Lollard Writing After Wyclif.* London: Cornell University Press.

Strohm, P. 1998. *England's Empty Throne: Usurpation and the Language of Legitimation 1399–1422.* London: University of Notre Dame Press.

Thomson, J. A. F. 1960. *Clergy and Laity in London, 1376–1531.* Unpublished PhD thesis: University of Oxford.

Thomson, J. A. F. 1965. *The Later Lollards, 1414–1520.* London: Oxford University Press.

Wilson, M. 2014. *Community, Kinship and Piety: Lincoln Cathedral Close c. 1450–1500.* Unpublished PhD thesis: University of Nottingham.

Index